COMMUNICATING PROCESS ARCHITECTURES 2013

Communicating Process Architectures 2013

WoTUG-35

Edited by

Peter H. Welch
University of Kent, UK

Frederick R.M. Barnes
University of Kent, UK

Jan F. Broenink
University of Twente, The Netherlands

Kevin Chalmers
Edinburgh Napier University, UK

Jan Bækgaard Pedersen
University of Nevada, Las Vegas, USA

and

Adam Sampson
University of Abertay Dundee, UK

Proceedings of the 35th WoTUG Technical Meeting,
25–28 August 2013, Edinburgh Napier University, UK

Communicating Process Architectures 2013
Proceedings of the 35th WoTUG Technical Meeting

ISBN : 978-0-9565409-7-3

www.openchannelpublishing.com

Communicating Process Architectures 2013
P.H. Welch et al. (Eds.)
Open Channel Publishing Ltd., 2013

Preface

Communicating Process Architectures 2013 is the thirty-fifth in the WoTUG series of conferences, and takes place from Sunday August 25th to Wednesday August 28th 2013. The conference is hosted by the School of Computing at Edinburgh Napier University, in Edinburgh, Scotland. Last year, we enjoyed the Scotland experience so much – thank you Abertay Dundee – we insisted on an immediate return!

As normal, paper submission came from around the world, covering a wide range of themes concerning concurrency and computer science. Seventeen papers were accepted for presentation and publication in these Proceedings, following the very strong editorial process developed by this conference over many years. These papers include four that throw new light on ideas in process algebra, formal methods and model checking. Five report new developments in language/library design and run-time support for process oriented mechanisms derived from CSP and the π-calculus. Another seven look at specific concurrency support mechanisms and patterns (including debugging, networking, real-time, simulation, GPGPU acceleration and transactional memory), explicitly deriving their robustness and performance from underlying formal theory. Last, but not least, we are very pleased to see the range of problems facing *High Performance Computing* analysed in the remaining accepted paper and by our two Keynote speakers (Professor Greg Michaelson, Heriot-Watt University, and Dr. Andrew Turner, Edinburgh Parallel Computing Centre – to whom we give special thanks).

The conference also hosts two evening Fringe sessions, which are for reporting current work not yet ready for peer-reviewed publication and/or for floating wild ideas to see what happens. The subjects are hard to predict but we are pleased to be able to include seven abstracts, received just in time, for inclusion in these Proceedings.

Finally, we thank everyone who submitted papers, the reviewers for their conscientious and very hard work, the delegates attending for their contributions that make it all worthwhile and the secretaries and conference officers who just make it work. Thank you all very much.

Peter H. Welch (*University of Kent*),
Frederick R.M. Barnes (*University of Kent*),
Jan F. Broenink (*University of Twente*),
Kevin Chalmers (*Edinburgh Napier University*),
Jan Bækgaard Pedersen (*University of Nevada, Las Vegas*),
Adam Sampson (*University of Abertay Dundee*).

Editorial Board

Reviewing Committee

Alastair Allen, *University of Aberdeen, UK*
Rick Beton, *Big Bee Consultants Ltd., UK*
John Markus Bjørndalen, *University of Tromsø, Norway*
Jan F. Broenink, *University of Twente, The Netherlands*
Phil Brooke, *Teesside University, UK*
Kevin Chalmers, *Edinburgh Napier University, UK*
Barry Cook, *4Links Ltd., UK*
Ian East, *Open Channel Publishing Ltd., UK*
Martin Ellis, *University of Kent, UK*
Rune Møllegaard Friborg, *University of Aarhus, Denmark*
William B. Gardner, *University of Guelph, Canada*
Ruth Ivimey-Cook, *IvimeyCom, UK*
Matthew Jadud, *Berea College, USA*
Temesghen Kahsai, *NASA Ames Research Center, USA*
Brian Kauke, *University of Nevada, Las Vegas, USA*
Gavin Lowe, *University of Oxford, UK*
Jan Bækgaard Pedersen, *University of Nevada, Las Vegas, USA*
Carl G. Ritson, *University of Kent, UK*
Herman Roebbers, *TASS Technology Solutions B.V., The Netherlands*
Adam Sampson, *University of Abertay Dundee, UK*
Steve Schneider, *University of Surrey, UK*
Jon Simpson, *University of Kent, UK*
Marc L. Smith, *Vassar College, USA*
Brian Vinter, *University of Copenhagen, Denmark*
Alan Wagner, *University of British Columbia, Canada*
Peter H. Welch, *University of Kent, UK*

Contents

Communicating Process Architectures 2013
P.H. Welch et al. (Eds.)
Open Channel Publishing Ltd., 2013

Costing by Construction

Keynote Presentation

Gregory J. MICHAELSON

School of Mathematical and Computer Sciences, Heriot-Watt University, UK

G.Michaelson@hw.ac.uk

Abstract. Predicting the performance of sequential systems is hard and concurrency brings additional complexities of coordination and timing. However, current models of concurrency tend to be unitary and so conflate computation and coordination. In contrast, the Hume language is based on concurrent generalised finite state boxes linked by wires. Boxes are stateless with transitions driven by pattern matching to select actions in a full strength functional language. This explicit separation of coordination and computation greatly eases concurrent system modelling: classical inductive reasoning may be used to establish properties within boxes, while box coordination may be explored independently through the novel box calculus. This seminar gives an introduction to the Hume language, cost models for Hume, and the box calculus, and considers how they might be integrated in a system to support costing by construction, where the resource implications of design decisions are made manifest as a system evolves.

Keywords. Hume, concurrency, coordination, computation, box calculus, cost models, performance

Communicating Process Architectures 2013
P.H. Welch et al. (Eds.)
Open Channel Publishing Ltd., 2013

National HPC Facilities at EPCC: Exploiting Massively Parallel Architectures for Scientific Simulation

Keynote Presentation

Andrew TURNER
Edinburgh Parallel Computing Centre, University of Edinburgh, UK
aturner@epcc.ed.ac.uk

Abstract. This presentation gives an overview of the challenges facing scientific software developers in exploiting current and upcoming massively parallel HPC facilities. The different levels of parallelism available in the architectures will be discussed, together with their impact for software design and what the trend is for the future. The discussion will be illustrated with examples from the two national facilities currently hosted at EPCC: HECToR (a Cray XE6) and the DiRAC IBM BlueGene/Q.

Keywords. massively parallel, architecture levels, scientific software, software design, Cray XE6, IBM BlueGene/Q

Communicating Process Architectures 2013
P.H. Welch et al. (Eds.)
Open Channel Publishing Ltd., 2013

The Meaning and Implementation of SKIP in CSP

Thomas GIBSON-ROBINSON and Michael GOLDSMITH

Department of Computer Science, University of Oxford, UK

thomas.gibson-robinson@cs.ox.ac.uk
michael.goldsmith@cs.ox.ac.uk

Abstract. The CSP model checker FDR has long supported Hoare's termination semantics for CSP, but has not supported the more theoretically complete construction of Roscoe's, largely due to the complexity of adding a second termination semantics. In this paper we provide a method of simulating Roscoe's termination semantics using the Hoare termination semantics and then prove the equivalence of the two different approaches. We also ensure that FDR can support the simulation reasonably efficiently.

Keywords. CSP, FDR, model checking, termination

Introduction

The process algebra CSP [1,2,3] provides a way of composing processes *sequentially*, such that a second process is only started once the first process has *terminated*. There have been many different semantics proposed for termination in CSP. The first semantics was proposed by Hoare in [1]. This semantics is perhaps the most intuitive, and probably the most useful for constructing CSP systems in practice, since parallel compositions terminate only when all components do, providing support for *distributed* termination.

In [2], Roscoe proposes a second termination semantics that that has since become the de-facto formalisation of termination in CSP. Roscoe's primary motivation for defining this second semantics was to admit a correct treatment of the algebraic semantics of termination in CSP, which was known [4] to be problematic[1] under Hoare's termination semantics.

The CSP model checker, FDR [5], has only ever supported the Hoare termination semantics, which is particularly unfortunate as it means that those learning CSP are unable to experiment with the more usual theoretical formalisation of termination. Unfortunately, adding support for a second termination semantics in FDR imposes a large overhead. For efficiency reasons, FDR requires several different definitions of every operator, meaning that an increase in the number of operator variants quickly leads to the number of lines of code increasing. Further, the different termination semantics require different ways of calculating the denotational value of a process (in particular, the failures of a process), which requires changes deep inside FDR. Therefore, for the sake of efficiency and in order to reduce the number of lines of code required, it is highly desirable to develop a solution that allows Rosoce's termination semantics to be simulated using the already implemented termination semantics within FDR.

[1]For example, one would reasonably expect $P : SKIP = P$, but this is not always the case under Hoare's termination semantics. An example is given at the end of Section 1.3.

Contributions In this paper we propose a method of simulating Roscoe's termination semantics using the Hoare termination semantics that requires only minimal changes to the existing operator definitions. Further, we prove that the simulation is equivalent to the original formulation. Thus, this provides a way of implementing both termination semantics inside of FDR with minimal changes.

Outline of Paper In Section 1, we introduce the relevant fragment of CSP and define the two main termination semantics. In Section 2 we then develop and prove the correctness of a simulation of the Roscoe semantics in the Hoare semantics. Further, this simulation will be efficient, in that it requires minimal changes to the existing operators. We also discuss an optimisation to the transformation that will improve the performance of the simulation under FDR. In Section 3 we summarise the results of the paper.

1. CSP and its Termination Semantics

In this section we give a brief overview of CSP and detail the two standard termination semantics for CSP.

1.1. CSP

CSP [1,2,3] is a *process algebra* in which programs or *processes* that communicate events from a set Σ with an environment may be described. We sometimes structure events by sending them along a *channel* . For example, $c.3$ denotes the value 3 being sent along the channel c. Further, given a channel c the set $\{|c|\} \subseteq \Sigma$ contains those events of the form $c.x$.

The simplest CSP process is the process $STOP$, that can perform no events and thus represents a deadlocked process. The process $a \rightarrow P$ offers the environment the event $a \in \Sigma$ and then when it is performed, behaves like P. The process $P \square Q$ offers the environment the choice of the events offered by P and by Q. Alternatively, the process $P \sqcap Q$, non-deterministically chooses which of P or Q to behave like. Note that the environment cannot influence the choice, the process chooses *internally*. $P \triangleright Q$ initially offers the choice of the events of P but can *timeout* and then behaves as Q.

The process $P\ {}_A\|_B\ Q$ allows P and Q to perform only events from A and B respectively and forces P and Q to synchronise on events in $A \cap B$. The process $P \underset{A}{\|} Q$ allows P and Q to run in parallel, forcing synchronisation on events in A and allowing arbitrary interleaving of events not in A. The *interleaving* of two processes, denoted $P \mathbin{|||} Q$, runs P and Q in parallel but enforces no synchronisation. The process $P \setminus A$ behaves as P but hides any events from A by transforming them into a special internal event, τ. This event does not synchronise with the environment and thus can always occur. The process $P[\![R]\!]$, behaves as P but renames the events according to the relation R. Hence, if P can perform a, then $P[\![R]\!]$ can perform each b such that $(a, b) \in R$, where the choice (if more than one such b) is left to the environment (like $\square$). The process $P \triangle Q$ initially behaves like P but allows Q to *interrupt* at any point and perform an event, at which point P is discarded and the process behaves like Q. The process $P \Theta_A Q$ initially behaves like P, but if P ever performs an event from A (the *exception* set), P is discarded and $P \Theta_A Q$ behaves like Q.

Recursive processes can be defined either equationally or using the notation $\mu X \cdot P$. In the latter, every occurrence of X within P represents a recursive call to $\mu X \cdot P$.

Non-determinism can arise in a variety of ways in CSP processes. For example, non-determinism can be explicitly introduced via operators such as the internal choice operator (providing the arguments are semantically distinct). Further (and more subtly), other operators can introduce non-determinism when combined in certain ways. For example, $a \rightarrow STOP \square a \rightarrow b \rightarrow STOP$ is non-deterministic since b can be both accepted and refused after performing an a. For a more thorough description of non-determinism see, e.g. [3].

There are a number of ways of giving meaning to CSP processes. The simplest approach is to give an operational semantics. The operational semantics of a CSP process naturally creates a labelled transition system (LTS) where the edges are labelled by events from $\Sigma \cup \{\tau\}$ and the nodes are process states. The usual way of defining the operational semantics of CSP processes is by presenting *Structured Operational Semantics* (SOS) style rules in order to define the transition relation $\xrightarrow{e}$ for $e \in \Sigma \cup \{\tau\}$. For instance, the operational semantics of the exception operator can be defined by the following inductive rules:

$$\frac{P \xrightarrow{a} P'}{P \Theta_A Q \xrightarrow{a} Q} a \in A \qquad \frac{P \xrightarrow{b} P'}{P \Theta_A Q \xrightarrow{b} P' \Theta_A Q} b \notin A \qquad \frac{P \xrightarrow{\tau} P'}{P \Theta_A Q \xrightarrow{\tau} P' \Theta_A Q}$$

The interesting rule is the first, which specifies that if P performs an event $a \in A$, then $P \Theta_A Q$ can perform the event a and then behave like Q (i.e. the exception has been *thrown*). The last rule is known as a *tau-promotion* rule as it promotes any τ performed by a component (in this case P) into a τ performed by the operator. The justification for this rule is that τ is an unobservable event, and therefore the environment cannot prevent P from performing the event. Note that τs from Q are not promoted, since Q is not active. Formally, an argument P of a CSP operator Op is **on** iff it can perform an event, i.e. there exists a SOS rule of the form:

$$\frac{P \xrightarrow{a} P'}{Op(\ldots, P, \ldots) \xrightarrow{a} Op(\ldots)}$$

P is **off** iff no such rule exists. For example, the left argument of the exception operator is **on**, whilst the right argument is **off**. Also, given that the SOS rules for internal choice are:

$$\frac{}{P \sqcap Q \xrightarrow{\tau} P} \qquad \frac{}{P \sqcap Q \xrightarrow{\tau} Q}$$

it follows that both arguments of $\sqcap$ are **off**. Conversely, both arguments of $\Box$ are **on**.

CSP also has a number of *denotational models*, such as the traces, failures and failures-divergences models. In these models, each process is represented by a set of behaviours. Two processes are equal in a denotational model iff they have the same set of behaviours. In this paper we consider only the traces and failures models. In the traces model, a process is modelled by the set of finite traces (which are sequences of events from Σ) that it can perform. In the failures model, a process is represented by a set of pairs, each consisting of a trace and a set of events that the process can stably (i.e. the process must not be able to perform a τ) refuse to perform after the trace.

Notation Given a sequence $tr \in A^*$ and $X \subseteq A$, the *restriction of tr to X*, denoted $tr \upharpoonright X$, is inductively defined by removing events not in X, as follows:

$$\langle\rangle \upharpoonright X = \langle\rangle$$

$$(\langle x \rangle ^\frown xs) \upharpoonright X = \begin{cases} \langle x \rangle ^\frown (xs \upharpoonright X) & \text{if } x \in X \\ xs \upharpoonright X & \text{otherwise} \end{cases}$$

If $tr \in (\Sigma \cup \{\tau\})^+$, $P \overset{tr}{\longmapsto} Q$ iff there exist $P_1, \ldots, P_N = Q$, where $N = |tr|$, $tr = \langle a_1, \ldots, a_N \rangle$ and such that $P_1 \xrightarrow{a_1} P_2 \xrightarrow{a_2} \ldots \xrightarrow{a_N} P_N$. If $tr \in \Sigma^+$, $P \overset{tr}{\Longrightarrow} Q$ iff there exists tr' such that $tr' \upharpoonright \Sigma = tr$ and $P \overset{tr'}{\longmapsto} Q$. $P \not\xrightarrow{e}$ denotes that there does not exist a process P' such that $P \xrightarrow{e} P'$.

1.2. Termination in CSP

CSP also allows processes to be composed sequentially. The *sequential composition of P and Q*, denoted $P\,;Q$, runs P until it terminates, at which point Q is run. CSP defines the process $SKIP$ as the process that immediately terminates. Thus, the process $P = a \to SKIP$ is the process that terminates after performing an a, whilst the process $Q = P\,;Q$ is a process that never terminates, but performs an infinite number of a's (it is equivalent to $R \mathrel{\widehat{=}} a \to R$). The process Ω represents a process that has terminated. Note that whilst this is (operationally) equivalent to $STOP$, we require a syntactically distinct process in order to define Roscoe's termination semantics. The reason that this is required will become clear when the operational semantics of $|||$ are defined under the Roscoe termination semantics in Section 1.4.

In order to define formally the semantics of termination, the event $\checkmark \notin \Sigma$ is added and $SKIP$ is defined as $\checkmark \to \Omega$. The operational semantics of sequential composition are defined as:

$$\frac{P \xrightarrow{a} P'}{P\,;Q \xrightarrow{a} P'\,;Q}\; a \in \Sigma \cup \{\tau\} \qquad \frac{P \xrightarrow{\checkmark} \Omega}{P\,;Q \xrightarrow{\tau} Q}$$

Note that $\checkmark$, since it represents termination, should only ever appear at the end of a trace. Clearly, the above definition of sequential composition respects this.

In order to define termination fully in CSP, we need to define how each of the CSP operators treat a $\checkmark$ performed by one of their arguments. If an operator has no **on** arguments (e.g. $\sqcap$ and $\to$), then these processes are simply defined as non-terminating. If an operator other than sequential composition (which has a special definition, due to its central role in termination) has one **on** argument (e.g. $\Theta_\cdot$, $\setminus \cdot$, $[\![\cdot]\!]$ and $\rhd$), then the operator terminates precisely when its **on** argument does (thus termination is *promoted*). The remaining CSP operators, which have two **on** arguments, are grouped into one of the following two categories with regards to how they treat termination:

Independent Operators that treat termination *independently* terminate whenever *any* of their **on** arguments terminate (e.g. $\Box$ and $\triangle$);

Synchronising Operators that *synchronise* termination terminate only when *all* of their **on** arguments terminate (e.g. $|||$, $\underset{A}{\|}$ and ${}_A\|_B$).

The semantics we consider in this paper treat $\Box$ and $\triangle$ independently. This makes sense intuitively: since neither of these operators actually runs its arguments in parallel, but instead allows actions from one to occur and then discards the other, at some point.

Independent operators, by definition, simply terminate when either of their arguments do. Therefore, the termination operational semantics of independent operators do not vary between different termination semantics, and we can therefore define them now. We formally define the semantics for $\Box$: the operational semantics for other independent operators can be defined analogously:

$$\frac{P \xrightarrow{\checkmark} \Omega}{P \Box Q \xrightarrow{\checkmark} \Omega} \qquad \frac{Q \xrightarrow{\checkmark} \Omega}{P \Box Q \xrightarrow{\checkmark} \Omega}$$

All of the CSP parallel operators (i.e. $|||$, $\|$ etc.) synchronise termination of their arguments[2]. Thus, a parallel composition terminates only when all of its arguments do, which can be useful when constructing systems since it corresponds to distributed termination. In the following sections we describe how the two different termination semantics define the semantics of the synchronising operators.

[2]Whilst this is true of the standard CSP semantics, in [6] Davies defines a version of interleave that does not synchronise on termination, but instead treats it independently. The results of this paper could equally be applied to treat interleave's termination independently instead, but we concentrate on the more usual semantics.

1.3. ✓-as-Refusable

The Hoare semantics, as used by FDR, treats ✓ as a *refusable* event. Thus, henceforth we refer to this as the the ***✓-as-Refusable*** semantics. The motivation for this semantics is that a process should be able to offer to the environment the option of terminating, but the environment should be able to decide whether the process terminates or not. For example, consider the process $SKIPChoice_a \mathrel{\widehat{=}} SKIP \mathbin{\Box} a \rightarrow STOP$ (which we use as a running example): under the ✓-as-Refusable semantics, this process offers the environment the choice of either terminating, or performing the event a.

Under the ✓-as-Refusable semantics, synchronising operators only terminate when both of their arguments are ready to terminate, and the termination is performed in lock-step. The operational semantics for termination of ||| is defined as follows. Again, the operational semantics for other synchronising operators can be defined analogously.

$$\frac{P \xrightarrow{\checkmark} \Omega \wedge Q \xrightarrow{\checkmark} \Omega}{P \mathbin{|||} Q \xrightarrow{\checkmark} \Omega}$$

For example, the above process $SKIPChoice_a$ can terminate immediately. Equally, $SKIPChoice_a \mathbin{|||} STOP = a \rightarrow STOP$, since the right-hand argument can never terminate and thus blocks $SKIPChoice_a$ from terminating. Conversely, $SKIPChoice_a \underset{\{a\}}{\parallel} SKIP = SKIP$, since the right-hand argument cannot perform an a and thus blocks the left-hand argument from performing it.

The value of a process in the denotational models is also affected by the termination semantics in use. Under any of the termination semantics, the traces of a process can be easily extracted using:

$$traces(P) \mathrel{\widehat{=}} \{tr \mid \exists P' \cdot P \stackrel{tr}{\Longrightarrow} P'\}.$$

The failures of a process P under the ✓-as-Refusable semantics, denoted $failures^{r}(P)$, can easily be extracted from the operational semantics:

$$failures^{r}(P) \mathrel{\widehat{=}} \{(tr, X) \mid \exists Q \cdot P \stackrel{tr}{\Longrightarrow} Q \wedge X \subseteq \Sigma \cup \{\checkmark\} \wedge Q \,\mathbf{ref}\, X\}.$$

where $Q \,\mathbf{ref}\, X$ iff Q is stable (i.e. $Q \stackrel{\tau}{\not\longrightarrow}$), and, $\forall x \in X \cdot Q \stackrel{x}{\not\longrightarrow}$.

For example, considering $P \mathrel{\widehat{=}} a \rightarrow STOP \sqcap b \rightarrow STOP$, $(\langle\rangle, \{a\}) \in failures^{r}(P)$ because, although the initial state of P is unstable, $P \xrightarrow{\tau} (b \rightarrow STOP)$. Further, since $b \rightarrow STOP$ is stable it follows that $(\langle\rangle, \{a\}) \in failures^{r}(b \rightarrow STOP)$ and hence $(\langle\rangle, \{a\}) \in failures^{r}(P)$ since $P \stackrel{\langle\rangle}{\Longrightarrow} b \rightarrow STOP$.

A direct consequence of the above definitions is that P is not necessarily equal to $P \,; SKIP$, which might reasonably be expected. For example, $SKIPChoice_a \,; SKIP$ is equal to $a \rightarrow STOP \rhd SKIP$, which is not equal to $SKIPChoice_a$. This is because:

$$(SKIPChoice_a \,; SKIP) \xrightarrow{\tau} SKIP$$

since $SKIPChoice_a$ can perform a ✓ which is converted into a τ by ;. Hence $(\langle\rangle, \{a\}) \in failures^{r}(SKIPChoice_a \,; SKIP)$ since $SKIPChoice_a \,; SKIP \stackrel{\langle\rangle}{\Longrightarrow} SKIP$ and $(\langle\rangle, \{x\}) \in failures^{r}(SKIP)$ for all $x \in \Sigma$. However, $(\langle\rangle, \{a\}) \notin failures^{r}(SKIPChoice_a)$ since $SKIPChoice_a$ is stable and explicitly offers an a, meaning the two sides are not equal.

1.4. ✓-as-Signal

Roscoe's termination semantics for CSP_M, known as the ***✓-as-Signal*** semantics, instead treats ✓ as an event that cannot be refused by the environment. Thus, when a process decides that it is terminating, the environment is not allowed to prevent termination: instead the process immediately terminates and signals this to the environment via the ✓ event. Thus, in the ✓-as-Refusable semantics, the offer of a ✓ can be thought of as a communication that the process can terminate if desired, whilst in the ✓-as-Signal semantics, the offer of a ✓ means the process can terminate on its own accord.

The operational semantics for termination of $|||$ is defined as follows. Again, the operational semantics for other synchronising operators can be defined analogously.

$$\frac{P \xrightarrow{\checkmark} \Omega}{P \mathbin{|||} Q \xrightarrow{\tau} \Omega \mathbin{|||} Q} \qquad \frac{Q \xrightarrow{\checkmark} \Omega}{P \mathbin{|||} Q \xrightarrow{\tau} P \mathbin{|||} \Omega} \qquad \frac{Q \xrightarrow{\checkmark} \Omega}{\Omega \mathbin{|||} Q \xrightarrow{\tau} \mathit{SKIP}} \qquad \frac{P \xrightarrow{\checkmark} \Omega}{P \mathbin{|||} \Omega \xrightarrow{\tau} \mathit{SKIP}}$$

In the above, the first two clauses ensure that as soon as an argument wishes to terminate, it is allowed to do so. This is in contrast to the ✓-as-Refusable semantics, in which an argument is not allowed to terminate until the other argument is ready to do so. The second two clauses specify that as soon as one argument has terminated, if the other argument wishes to terminate then the whole operator becomes SKIP, which can immediately terminate.

The above rules also illustrate the need for Ω and $STOP$ to be syntactically distinct. Clearly, $STOP \mathbin{|||} \mathit{SKIP}$ should not be allowed to terminate since $STOP$ cannot terminate. However, if the last two rules above used $STOP$ rather than Ω, then this would allow the interleave to erroneously terminate. Thus, we need $STOP$ and Ω to be syntactically distinct so that we can differentiate between a process that has terminated via a tick and a process that has just deadlocked, so we can define the termination semantics of synchronising operators.

The most important difference between the two semantics is how the denotational values are extracted from the operational semantics. As stated above, Roscoe views termination as something a process tells the environment it is going to do, rather than something the environment can choose. Therefore, after a trace, if a process can terminate (i.e. a ✓ is available), then it should be able to refuse to communicate anything apart from the ✓ (i.e. all events from Σ). Thus, the failures of a process P under the ✓-as-Signal semantics, denoted $\mathit{failures}^s(P)$, contains an extra clause versus $\mathit{failures}^r(P)$ in order to add these extra failures:

$$\begin{aligned} \mathit{failures}^s(P) \mathrel{\widehat{=}} \{(tr, X) \mid \exists Q \cdot P \stackrel{tr}{\Longrightarrow} Q \wedge X \subseteq \Sigma \cup \{\checkmark\} \wedge Q \mathbf{\ ref\ } X\} \\ \cup \{(tr, X) \mid P \stackrel{tr ^\frown \langle\checkmark\rangle}{\Longrightarrow} \Omega, X \subseteq \Sigma\} \end{aligned} \tag{1.1}$$

However, as noted above, the traces extracted from the two semantics are identical.

The addition of the above failures has a number of interesting consequences for the behaviour of processes that involve choices between ✓ and visible events. For example, consider the process $\mathit{SKIPChoice}_a$ defined in Section 1.3. Under the ✓-as-Signal semantics this now has the failure $(\langle\rangle, \{a\})$ (since it can perform a ✓ immediately) which was not a failure under the ✓-as-Refusable semantics. Thus, $\mathit{SKIPChoice}_a \underset{\{a\}}{\parallel} a \rightarrow STOP = a \rightarrow STOP \rhd STOP$ under the ✓-as-Signal semantics, since $\mathit{SKIPChoice}_a$ may possibly refuse the a which would cause a deadlock. Under the ✓-as-Refusable semantics $\mathit{SKIPChoice}_a \underset{\{a\}}{\parallel} a \rightarrow STOP = a \rightarrow STOP$, since the ✓ can just be ignored. Equally, $\mathit{SKIPChoice}_a \underset{\{a\}}{\parallel} \mathit{SKIP} = \mathit{SKIP}$ under both semantics since environmental control over the a is retained. This suggests the most important difference between the two semantics: under the ✓-as-Signal semantics the environment is unable to stop ✓ from occurring whilst under the ✓-as-Refusable semantics the environment is free to choose ✓ just like any other visible event.

Recall that under the ✓-as-Refusable semantics, it is not necessarily true that $P = P \mathbin{;} \mathit{SKIP}$ and, in particular, that $\mathit{SKIPChoice}_a \mathbin{;} \mathit{SKIP}$ is actually equal to $a \rightarrow STOP \rhd \mathit{SKIP}$

and not $SKIPChoice_a$. However, under the ✓-as-Signal semantics $a \rightarrow STOP \triangleright SKIP = SKIPChoice_a$, since the extra failures that $SKIPChoice_a$ has (as per Equation 1.1) mean that the a can be refused, as it can in $a \rightarrow STOP \triangleright SKIP$. More generally, $P = P \mathbin{;} SKIP$ since in any state where the ✓ can be performed by P, all events other than ✓ can already be refused.

2. Simulating ✓-as-Signal

We now consider how to simulate the ✓-as-Signal semantics under the ✓-as-Refusable semantics. Intuitively, the difference between the two different semantics is that the ✓-as-Refusable semantics just treats ✓ as a regular event and therefore, in $SKIPChoice_a = SKIP \mathbin{\Box} a \rightarrow STOP$, offers the deterministic choice between the ✓ and the a. In the ✓-as-Signal semantics, whilst the choice is still offered, it is no longer a deterministic choice since P can refuse the a (cf. Equation 1.1).

In order to define our translation, we firstly consider how operators with Independent termination semantics are affected. Firstly, consider the process: $SKIPChoice_a \mathbin{|||} STOP$. The LTSs for this process under the two different semantics are given in Figure 1.

(a) ✓-as-Refusable (b) ✓-as-Signal

Figure 1. The LTS of the process $SKIPChoice_a \mathbin{|||} STOP$ under different termination semantics.

Note, in particular, that under the ✓-as-Signal semantics, this process can refuse to do an a, but under the ✓-as-Refusable semantics, it cannot. Thus, it follows that to simulate this process correctly in the ✓-as-Signal semantics, somehow the process must be altered to allow the a to be refused. In the above example, the only way that the a can be refused is if the external choice is resolved[3]. In order to solve this we add a new *resolving tau* event, denoted τ_r, that is treated by all operators precisely as they treat ✓. Thus, in the above example, a τ_r would cause the $\Box$ to be resolved, leading to a state in which only ✓ is offered and, in particular, refuses the a as required. In our translation, if we arrange that τ_r always (and only) occurs directly before a ✓, then it follows that we can correctly resolve $\Box$ (and other operators with Independent termination semantics) and introduce the failures required by Equation 1.1.

We now discuss how to simulate correctly operators that have synchronising termination semantics. In Section 1.4, we saw that $SKIPChoice_a \mathbin{|||} STOP$ is equivalent under ✓-as-Refusable semantics to $a \rightarrow STOP$, whilst under ✓-as-Signal semantics it is equivalent to $a \rightarrow STOP \mathbin{\triangleright} STOP$ (since $SKIPChoice_a$ can decide to terminate independently). We can achieve the same effect by simulating the left hand side as $SKIPChoice_a \mathbin{;} SKIP$, thus allowing the termination to resolve the external choice and introduce the failures required by Equation 1.1. Note that this is essentially taking advantage of the equivalence $P = P \mathbin{;} SKIP$ that holds under the ✓-as-Signal semantics, but not in the ✓-as-Refusable semantics.

Outline In Section 2.1 we formalise the translation that we have sketched above. We then prove the equivalence of the translation in Section 2.2 and then discuss the efficency of the translation in the context of FDR in Section 2.3.

[3]An alternative method of solving this problem is to simulate each external choice $P \mathbin{\Box} Q$ as $(P \mathbin{\Box} Q); SKIP$. The problem with this simulation is that it produces processes that are much larger than the original process, and more problematically, will cause FDR to be unable to compile many recursive definitions.

2.1. Formalising The Translation

We now formalise the above translation, as follows. Firstly, we define how the new resolving tau τ_r is treated by the existing operators (by assumption, $\tau_r \notin \Sigma$, so none of the existing operational semantic rules apply). We define $BSkip \mathrel{\widehat{=}} \tau_r \rightarrow \checkmark \rightarrow \Omega$. In order to ensure that a τ_r occurs *only* directly before a $\checkmark$, we add the following additional rule to $P \mathbin{;} Q$:

$$\frac{P \xrightarrow{\tau_r} P'}{P \mathbin{;} Q \xrightarrow{\tau} P' \mathbin{;} Q}$$

As a result of this rule, note that $P \mathbin{;} Q$ can only perform a τ_r if Q can after P has terminated (since any τ_r of P is converted into a τ). Thus, assuming Q only performs a τ_r directly before a $\checkmark$, the desired property immediately follows. Further, to each other operator we add an extra rule to promote τ_r exactly as τ is promoted. Note that if none of the arguments of some CSP operator perform a τ_r, then the above changes do not alter the semantics of the operator at all. This should simplify the implementation of these rules.

In order to simplify the formalisation of the transformation, we consider only a restricted subset of CSP that includes $STOP$, $STOP$, $\rightarrow$, $;$, $\Box$ and $|||$. Note that this includes all categories of operators (in terms of their termination semantics), and thus the proof and transformation could easily be generalised to all CSP operators.

Definition 2.1. The $\checkmark$*-as-Signal translation function*, Sig, is defined on CSP processes as follows:

$$\begin{aligned}
Sig(SKIP) &\mathrel{\widehat{=}} BSkip \\
Sig(STOP) &\mathrel{\widehat{=}} STOP \\
Sig(a \rightarrow P) &\mathrel{\widehat{=}} a \rightarrow Sig(P) \\
Sig(P \Box Q) &\mathrel{\widehat{=}} Sig(P) \Box Sig(Q) \\
Sig(P \mathbin{;} Q) &\mathrel{\widehat{=}} Sig(P) \mathbin{;} Sig(Q) \\
Sig(P \mathbin{|||} Q) &\mathrel{\widehat{=}} (Sig(P) \mathbin{;} BSkip) \underset{\{\tau_r\}}{\|} (Sig(Q) \mathbin{;} BSkip)
\end{aligned}$$

As discussed informally, we ensure that a τ_r occurs exactly once before a $\checkmark$ by redefining $SKIP$ in the obvious way. We then ensure that $|||$ synchronises on τ_r to ensure that only one τ_r can occur. We prove that this holds in Lemma 2.3.

Notation In the following we use $\Sigma^{\checkmark}$ to denote $\Sigma \cup \{\checkmark\}$, Σ^{τ_r} to denote $\Sigma \cup \{\tau_r\}$ and $\Sigma^{\checkmark,\tau_r}$ to denote $\Sigma \cup \{\checkmark, \tau_r\}$.

2.2. Proving Equivalence

We now prove that our translation gives the correct result. In particular, since we are interested in using our translation in the context of FDR, we check that the denotational value of a process P under the $\checkmark$-as-Signal semantics is the same as the value of $Sig(P) \setminus \{\tau_r\}$[4] under the $\checkmark$-as-Refusable semantics.

In order to prove that the translation produces the correct denotational values, we firstly prove a couple of results about the operational semantics. Together, these essentially prove that τ_r functions as discussed above. Firstly we prove that τ_r's always occur directly before $\checkmark$'s.

[4]Clearly we need to hide the newly introduced τ_r event, otherwise we will obtain extra traces.

Lemma 2.2. $(Sig(P) \setminus \{\tau_r\}) \stackrel{\langle\checkmark\rangle}{\Longrightarrow} \Omega$ iff $Sig(P) \stackrel{\langle\tau_r,\checkmark\rangle}{\Longrightarrow} \Omega$.

Proof (Sketch). This follows by a trivial structural induction over P, using the definitions of the modified operational semantics and the definition of Sig. In particular, note that $BSkip$ always performs a τ_r immediately before a $\checkmark$. Otherwise, if an operator performs a $\checkmark$, then it must not be ; (since ; cannot perform a $\checkmark$), and thus it must be because an argument performs a $\checkmark$. Thus the inductive hypothesis applies, noting that all operators promote τ_r. □

We now prove a second operational result, and prove that if a process can do a τ_r, then the resulting state must be operationally equivalent to $SKIP$ (i.e. it can do a $\checkmark$, but can perform no other event).

Lemma 2.3. $Sig(P) \stackrel{\langle\tau_r\rangle}{\Longrightarrow} X$ iff $X \stackrel{\checkmark}{\longrightarrow} \Omega$ and, for all $a \in \Sigma^{\tau_r}$, $X \stackrel{a}{\not\longrightarrow} X'$, for some X'.

Proof (Sketch). This follows by a trivial induction over P, noting that the translation ensures that whenever τ_r occurs, the resulting state is equivalent to $SKIP$. □

Using the above we can now prove that our translation produces the correct traces.

Theorem 2.4. $traces(P) = traces(Sig(P) \setminus \{\tau_r\})$

Proof. This follows by a trivial induction on P, noting that $traces(SKIP) = traces(BSkip \setminus \{\tau_r\})$. □

We can now prove our that the translation produces the correct failures.

Theorem 2.5. $failures^s(P) = failures^r(Sig(P) \setminus \{\tau_r\})$.

Proof. We prove the lemma by structural induction on P. We elide the case for $STOP$ since it is trivial.

$P = SKIP$:

$$\begin{aligned} & failures^r(Sig(SKIP) \setminus \{\tau_r\}) \\ &= failures^r(BSkip \setminus \{\tau_r\}) \\ &= failures^r(SKIP \setminus \{\tau_r\}) \\ &= failures^r(SKIP) \\ &= failures^s(SKIP) \end{aligned}$$

$P = a \rightarrow Q$:

$$\begin{aligned} & failures^s(a \rightarrow Q) \\ &= \{(\langle\rangle, X) \mid X \subseteq \Sigma^{\checkmark,\tau_r}, a \notin X\} \\ &\qquad \cup \{(\langle a\rangle^\frown tr, X) \mid (tr, X) \in failures^s(Q)\} \\ &= \{(\langle\rangle, X) \mid X \subseteq \Sigma^{\checkmark,\tau_r}, a \notin X\} \\ &\qquad \cup \{(\langle a\rangle^\frown tr, X) \mid (tr, X) \in failures^r(Sig(Q) \setminus \{\tau_r\})\} && \langle \text{IH} \rangle \\ &= failures^r(a \rightarrow (Sig(Q)) \setminus \{\tau_r\}) \\ &= failures^r(Sig(a \rightarrow Q) \setminus \{\tau_r\}) \end{aligned}$$

$P = Q \Box R$:

$$
\begin{aligned}
& \mathit{failures}^s(Q \Box R) \\
&= \{(\langle\rangle, X) \mid (\langle\rangle, X) \in \mathit{failures}^s(Q) \cap \mathit{failures}^s(R)\} \\
&\quad \cup \{(tr, X) \mid (tr, X) \in \mathit{failures}^s(Q) \\
&\qquad\qquad \cup \mathit{failures}^s(R), tr \neq \langle\rangle\} \\
&\quad \cup \{(\langle\rangle, X) \mid X \subseteq \Sigma^{\tau_r} \wedge \langle\checkmark\rangle \in traces(Q) \cup traces(R)\} \\
&= \{(\langle\rangle, X) \mid (\langle\rangle, X) \in \mathit{failures}^r(Sig(Q) \setminus \{\tau_r\}) \\
&\qquad\qquad \cap \mathit{failures}^r(Sig(R) \setminus \{\tau_r\})\} \\
&\quad \cup \{(tr, X) \mid (tr, X) \in \mathit{failures}^r(Sig(Q) \setminus \{\tau_r\}) \\
&\qquad\qquad \cup \mathit{failures}^r(Sig(R) \setminus \{\tau_r\}), tr \neq \langle\rangle\} \\
&\quad \cup \{(\langle\rangle, X) \mid X \subseteq \Sigma^{\tau_r} \wedge \langle\checkmark\rangle \in traces(Sig(Q) \setminus \{\tau_r\}) \\
&\qquad\qquad \cup traces(Sig(R) \setminus \{\tau_r\})\} && \langle \text{IH, Theorem 2.4} \rangle \\
&= \{(\langle\rangle, X) \mid tr \neq \langle\rangle \wedge \exists\, tr \in \{\tau_r\}^* \cdot \\
&\qquad (tr, X \cup \{\tau_r\}) \in \mathit{failures}^r(Sig(Q)) \\
&\qquad\qquad \cap \mathit{failures}^r(Sig(R))\} \\
&\quad \cup \{(tr, X) \mid \exists\, tr' \cdot tr' \upharpoonright \Sigma^{\checkmark} = tr \\
&\qquad\qquad \wedge (tr', X \cup \{\tau_r\}) \in \mathit{failures}^r(Sig(Q)) \\
&\qquad\qquad\qquad \cup \mathit{failures}^r(Sig(R))\} \\
&\quad \cup \{(\langle\rangle, X) \mid X \subseteq \Sigma^{\tau_r} \wedge \langle \tau_r, \checkmark\rangle \in traces(Sig(Q)) \\
&\qquad\qquad\qquad \cup traces(Sig(R)\} && \langle \text{Lemma 2.2} \rangle \\
&= \{(\langle\rangle, X) \mid \exists\, tr \in \{\tau_r\}^* \cdot \\
&\qquad (tr, X \cup \{\tau_r\}) \in \mathit{failures}^r(Sig(Q)) \\
&\qquad\qquad \cap \mathit{failures}^r(Sig(R))\} \\
&\quad \cup \{(tr \upharpoonright \Sigma^{\checkmark}, X) \mid (tr, X \cup \{\tau_r\}) \in \mathit{failures}^r(Sig(Q)) \\
&\qquad\qquad \cup \mathit{failures}^r(Sig(R)) \wedge tr \neq \langle\rangle\} && \langle \dagger \rangle \\
&= \{(tr \upharpoonright \Sigma^{\checkmark}, X) \mid (tr, X \cup \{\tau_r\}) \in \mathit{failures}^r(Sig(Q) \Box (R))\} \\
&= \mathit{failures}^r((Sig(Q) \Box Sig(R)) \setminus \{\tau_r\}) \\
&= \mathit{failures}^r(Sig(Q \Box R) \setminus \{\tau_r\})
\end{aligned}
$$

We can prove that the step marked † holds by proving the following equivalence:

$$
\begin{aligned}
& \{(tr, X) \mid tr \neq \langle\rangle \wedge \exists\, tr' \cdot tr' \upharpoonright \Sigma^{\checkmark} = tr \wedge \\
&\qquad (tr', X \cup \{\tau_r\}) \in \mathit{failures}^r(Sig(Q)) \cup \mathit{failures}^r(Sig(R))\} \\
&\quad \cup \{(\langle\rangle, X) \mid X \subseteq \Sigma^{\tau_r} \wedge \langle \tau_r, \checkmark\rangle \in traces(Sig(Q)) \cup traces(Sig(R))\} \\
&= \{(tr \upharpoonright \Sigma^{\checkmark}, X) \mid (tr, X \cup \{\tau_r\}) \in \mathit{failures}^r(Sig(Q)) \cup \mathit{failures}^r(Sig(R)) \\
&\qquad \wedge tr \neq \langle\rangle\}
\end{aligned}
$$

Firstly, suppose (tr, X) is a member of the left-hand equation. Then, if (tr, X) is a member of the first clause, $tr \neq \langle\rangle$ and there exists tr' such that $tr' \upharpoonright \Sigma^{\checkmark} = tr$ and

$(tr', X \cup \{\tau_r\}) \in failures^r(Sig(Q))$. Hence, (tr, X) is a member of the right-hand equation, as required. Otherwise, (tr, X) must be a member of the second clause, and hence $tr = \langle\rangle$, $X \subseteq \Sigma^{\tau_r}$ and $\langle\tau_r, \checkmark\rangle \in traces(Sig(Q)) \cup traces(Sig(R))$. WLOG, assume there exists S such that $Sig(Q) \stackrel{\langle\tau_r\rangle}{\Longrightarrow} S \stackrel{\langle\checkmark\rangle}{\Longrightarrow} \Omega$ (note that the only state that can occur after $\checkmark$ is Ω). Hence, by Lemma 2.3, $(\langle\tau_r\rangle, Y) \in failures^r(S)$ for any $Y \subseteq \Sigma^{\tau_r}$. Hence, as $X \subseteq \Sigma^{\tau_r}$, it follows that $(\langle\tau_r\rangle, X \cup \{\tau_r\}) \in failures^r(S)$, and thus $(\langle\rangle, X)$ is a failure of the right-hand equation, as required
Otherwise, suppose $(tr \restriction \Sigma^{\checkmark}, X)$ is a member of the right-hand equation. Thus, $tr \neq \langle\rangle$ and $(tr, X \cup \{\tau_r\}) \in failures^r(Sig(Q)) \cup failures^r(Sig(R))$. WLOG, assume $(tr, X \cup \{\tau_r\}) \in failures^r(Sig(Q))$. There are two cases to consider. Firstly, assume $tr \restriction \Sigma^{\checkmark} \neq \langle\rangle$. Hence, trivially it follows that $(tr \restriction \Sigma^{\checkmark}, X)$ is a member of the first clause of the left-hand equation. Otherwise, $tr \restriction \Sigma^{\checkmark} = \langle\rangle$, and hence tr must consist of one or more τ_r's. By Lemma 2.3, $tr = \langle\tau_r\rangle$, and, letting S be such that $Sig(Q) \stackrel{\langle\tau_r\rangle}{\Longrightarrow} S$, $S \stackrel{\checkmark}{\longrightarrow} \Omega$. Hence, $\langle\tau_r, \checkmark\rangle \in traces(Sig(Q))$. Further, since the above holds for any such S, it follows $\checkmark$ cannot be refused after τ_r and thus $X \subseteq \Sigma^{\tau_r}$. Hence, $(\langle\rangle, X)$ is a member of the second clause of the left-hand equation, as required.

$P = Q \,;\, R$:

$$\begin{aligned}
& failures^s(Q \,;\, R) \\
& = \{(tr, X) \mid (tr, X \cup \{\checkmark\}) \in failures^s(Q)\} \\
& \quad \cup \{(tr^\frown tr', X) \mid tr^\frown\langle\checkmark\rangle \in traces(Q) \\
& \qquad\qquad \wedge (tr', X) \in failures^s(R)\} \\
& = \{(tr, X) \mid (tr, X \cup \{\checkmark\}) \in failures^r(Sig(Q) \setminus \{\tau_r\})\} \\
& \quad \cup \{(tr^\frown tr', X) \mid tr^\frown\langle\checkmark\rangle \in traces(Sig(Q) \setminus \{\tau_r\}) \\
& \qquad\qquad \wedge (tr', X) \in failures^s(Sig(R) \setminus \{\tau_r\})\} \quad \langle\text{IH, Theorem 2.4}\rangle \\
& = failures^r((Sig(Q) \setminus \{\tau_r\}) \,;\, (Sig(R) \setminus \{\tau_r\})) \\
& = failures^r(Sig(Q \,;\, R) \setminus \{\tau_r\}).
\end{aligned}$$

$P = Q \,|||\, R$:

In order to prove this case we firstly define, using the definition from [2], the set of all interleavings of two traces s and t, denoted $s \,|||\, t$:

$$\begin{aligned}
s^\frown\langle\checkmark\rangle \,|||\, t^\frown\langle\checkmark\rangle &= \{u^\frown\langle\checkmark\rangle \mid u \in s \,|||\, t\} \\
s^\frown\langle\checkmark\rangle \,|||\, t &= s \,|||\, t \\
s \,|||\, t^\frown\langle\checkmark\rangle &= s \,|||\, t \\
s \,|||\, \langle\rangle &= \{s\} \\
\langle\rangle \,|||\, t &= \{t\} \\
\langle x\rangle^\frown xs \,|||\, \langle y\rangle^\frown ys &= \{\langle x\rangle^\frown u \mid u \in xs \,|||\, \langle y\rangle^\frown ys\} \cup \{\langle y\rangle^\frown u \mid u \in ys \,|||\, \langle x\rangle^\frown xs\}
\end{aligned}$$

Using the above, we can now prove the required result as follows:

$$\begin{aligned}
& failures^s(Q \,|||\, R) \\
& = failures^s((Q \,;\, SKIP) \,|||\, (R \,;\, SKIP))
\end{aligned}$$

$$
\begin{aligned}
&= \{(tr, Y \cup Z) \mid Y \setminus \{\checkmark\} = Z \setminus \{\checkmark\} \\
&\qquad \wedge \exists\, s, t \cdot tr \in s \;|||\; t \wedge (s, Y) \in \mathit{failures}^s(Q\,;\mathit{SKIP}) \\
&\qquad\qquad \wedge (t, Z) \in \mathit{failures}^s(R\,;\mathit{SKIP})\} \\
&= \{(tr, Y \cup Z) \mid Y \setminus \{\checkmark\} = Z \setminus \{\checkmark\} \\
&\qquad \wedge \exists\, s, t \cdot tr \in s \;|||\; t \\
&\qquad\qquad \wedge (s, Y) \in \mathit{failures}^r((\mathit{Sig}(Q)\,;\mathit{BSkip}) \setminus \{\tau_r\}) \\
&\qquad\qquad \wedge (t, Z) \in \mathit{failures}^r((\mathit{Sig}(R)\,;\mathit{BSkip}) \setminus \{\tau_r\})\} && \langle \mathrm{IH} \rangle \\
&= \mathit{failures}^r(((\mathit{Sig}(Q)\,;\mathit{BSkip}) \setminus \{\tau_r\}) \;|||\; ((\mathit{Sig}(R)\,;\mathit{BSkip}) \setminus \{\tau_r\})) \\
&= \mathit{failures}^r(((\mathit{Sig}(Q)\,;\mathit{BSkip}) \;|||\; (\mathit{Sig}(R)\,;\mathit{BSkip})) \setminus \{\tau_r\}) \\
&= \mathit{failures}^r(((\mathit{Sig}(Q)\,;\mathit{BSkip}) \underset{\{\tau_r\}}{\|} (\mathit{Sig}(R)\,;\mathit{BSkip})) \setminus \{\tau_r\}) && \langle \dagger \rangle \\
&= \mathit{failures}^r(\mathit{Sig}(Q \;|||\; R) \setminus \{\tau_r\})
\end{aligned}
$$

The step marked † us justified by observing that forcing synchronisation on τ_r does not alter the failures. This is because, in any state in which the first equation could perform a τ_r, it must be because a $BSkip$ is doing so. However, note that it does not matter if a $BSkip$ is blocked from proceeding by a synchronisation on a ✓ or a τ_r; due to the hiding involved, in such a state both equations can refuse the whole of $\Sigma^{\checkmark,\tau_r}$, as required.

Hence, as this covers all possible cases for P, the required result follows. □

2.3. *Efficiency*

The solution sketched above is certainly efficient with regards to implementation, since it requires only a straightforward substitution to be performed on a process. However, unfortunately, the transformation would cause FDR to produce less efficient representations of the various LTSs, thus slowing down the refinement checking process and causing more memory to be consumed. In order to explain the issue, we firstly briefly review how FDR internally represents labelled transition systems (LTSs).

FDR has two main internal representations of LTSs. The simplest, known as a *low-level* machine, is an explicit graph. FDR's most useful representation is known as a *supercombinator*. A supercombinator takes a number of other LTSs and a set of rules that say how the transitions of the component machines combine together to produce the transitions of the overall machine. For example, a supercombinator for $P \;|||\; Q$, assuming P and Q are represented as explicit graphs, would have: two component machines, P and Q; a rule for each non-✓ event that P or Q can perform to promote the event; a rule that causes the machine to perform a ✓ when both P and Q do. Further, in order to support operators such as $P\,;Q$, these rules are arranged into *formats*. For example, the supercombinator for $P\,;Q$ has two *formats*. The first format represents the rules that are active before P has performed a ✓, whilst the second format is active after P has performed a ✓ (which will have been converted into a τ) and represents how the transitions of Q are promoted. Thus rules can also specify which format they transition into. Further, supercombinators are actually constructed recursively (this is the key to efficiency). For example, assuming that each P_i is a low-level machine, then a single supercombinator is constructed for $(P_1\,;P_2) \;|||\; (P_3\,;P_4)$, rather than three different supercombinators. This is constructed by combining together the rules for the ; and $|||$ operators.

In our particular simulation it is this recursive construction that is the source of the inefficiency. For example, consider $Q = |||_{i:\{1..N\}}(P_i \,;\, SKIP)$: assuming that each of the P_i is represented as a low-level machine, this machine will have 2^N formats since the supercombinator has to have a format that represents Q being in any possible combination of the formats of each $P_i;SKIP$ process. Since it is not possible (or at least, certainly not desirable), to lazily construct formats, this could add significant time and memory consumption to the check.

Recall that the only difference in the denotational semantics between the ✓-as-Signal and ✓-as-Refusable semantics is that the former adds extra failures. In particular, according to Equation 1.1, $failures^s(P)$ contains the extra failures:

$$\{(tr, X) \mid P \xRightarrow{tr^\frown\langle\checkmark\rangle} \Omega, X \subseteq \Sigma\}$$

Note that any process that just offers a ✓ will already have all of the above failures. Therefore, the only processes that the above will add failures to are those that contain a choice between a visible event and a ✓. For example, $SKIPChoice_a$ and $SKIP \;|||\; SKIPChoice_a$ both contain choices between visible events and ✓'s, and thus have extra failures added. However, as processes such as $P \;\Box\; SKIP$ are sufficiently unusual[5], if we only apply the transformation to the relevant portions of a process the increase in cost will be negligible. We formalise this as follows.

Definition 2.6. A process P is *operationally discretionary* iff $P \xrightarrow{\checkmark} \Omega$ and there exists $a \in \Sigma$ such that $P \xrightarrow{a} P'$.

Theorem 2.7. Let P be a process such that for all P' such that $P \xRightarrow{tr} P'$, P' is not operationally discretionary. Then $failures^s(P) = failures^r(P)$.

Proof (Sketch). This can be proven by induction over the process, noting the above observation regarding the extra failures in $failures^s(P)$ as opposed to $failures^r(P)$. □

Checking if every state has a ✓ alongside a visible event available is clearly not feasible due to the time required to check such a property. Therefore, we need to compute a sound over-approximation of the operationally-discretionary property. We can do so by noting that an Independent operator only yields an operationally discretionary process if it has an argument that is $SKIP$ and another argument that is a visible event. Further, a synchronising operator only yields an operationally discretionary process if one of its arguments is operationally discretionary, and the remaining arguments can terminate.

Whilst the above works for a large class of operators, incorporating hiding requires a little more complexity. In particular, consider the processes $P \;\hat{=}\; a \rightarrow SKIP$ and $P \setminus \{Y\} \;\Box\; b \rightarrow STOP$. Note that this process has an operationally-discretionary argument iff $a \in Y$. Hence, to detect the above, we also need to track the set of events that are hidden and thus could result in a $SKIP$ being reached by only τ events. We formalise this in the following definition.

Definition 2.8. A process P has a X*-guarded SKIP* iff: $P \xRightarrow{s^\frown\langle\checkmark\rangle} \Omega$ for $s \in X^*$ iff P is one of the following forms:

1. $P = SKIP$;
2. $P = a \rightarrow Q$, $a \in X$ and Q has an X-guarded $SKIP$;
3. P is either an Independent operator, is a non-deterministic choice, or is a sequential composition, and P has at least one argument that has an X-guarded $SKIP$;
4. P is a synchronising operator such that all of its arguments have X-guarded $SKIP$'s;

[5] As Roscoe notes in [2], Hoare actually banned this process because it appears so unnatural. Thus, whilst our simulation is less efficient on such processes, this is arguably a reasonable price for perversity!

5. $P = Q[\![R]\!]$ and Q has a Y-guarded $SKIP$ where Y is the pre-image of X under R;
6. $P = Q \setminus Y$ and Q has an $(X \cup Y)$-guarded $SKIP$.

A process P has an *unguarded SKIP* iff it has a $\{\}$-guarded $SKIP$. A process P is *problematic* iff either:

1. P is an Independent operator, one argument of P has an unguarded $SKIP$, and there exists an **on** argument P' of P such that $P' \overset{\langle a \rangle}{\Longrightarrow}$ for some $a \in \Sigma$; or
2. P has a problematic process argument.

For example, consider the process $P = a \rightarrow SKIP \square SKIP$. By the above definition, this is rightly considered problematic since P is an independent operator and has an argument with an unguarded $SKIP$. However, $P \setminus \{a\} = SKIP$ and is thus intuitively unproblematic. It is also formally unproblematic since $a \rightarrow SKIP \square SKIP$ has a $\{a\}$-guarded $SKIP$.

Note that the above is indeed an over-approximation. For example, consider the process $P = a \rightarrow SKIP \square b \rightarrow SKIP$. Whilst P is unproblematic, $P \setminus \{a\}$ is identified as problematic, even though $P \setminus \{a\} = b \rightarrow SKIP \rhd SKIP$, which is clearly unproblematic. We believe that such processes should be sufficiently uncommon to ensure that the over-approximation is a useful one.

We now prove that the above computes a sound approximation, as follows.

Theorem 2.9. If P is not problematic then P contains no operationally-discretionary subprocess.

Proof (Sketch). This can be proven by a structural induction over P, noting that in the definition of X-guarded, X is a sound approximation to the set of events that can be τ's. Hence, since a non-problematic process has no Independent operators that have **on** arguments that have unguarded $SKIP$'s and visible events, the required result immediately follows. □

3. Conclusions

In this paper we have discussed the difference between the ✓-as-Refusable and ✓-as-Signal semantics. In Section 2.1, we specified a way of simulating the ✓-as-Signal semantics within the ✓-as-Refusable semantics before proving, in Section 2.2, that the simulation is correct, in that it produces the same denotational value in the ✓-as-Refusable semantics as the ✓-as-Signal semantics. Lastly, in Section 2.3, we discuss how the simulation can be inefficient when considering how FDR internally represents LTSs and proposed a solution that reduces the overhead for all but the most unusual of processes.

We believe that the technique that we have presented in this paper provides a way of simulating the ✓-as-Signal semantics within FDR with a relatively minimal number of changes. We hope to implement the simulation in the context of FDR3, which will be released towards of the end of the year. This will be the first time that the FDR refinement checker has supported checking refinements under different termination semantics, which will hopefully be of wider interest to those who are interested in studying termination in a more theoretical context.

Acknowledgements

We would like to thank Bill Roscoe for several interesting discussions on this work. We would also like to thank the anonymous reviewers for providing many useful comments and interesting discussions.

References

[1] C. A. R. Hoare. *Communicating Sequential Processes*. Prentice-Hall, Inc., Upper Saddle River, NJ, USA, 1985.

[2] A. W. Roscoe. *The Theory and Practice of Concurrency*. Prentice Hall, 1997.

[3] A. W. Roscoe. *Understanding Concurrent Systems*. Springer, 2010.

[4] H. Tej and B. Wolff. A Corrected Failure-Divergence Model for CSP in Isabelle/HOL. In *FME '97: Industrial Applications and Strengthened Foundations of Formal Methods*, volume 1313 of *Lecture Notes in Computer Science*, pages 318–337. Springer, 1997.

[5] Formal Systems (Europe) Ltd. *Failures-Divergence Refinement—FDR 2 User Manual*, 2011.

[6] J. Davies. *Specification and Proof in Real Time CSP*. Distinguished Dissertations in Computer Science. Cambridge University Press, 1993.

Communicating Process Architectures 2013
P.H. Welch et al. (Eds.)
Open Channel Publishing Ltd., 2013

Successful Termination in Timed CSP

Paul HOWELLS [a,1] and Mark D'INVERNO [b]
[a] *Department of Computer Science & Software Engineering, University of Westminster, UK*
[b] *Department of Computing, Goldsmiths, University of London, UK*

Abstract. In previous work the authors investigated the inconsistencies of how successful termination was modelled in Hoare, Brookes and Roscoe's original CSP. This led to the definition of a variant of CSP, called CSP_T. CSP_T presents a solution to these problems by means of adding a termination axiom to the original process axioms. In this paper we investigate how successful process termination is modelled in Reed and Roscoe's Timed CSP, which is the temporal version of Hoare's original untimed CSP. We discuss the issues that need to be considered when selecting termination axioms for Timed CSP, based on our experiences in defining CSP_T. The outcome of this investigation and discussion is a collection of candidate successful termination axioms that could be added to the existing Timed CSP models, leading to an improved treatment of successful termination within the Timed CSP framework. We outline how these termination axioms would be added to the family of semantic models for Timed CSP. Finally, we outline what further work needs to be done once these new models for Timed CSP have been defined. For example, it would then be possible to define timed versions of the new more flexible parallel operators introduced in CSP_T.

Keywords. concurrency, CSP, Timed CSP, CSP_T, process termination

Introduction

In the original *failure-divergence* semantic model for *Communicating Sequential Processes* (CSP) [1,2,3] the incomplete treatment of successful process termination, and in particular parallel termination, permitted intuitively contradictory processes to be defined. For instance, it is possible to define a parallel process that appears to terminate several times before actually doing so – see Section 1 for an example.

Several alternative solutions have been proposed for these problems associated with modelling termination. The solution due to Roscoe [4,5] is seen as the "standard" solution. However, his solution requires termination to be viewed as a new kind of event, known as a "signal", distinct from other events; and only allows asynchronous parallel termination semantics.

Our response was to develop an alternative solution, in the form of a variant of CSP called CSP_T [6,7]. CSP_T solves the original termination problems and introduces three distinct, but related, parallel operators that between them provide a transparent and intuitive means for specifying a range of termination behaviours for networks of processes. Our aim for these operators was to remain as consistent as possible with the original model defined for CSP, whilst giving the system designer *precision* over the extent of parallel interaction (by means of a parametrised synchronisation set) and *flexibility* over the type of parallel termination.

[1]Corresponding Author: *Paul Howells, Department of Computer Science & Software Engineering, University of Westminster, 115 New Cavendish St., London, W1W 6RU, UK.* E-mail: `P.Howells@westminster.ac.uk`.

In this paper, we continue our investigation of the successful termination of processes within the CSP framework, by considering how successful termination is or should be modelled within the timed version of CSP, known as Timed CSP. We begin this process by considering the issues that influence the choice of a process axiom to capture the notion of successful termination within both the untimed and timed CSP worlds.

Following on from this discussion, we present our proposals for how an improved treatment of successful termination can be incorporated into Timed CSP. To achieve this, we shall adopt a similar approach to that taken in formulating our solution to the problems associated with modelling termination in CSP, which led to the definition of CSP_T and introduction of a termination axiom. Since there are several semantic models for Timed CSP (Reed and Roscoe [8,9,10,11]), this requires the definition of a termination axiom for each model. Once this task is complete, our intention is to replace the collection of parallel operators available in these timed versions of CSP with timed versions of those introduced in CSP_T.

Outline of the Rest of the Paper

In Section 1, we provide a brief introduction to CSP_T. We provide an overview of Timed CSP in Section 2. In Section 3 we review the issues that influenced the choice of a termination axiom in CSP_T and present a collection of candidate termination axioms that we propose to add to the existing models of Timed CSP. Related work on improving the modelling of termination in Timed CSP is discussed in Section 4. Finally, we present our conclusions and suggestions for future work in Section 5. (The Appendix contains a summary of the notation and definitions used throughout the paper.)

1. An Introduction to CSP_T[1]

In this section[2],we provide an overview of CSP_T, for a complete description please see [6, 7,13]. (For a summary of the CSP and CSP_T notation and definitions used throughout the paper, see the Appendix.)

In Hoare, Brookes and Roscoe's original CSP [1,2,3,14] there was a well known problem concerning the incomplete treatment of parallel termination by two of original parallel operators: alphabetised $({}_A\|_B)$[3] and asynchronous/interleaving ($|||$). In particular, the type of termination that occurred using these operators was inconsistent and could vary depending on the termination of the processes being combined.

We shall now give an example to illustrate the type of problems that could arise. First recall, that the *successful termination* of a process in CSP (and CSP_T) is modelled by the event *tick* ($\checkmark$). Now consider the following process equivalence derivable within the original CSP:

$$\begin{aligned}(a \to SKIP)|||(b \to SKIP) \equiv \quad & (a \to ((\checkmark \to b \to SKIP) \,\Box\, (b \to \checkmark \to SKIP))) \\ & \Box\, (b \to ((a \to \checkmark \to SKIP) \,\Box\, (\checkmark \to a \to SKIP)))\end{aligned}$$

Clearly the $\checkmark$s on the right hand side cannot be interpreted as the successful termination of the process $(a \to SKIP)|||(b \to SKIP)$, since it continues to perform a, b and $\checkmark$ events. This

[1] This section contains material extracted from Sections 3.1, 5.2 and 5.6 of [6] and Sections 3.2, 3.3 and 3.4 of [7] by permission of Springer and Elsevier, respectively.

[2] This section is identical to Section 1 of [12], a companion paper in these same Proceedings (CPA 2013). This duplication, by permission of the Editors, is to let both papers be self-contained.

[3] Here and henceforth, A and B represent the alphabets of the processes being composed. For example, as in $P_A\|_B Q$, where $A = \alpha(P)$ and $B = \alpha(Q)$ respectively.

illustrates how the original semantics for CSP could gave rise to these obviously undesirable and intuitively contradictory processes.

Several solutions have been proposed to this problem, including Tej and Wolf [15], Roscoe [4,5], Hoare and He [16] and the authors's own solution CSP_T [6]. For a detailed comparison of these solutions the interested reader is referred to [6].

Our starting point in defining CSP_T was the original failure-divergence model developed by Hoare, Brookes and Roscoe [2]. Our aim in modifying this model was to provide a more robust treatment of termination through the consistent and special handling of ✓ by the language (processes and operators) and semantics (failures and divergences).

For CSP_T, this was achieved by defining a new process axiom that captured our view of termination:

$$t \neq \langle\rangle \ \wedge \ (s^\frown\langle\checkmark\rangle^\frown t, \varnothing) \in F \ \Rightarrow \ s \in D \qquad \text{(T1)}$$

where s and t are traces, F and D are the failure and divergence sets respectively of a process. This axiom means that if a process indicates that it has terminated (by means of the ✓) but continues to perform events (t), then it must have started diverging before it performed the ✓ (i.e. $s \in D$). For the rational behind this axiom, see Section 3.1.

This new termination axiom resolves the termination issues of the original semantics, and it is added to the existing CSP process axioms (D1) to (N5), see Appendix, to define CSP_T. Note that our view of tick (✓) is consistent with Hoare's, i.e. that it is a normal event, and have not adopted Roscoe's view that it is a special *signal* event. In doing this, we have defined a sub-model of the original failure-divergence model N_T such that, within this sub-model, all processes are well-behaved with respect to termination. In addition, three new forms of parallel operators were defined for CSP_T, each with a different form of termination semantics, as replacements for the original ones. We now introduce the language and model for CSP_T; we begin by introducing the three new parallel operators.

1.1. Parallel Operators of CSP_T

Our three new parallel operators are defined to be used as replacements for the original synchronous ($\|$), interleaving ($|||$) and alphabetised (${}_A\|_B$) parallel operators. It is necessary to define replacements for $|||$ and ${}_A\|_B$, as they do not satisfy (T1). These new parallel operators are *generalised* (or *interface*) style parallel operators, i.e. are parameterised by the set of events the processes are required to synchronise on. Each of these three operators has a distinct type of parallel termination semantics, and thus are distinct operators, see [7] for details. We call them *synchronous*, *asynchronous* and *race*, that we define here.

Synchronous: requires the successful termination of both P and Q; and the synchronisation of their termination, that is, ✓.

Asynchronous: requires the successful termination of both P and Q; and P and Q terminated asynchronously, i.e. they do not synchronise on ✓. (Roscoe [4,5] refers to this type of parallel termination semantics as *distributed termination*.)

Race: requires the successful termination of either P or Q asynchronously. Successful termination fails to occur only if both P and Q fail to terminate. Unlike synchronous and asynchronous termination where the environment observes a single ✓, under *race* termination semantics it can observe any ✓ that is performed by P or by Q. Consequently, the first ✓ the environment observes is taken as representing the termination of the parallel composition and whichever of P or Q did not terminate, i.e. did not preform the ✓, is aborted. Hence, with this type of termination semantics termination occurs as soon as either P or Q does so.

Note that in [4,5] Roscoe chooses to reject race termination semantics, due to the problems of dealing with the non-terminated process, e.g. the need for some powerful mechanism to manage its termination. We however, believe that it is preferable to at least offer system specifiers the choice of using this type of termination semantics and let them resolve these issues, rather than ban it outright. We believe that this is not against the spirit of CSP, since there are other features available in CSP that also raise similar implementation issues, e.g. the various interrupt style operators.

The CSP_T parallel operators with these different forms of parallel termination semantics, are given in Table 1. Each operator is parametrised by a *synchronisation set* (Ω, Δ, Θ), that is the set of events on which the combined processes are required to synchronise. Events which are not in the synchronisation set but that can be performed by either P or Q or both are asynchronous events.

Table 1. CSP_T parallel operators.

Termination Semantics	Operator	Synchronisation Set	Notes
Generalised	$P\|_\Omega Q$	$\varnothing \subseteq \Omega \subseteq \Sigma$	
Synchronous	$P\|_\Delta Q$	$\{\checkmark\} \subseteq \Delta \subseteq \Sigma$	$\checkmark \in \Delta$
Asynchronous	$P\|\|_\Theta Q$	$\varnothing \subseteq \Theta \subseteq \Sigma - \{\checkmark\}$	$\checkmark \notin \Theta$
Race	$P\|_\Theta Q$	$\varnothing \subseteq \Theta \subseteq \Sigma - \{\checkmark\}$	$\checkmark \notin \Theta$

We define a generalised parallel operator (denoted $P\|_\Omega Q$) that is used to define the operators with synchronous and race termination semantics. To distinguish this from the synchronous termination operator we denote the synchronisation set by Ω rather than Δ. The synchronous parallel operator $\|_\Delta$ is simply the generalised one $\|_\Omega$, with the constraint that $\checkmark \in \Delta$, thus ensuring synchronous termination. The race termination operator $|_\Theta$ can also be defined using $\|_\Omega$ and *SKIP* as follows:

$$P|_\Theta Q \;\widehat{=}\; (P\|_\Theta Q); SKIP \qquad [\varnothing \subseteq \Theta \subseteq \Sigma - \{\checkmark\}]$$

where Σ is the set of all events and, for CSP_T, includes $\checkmark$. Note that CSP_T does not have the law $(P; SKIP = P)$, otherwise the above would mean the race and generalised operators were the same. To illustrate how this definition behaves, consider the process $(P\|_\Theta Q); R$, with $\Theta \subseteq \Sigma - \{\checkmark\}$. Whichever of the two $\checkmark$s R (the environment) observes first is taken as representing the termination of $P\|_\Theta Q$ and, hence, R proceeds to execute; whichever of P or Q did not terminate is aborted.

The asynchronous parallel operator $|||_\Theta$ is defined independently, as its form of termination semantics is not compatible with $\|_\Omega$, i.e. it cannot be used to define $|||_\Theta$. Both process operands must terminate for $|||_\Theta$ to terminate, so it is not the same as $\|_\Theta$ (see previous paragraph). Both processes must terminate for $\|_\Delta$, if $\Delta = \Theta \cup \{\checkmark\}$, but they *synchronise* on their $\checkmark$s (i.e. a non-terminating process can prevent the other operand from terminating, which is *not* the case for $|||_\Theta$).

Here are examples of the three parallel operators of CSP_T, that illustrate the difference between $|_\Theta$ and the two others (which for these examples, $|||_\varnothing$ and $\|_\varnothing$, are the same):

$$\begin{aligned} a \to SKIP\|_\varnothing b \to SKIP &\equiv a \to SKIP|||_\varnothing b \to SKIP \\ &\equiv (a \to b \to SKIP) \,\square\, (b \to a \to SKIP) \\ a \to SKIP|_\varnothing b \to SKIP &\equiv (a \to (SKIP \sqcap (SKIP \,\square\, b \to SKIP))) \\ &\quad \square\, (b \to (SKIP \sqcap (SKIP \,\square\, a \to SKIP))) \end{aligned}$$

It is important to note that we do not include the generalised parallel operator $\|_\Omega$ in the language of CSP_T, since this would defeat the purpose of the whole exercise. Since, it would again result in inconsistent processes, similar to the one given above in Section 1, with multiple ticks in its traces. However, by restricting its use to defining $\|_\Delta$ and $|_\Theta$ we ensure these forms of processes do not occur. Full details, including the operational semantics and semantic functions, can be found in [6,7].

1.2. The Model and Language of CSP_T

The model for CSP_T, N_T is defined by adding the Termination axiom (T1) to the process axioms (D1) to (N5) of the original CSP model N, given in the Appendix. Hence, in N_T the failure and divergence sets of a process are defined as for N, except that they also satisfy the process axiom (T1) as well as (D1) to (N5). The semantic functions $\mathcal{F}$ and $\mathcal{D}$ for N_T are the same as for N, and are given in [6].

The language for CSP_T is the same as that of the original CSP, but uses the more recent form of relational renaming instead of functional renaming and uses the three new parallel operators as replacements for $\|$, $|||$ and ${}_A\|_B$, see Table 2. It is defined as follows:

$$P ::= \perp \mid STOP \mid SKIP \mid a \rightarrow P \mid P \sqcap P \mid P \Box P \mid P;P \mid P \backslash a$$
$$\mid P[\![R]\!] \mid \mu p.F(p) \mid p \mid P \|_\Delta P \mid P |||_\Theta P \mid P|_\Theta P$$

where $a \in \Sigma - \{\checkmark\}$. $\perp$ is the divergent process (can be defined as $\mu p.p$). $STOP$ is the deadlocked process and $SKIP$ is the successfully terminating process. $a \rightarrow P$ is action prefix. $P \sqcap Q$ is nondeterministic choice and $P \Box Q$ is deterministic choice. $P;Q$ is sequential composition. $P \backslash a$ is event hiding. $P[\![R]\!]$ is action (relational) renaming, with the usual constraint [4,5] applying to the renaming relation R with respect to $\checkmark$, i.e. that no other event is mapped to it or that it is mapped to another event. p is a process variable, $\mu p.F(p)$ is recursion and in the definition of $F(p)$, only the above processes and operators can be used. $P \|_\Delta Q$, $P |||_\Theta Q$ and $P|_\Theta Q$ are the generalised synchronous, asynchronous and race parallel operators respectively. The processes and operators of CSP_T are *well-defined* and *well-behaved* in N_T.

Table 2. CSP_T replacements for $\|$, $|||$ and ${}_A\|_B$.

Termination Semantics	$\vert\vert\vert$	${}_A\vert\vert_B$	$\vert\vert$
Synchronous ($\vert\vert_\Delta$)	$\Delta = \{\checkmark\}$	$\Delta = (A \cap B) \cup \{\checkmark\}$	$\Delta = (A \cup B) \cup \{\checkmark\}$
Asynchronous ($\vert\vert\vert_\Theta$)	$\Theta = \varnothing$	$\Theta = (A \cap B) - \{\checkmark\}$	$\Theta = (A \cup B) - \{\checkmark\}$
Race ($\vert_\Theta$)	$\Theta = \varnothing$	$\Theta = (A \cap B) - \{\checkmark\}$	$\Theta = (A \cup B) - \{\checkmark\}$

1.3. Summary of Differences between CSP and CSP_T

The essential differences between Hoare, Brookes and Roscoe's original CSP and CSP_T can be summarised as follows:

- Termination axiom (T1) has been added to the original axioms (D1) - (N5). See Section 3.1 and the Appendix for details and explanation.
- Use relational renaming ($P[\![R]\!]$), instead of functional and inverse renaming.
- Replaced the interleaving ($|||$), synchronous ($\|$) and alphabetised (${}_A\|_B$) parallel operators with generalised synchronous ($\|_\Delta$), asynchronous ($|||_\Theta$) and race ($|_\Theta$) operators.

For a more detailed account of the development of CSP_T, the interested reader is referred to our previous work [6,7]. In the next section we provide an overview of the new concepts introduced for Timed CSP.

2. An Overview of Timed CSP

In this section we provide a brief introduction to the new notions introduced by moving from untimed to Timed CSP. We begin by outlining the different models for Timed CSP (Section 2.1), the timed versions of traces, refusals and failures and the new notion of stability (Section 2.2). Finally, we discuss the different forms of parallel operator used in the existing versions of Timed CSP and their termination behaviour (Section 2.3).

2.1. Timed CSP Models

Timed CSP was first introduced by Reed and Roscoe [8] and is the timed extension of Hoare's CSP [3]. The only extra component added to the original language of CSP was the delay process $WAIT\ t$, this process introduces an explicit delay of $t \geq 0$ time units (usually seconds). $WAIT\ t$ is a delayed form of *SKIP*, which successfully terminates after time t, where $WAIT\ 0 = SKIP$.

Since the introduction of Timed CSP, it has been developed by the addition of different semantic models [9], additional operators [17], proof systems [18,19,20,21,22] and the notion of timewise refinement [23] and operational semantics [24]. For general introductions to Timed CSP see the work of Davies and Schneider [17,25,26,27].

There are several semantic models for Timed CSP, developed by Roscoe and Reed [8,9, 10,28,29,11]. The models form a hierarchy, see Figure 1, ordered by the information content of the observations that can be made in each model. The untimed models occupy the lowest position (the inner diamond of Figure 1) in the hierarchy; above them (the outer diamond) are the timed models. The lowest model is the Untimed Trace model M_T, and the highest (most complex and comprehensive) is the Timed Failure-Stability model TM_{FS}. The mathematics for the semantic models for untimed CSP have largely been based on lattice theory, whereas for Timed CSP it is topology [9,30], and in particular complete (ultra) metric spaces.

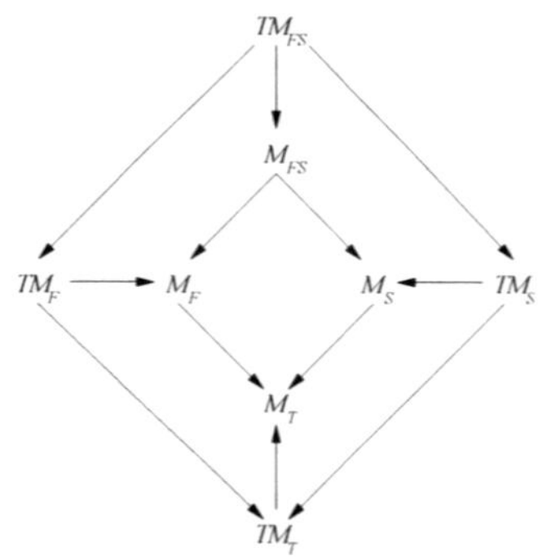

Figure 1. Reed's Hierarchy of Timed and Untimed CSP models. The suffices T, S and F stand for Traces, Stability and Failures respectively.

2.2. Timed Traces, Refusals, Failures and Stability

There are several new notions that are central to the semantics of Timed CSP: *timed events*, *timed traces*, *timed refusal sets* and *stability values*, which we now describe.

2.2.1. Timed Events and Traces

A *timed trace* is a finite sequence of observable events performed by a process, such that each event is labelled with the time at which it occurs. The events in the sequence are ordered chronologically. There are two forms of timed traces, one for the TM_F and TM_{FS} models, and one for the TM_T and TM_S models.

The set of all observable events is Σ. The time domain is the non-negative reals $TIME = [0, \infty)$. A *timed event* is an ordered pair (t, a), where $a \in \Sigma$ and $t \in TIME$. The set of all timed events is $T\Sigma = TIME \times \Sigma$. The set of all timed traces for the TM_F and TM_{FS} models is defined as follows:

$$(T\,\Sigma)^*_{\leq} = \{s \in T\,\Sigma^* \mid \langle (t_1, a_1), (t_2, a_2)\rangle \preccurlyeq s \Rightarrow t_1 \leq t_2\} \qquad \text{(Timed Traces)}$$

where $s_1 \preccurlyeq s_2$ iff s_1 is a subsequence of s_2.

In addition $\Sigma(s)$ is the set of events occurring in s. If $s, w \in (T\,\Sigma)^*_{\leq}$ then define $s \cong w$ if and only if s and w are the same, *except* that the order of events at the same time may differ.

In both of the TM_T and TM_S models it is necessary to record when an event is performed as soon as it becomes available. This is required to provide a correct treatment of hiding ($\backslash$) in these models, because a hidden event is required to occur as soon as is possible. (For the other timed models TM_F and TM_{FS} this information can be inferred from the timed refusals associated with the trace.) This information is indicated in a trace for these models by denoting the event with a circumflex, e.g. $\hat{a}$. This set of events is denoted by: $\hat{\Sigma} = \Sigma \cup \{\hat{a} \mid a \in \Sigma\}$. This is then used to define timed traces for these two models as follows:

$$(T\,\hat{\Sigma})^*_{\leq} = \{s \in T\,\hat{\Sigma}^* \mid \langle (t_1, a_1), (t_2, a_2)\rangle \preccurlyeq s \Rightarrow t_1 \leq t_2\} \qquad \text{(Timed Traces)}$$

In these models, it is also the case that when an $\hat{a}$ event occurs, it can also occur normally as simply an a event. To facilitate this, the function $\tilde{s}$ is defined over these types of traces and simply removes the circumflexes from all of the $\hat{a}$ events in s.

To illustrate this point consider the process $WAIT\ 1; (a \rightarrow P)$. For this process, if a occurs at time 1, then it has occurred as soon as it becomes available and this information is captured by the trace $\langle (1, \hat{a})\rangle$; it also results in the trace $\langle (1, a)\rangle$. However, if a occurs after time 1, say time 2, then it has not occurred when first available, so results in traces $\langle (1, \hat{a})\rangle$ and $\langle (2, a)\rangle$.

2.2.2. *Timed Refusals and Failures*

In the untimed CSP models that use *failures*, (s, X), to denote processes, the refusal set X represents the set of events that may be refused after the process has performed the trace s. In the timed models TM_F and TM_{FS} a *timed failure* not only represents what a process may refuse after performing a timed trace, but can also include a record of the refusals as the trace is being performed – for example, before the first event is performed, during the time between consecutive events or after the final event of the trace. Consequently, a timed *refusal set* is a (finite) union of "initial", "intermediate" and "final" sets of timed *refusal tokens* that each describe refusal information (with timings) at various stages during the execution of the associated timed trace.

An additional factor (introduced by Reed and Roscoe [9,11]) that affects the definition of timed refusal sets is the "realism" requirement: a process can only change state a finite number of times during a finite time period. This is known as *finite variability*. The effect of this requirement on timed refusal sets is that they can only be formed from a finite number of refusal tokens, as each one corresponds to a state change.

A *timed failure* is then straightforwardly defined as a timed trace combined with a timed refusal. This represents an observation of a process performing the timed trace while refusing sets of events during the time intervals described by the timed refusal.

These informal notions are captured by the following formal definitions:

$$I : TINT = \{\, [l(I), r(I)) \mid 0 \leq l(I) < r(I) < \infty \,\} \qquad \text{(Time Intervals)}$$

$$RT : RTOK = \{\, I \times X \mid I \in TINT \wedge X \in \mathbb{P}(\Sigma) \,\} \qquad \text{(Refusal Tokens)}$$

$$\aleph : RSET = \{\, \bigcup Z \mid Z \subseteq RTOK \wedge Z\, finite \,\} \qquad \text{(Refusal Sets)}$$

$$(s, \aleph) : TFAIL = \{\, (s, \aleph) \mid s \in (T\,\Sigma)^*_{\leq} \wedge \aleph \in RSET \,\} \qquad \text{(Time Failures)}$$

A *time interval* is a finite half-open time interval. The set of all time intervals is *TINT*[4]. A *refusal token RT* is a set product, $I \times X$, where I is a time interval and X is a set of events that

[4]In this definition $r(I)$ and $l(I)$ are the projection functions that return the two values for a time interval I.

can be refused *continuously* during the time interval. The set of all refusal tokens is *RTOK*. A *timed refusal* $\mathcal{N}$ is a finite union of refusal tokens. The set of all timed refusals is *RSET*. In summary, a *timed refusal* represents the times at which events may be refused during the execution of a timed trace, e.g. a set of (*time*, *event*) pairs. A *timed failure* $(s, \aleph)$ is a pair composed of a timed trace s and a timed refusal $\aleph$.

There are two standard functions (see [11]) defined on timed refusals: $I(\aleph)$ returns the time interval covered by the refusal set $\aleph$ and $\Sigma(\aleph)$ returns the set of events in $\aleph$. These two functions are defined as follows:

$$I(\aleph) = \{ t \in [0, \infty) \mid \exists a \in \Sigma.(t, a) \in \aleph \}$$
$$\Sigma(\aleph) = \{ a \in \Sigma \mid \exists t \in [0, \infty).(t, a) \in \aleph \}$$

2.2.3. Stability

In Timed CSP, *stability* is used to model the internal activity of a process. It is the dual of divergence, as is used in the failure-divergence model of CSP. A process is defined as being *stable* once it has ceased all internal activity. When a process is stable it cannot change state without performing an external event. The *stability value* associated with an observation (timed trace or failure) of a process is the earliest time by which all internal activity of the process is guaranteed to have stopped. It is formally defined to be the least upper bound of all the times when the process becomes stable given the observation. A process which diverges has a stability value of ∞. α is usually used to denote stability values as follows:

$$\alpha : TSTAB = TIME \cup \{\infty\} \qquad \text{(Stability Values)}$$

Stability values are not used in either the TM_T timed traces or TM_F timed failures models. However, in the TM_S stability model, a stability value is associated with every timed trace. For example, (s, α) where s is a timed trace and α is the stability value for the trace s, i.e. the time by which internal activity has ceased after performing s. Similarly, for the TM_{FS} failure-stability model, a stability value is associated with every timed failure. For example, $(s, \alpha, \aleph)$ where α is the stability value associated with the failure with trace s and refusal $\aleph$.

2.3. Termination and Parallel Operators in Timed CSP [5]

We now discuss the different forms of parallel operators used in the different versions of Timed CSP. Although the termination semantics of ||| can result in inconsistent processes in untimed CSP (as shown in Section 1), it has been used to advantage by Davies and Schneider [18,17,22] to define event interrupt and timed interrupt (*tireout*) operators in Timed CSP. These operators were originally defined using ||| and ";" in such a way to ensure that ✓s only occur at the end of traces. However, in Schneider's later treatment [27], these operators are defined directly. The problem of inconsistent termination semantics associated with the use of the original version of ||| in Timed CSP suggests that this form of ||| should not be used in CSP or Timed CSP explicitly, but at best only as a means to define other operators.

The use of a *generalised* or *interface* style parallel operator is now standard within the CSP community, examples in (untimed) CSP can be found in [5,15,27]. A version has also been defined for the timed failures model TM_F for Timed CSP by Davies [18] called the *communicating* parallel operator. However, there are slight differences between the definitions used by various authors. For example, whether ✓ is an element of the synchronisation set.

In their early work on Timed CSP, Davies and Schneider [26] also use synchronous termination semantics for all their parallel operators in Timed CSP. Their method is similar to

[5]This section contains material extracted from Section 6.4 of [6] by permission of Springer.

the one used by the authors in CSP_T. In particular, they use their generalised parallel operator $P[|X|]Q$ with $X = \{\checkmark\}$ to re-define $|||$. In the definition of ${}_A\|_B$ they require $\checkmark$ to be implicitly included in A and B. However, they still use the old version of $|||$ to define the event and time interrupt operators and have not added a termination axiom.

3. Factors Influencing the Definition of a Termination Axiom

In this section we begin by summarising the factors and issues the authors considered relevant when defining and selecting a termination axiom for CSP_T. This was done in relation to Brookes and Roscoe's [2] original CSP model N. (For full details see [6]). We shall then consider the factors that we believe should be considered when selecting a termination axiom for Timed CSP, in relation to the Reed and Roscoe Timed CSP models of Figure 1.

3.1. Factors in CSP_T[6]

It was our intention that in CSP_T a $\checkmark$ signifies the successful termination of any (non-divergent) process, whether it is a sequential or parallel process. This means that if a $\checkmark$ occurs at the end of the trace of a (non-divergent) parallel process, then it has *successfully terminated.* Our aim was to ensure that a $\checkmark$ signified not just sequential process termination but also parallel termination, which was not the case in the original CSP. This is captured by the following property on traces:

Definition 1: A process trace satisfies the *$\checkmark$-requirement* if a $\checkmark$ only occurs at the end of the trace.

We extend this definition to processes as follows:

Definition 2: A process satisfies the *$\checkmark$-requirement* if all its traces satisfy the *$\checkmark$-requirement.*

Next, we considered how to capture this intuitive idea by considering four candidate termination axioms. The selected one was added to the existing CSP process axioms (D1) to (N5), see Appendix. The consequence of adding our chosen axiom to the existing process axioms was to exclude from this new failure-divergence model all processes whose traces do not satisfy our $\checkmark$-requirement. The resulting CSP_T model (N_T) is a sub-model of the original CSP failure-divergence model (N), such that, within this new sub-model all processes satisfy the $\checkmark$-requirement.

In our search to define an axiom that captures our notion of successful termination, it became clear that we needed to decide what types of processes the axiom should be applied to. Thus, we needed to resolve the following questions.

- Should the $\checkmark$-requirement apply only to those processes which can never diverge (i.e. those with no divergent traces) or to all processes?
- If it does apply to all processes should it apply to all traces of the process or only to the non-divergent ones?

It is necessary to consider the $\checkmark$-requirement in relation to divergence because in the original CSP model N (see [2]), a diverging process can perform any and every trace; in particular it can perform traces which do not satisfy the $\checkmark$-requirement. Hence, it is necessary to decide to which processes and traces the $\checkmark$-requirement should be applied. We considered three main types of termination axioms:

[6]This section contains material extracted from Sections 4.1 and 4.2 of [6] and Section 3.1 of [7] by permission of Springer and Elsevier, respectively.

1. those that apply only to non-divergent processes;
2. those that apply to divergent and non-divergent processes;
3. those that apply only to the non-divergent traces of both divergent and non-divergent processes.

From these three cases we formalised the following four termination axioms: (TA1) corresponds to the first case, (TA2) and (TA3) are alternatives corresponding to the third case and (TA4) corresponds to the second case.

$$t \neq \langle\rangle \wedge D = \varnothing \wedge (s^\frown\langle\checkmark\rangle, \varnothing) \in F \;\Rightarrow\; (s^\frown\langle\checkmark\rangle^\frown t, \varnothing) \notin F \qquad \text{(TA1)}$$

If the divergent set of a process is empty and $(s^\frown\langle\checkmark\rangle, \varnothing) \in F$ (i.e. $s^\frown\langle\checkmark\rangle$ is a trace) then $s^\frown\langle\checkmark\rangle^\frown t$ is not a trace for any non-null extension t (i.e. no more events can be performed after a ✓ has occurred). However, if divergence is possible (i.e. $D \neq \varnothing$), there are no constraints on what can happen after a ✓ (even for non-divergent traces).

$$t \neq \langle\rangle \;\wedge\; (s^\frown\langle\checkmark\rangle^\frown t, \varnothing) \in F \;\Rightarrow\; s^\frown\langle\checkmark\rangle \in D \qquad \text{(TA2)}$$

If a process can perform the trace $s^\frown\langle\checkmark\rangle^\frown t$, where $t \neq \langle\rangle$, then the process must have started diverging no later than after performing the ✓.

$$t \neq \langle\rangle \;\wedge\; (s^\frown\langle\checkmark\rangle^\frown t, \varnothing) \in F \;\Rightarrow\; s \in D \qquad \text{(TA3)}$$

If a process can perform the trace $s^\frown\langle\checkmark\rangle^\frown t$, where $t \neq \langle\rangle$, then the process must have started diverging no later than after performing the s (i.e. before the ✓). A consequence of this is that after such a process has performed s, no matter what (if any) further events it performs, it can always refuse any set of events.

$$t \neq \langle\rangle \wedge (s^\frown\langle\checkmark\rangle, \varnothing) \in F \;\Rightarrow\; (s^\frown\langle\checkmark\rangle^\frown t, \varnothing) \notin F \qquad \text{(TA4)}$$

If $s^\frown\langle\checkmark\rangle$ is any trace, then the process cannot perform any further events after the ✓. Intuitively this means that once a ✓ has been performed the process stops, even if it was diverging at the time.

3.1.1. Comparison of the Four Termination Axioms

Having defined our four candidate termination axioms, we investigated their potential effect on various forms of both divergent and non-divergent parallel processes.

The first axiom, (TA1) reflects the view that if it is possible for a process to diverge then we do not expect it to be well-behaved with respect to successful termination. (TA2) and (TA3) capture the idea that if a process can diverge after some trace it is still required to be well-behaved with respect to successful termination as long as it is not actually diverging. However, when it is diverging it is allowed to *misbehave* with respect to successful termination. (TA4) captures the view that a process is still required to be well-behaved with respect to successful termination even when it is diverging.

Reflecting on these, (TA4) is immediately incompatible with Hoare's original axioms since it excludes divergent processes and we can abandon it. (Recall that a guiding principal is that we want to minimise any change to the original semantics.) However, (TA1), (TA2) and (TA3) are all compatible. One possible reason for adopting (TA1) is that it captures the view that if a process can diverge then we do not care whether it is well-behaved with respect to termination. At first, this seems appealing because we might argue that we do not care whether it violates the ✓-requirement given that it can diverge. However, the reason for not adopting it is that it is too weak to enforce the ✓-requirement over all processes. We therefore reject (TA1). However, this does not happen with (TA2) or (TA3) and they are both suitable for the following reasons.

- They enforce the ✓-requirement on all non-divergent traces of a process irrespective of whether it diverges or not, which (TA1) does not. This view seems to capture more closely our intuitive ideas which originally lead to the ✓-requirement and the search for a termination axiom.
- If a process can perform a trace of the form $s \frown \langle \checkmark \rangle \frown t$, where $t \neq \langle \rangle$, then we now know that it is diverging; whereas without this axiom this was not the case. So the only processes which can now exhibit the undesirable behaviour of not satisfying the ✓-requirement are those which do so because they are diverging, which is consistent with their interpretation as the most chaotic process.

The difference between (TA2) and (TA3) is at what point the process must have started diverging to produce the undesired trace. In (TA2) the process must have started no later than after the ✓ whereas in (TA3) it must have started before the ✓. In this sense we view (TA3) as being stronger than (TA2), and in fact we can prove (see [13]) that by adopting (TA3) in favour of (TA2), that all processes which are equivalent to, or contain as a subprocess the following process, would be excluded:

$$\checkmark \rightarrow \perp$$

So our final choice depends on whether we wish to eliminate processes of the form $\checkmark \rightarrow \perp$. The intuitive meaning of this process is that it first successfully terminates then diverges, which is at least intuitively contradictory, if nothing else. Therefore, on this basis, we choose to adopt the termination axiom (TA3) in favour of (TA2).

To summarise, the main reasons for choosing (TA3) (renamed T1) was that it solved the termination problems[7], most closely captured our intuitive views of termination and achieved this with only a minimal modification to the original semantics. In particular, (T1) reflects the view that even if a process can diverge after some trace it is still required to be well-behaved with respect to successful termination. That is to say it satisfies the ✓-requirement when it is not actually diverging; but when it is diverging it is allowed to *misbehave* with respect to successful termination. In this respect, (T1) also maintains the original view of Hoare, Brookes and Roscoe that a diverging process cannot recover by performing a ✓.

This concludes the outline of the process by which the authors selected a termination axiom for an untimed version of CSP, i.e. CSP_T. Once the (T1) axiom was chosen, it was necessary to prove that all of the processes and operators of CSP_T satisfied it and that this new model N_T was consistent and well-defined – see [6] for details. In the next section we discuss the issues that need to be considered in selecting termination axioms for each of the timed models for Timed CSP.

3.2. Factors in Timed CSP

Clearly the new notions of timed traces, timed refusals and stability values that are central to Timed CSP, need to be taken into account when defining suitable termination axioms to add to the various timed models. Of these, probably the most significant is the use of a stability value (the time at which a process ceases internal activity, i.e. becomes "stable") to model divergence, as compared to a divergence trace (a trace after which infinite internal activity is possible) used in CSP.

In addition, in the timed models that include stability values (Timed Stability TM_S and Timed Failures-Stability TM_{FS}), there appears to be an implicit notion of *"immediate stability at termination"*. That is the stability value for a process that performs a ✓ is the same as the time at which the ✓ was performed. This can be inferred from the stability values associated with traces that end in ✓, as defined in the semantic functions for *SKIP*, *WAIT t*

[7]For example, the intuitively contradictory processes described in Section 1 are eliminated.

and ";" (see [11]). This contrasts with the stability values defined for other traces, as defined in the semantic function for $\rightarrow$, where the delay constant "δ", associated with non-$\checkmark$ events, is added to the time of the event to produce the stability value. Thus, any termination axiom that is to be introduced must be consistent with these additional requirements as well as our earlier $\checkmark$-requirement.

Another constraint on the introduction of a termination axiom relates to the addition of various *timeout* and *interrupt* operators added to Timed CSP by Davies and Schneider [18, 17,22]. The definition of these operators [17] relies on the use of $|||$, thus the introduction of a termination axiom into the timed models would appear to prohibit the definition of these operators. However, it is the *race* termination semantics of $|||$ that is used to define these additional operators. Thus, if the race termination semantics version $|_{\varnothing}$ of $|||$ was introduced into the language of Timed CSP and used to define the timeout and interrupt operators, no inconsistency would arise.

However, recalling the following definition of $|_{\Theta}$:

$$P|_{\Theta}Q \mathrel{\widehat{=}} (P\|_{\Omega}Q); SKIP$$

It is important to note that this would only be the case in versions of Timed CSP where the *time postulates* [9], concerning the lower bound delay, δ, on the length of the time interval between any two consecutive events in the history of a sequential process, is zero when the first event is $\checkmark$, and that hidden events occur as soon as they become available. In the case of $P|_{\Theta}Q$, this is the $\checkmark$ of the first process in the sequential composition, i.e. $P\|_{\Omega}Q$. These two conditions ensure that the race termination semantics version of $|||$, i.e. $|_{\varnothing}$, will not introduce an additional delay into the timeout and interrupt operators.

It is currently not clear to the authors whether there are other factors, to those given above, which would need to be considered when defining termination axioms for the timed models. However, based on this preliminary analysis we now present *candidate* termination axioms for the four timed models that would be a starting point for any future research aimed at providing a consistent treatment of termination within these Timed CSP models (Section 5).

3.3. Adding Termination Axioms to the Timed Models

In this section, we outline how termination axioms could be added to the process axioms of the timed semantic models from Reed's hierarchy, shown in Figure 1. In the following sections, we present candidate termination axioms for each of them: *Timed Traces* (TM_T), *Timed Stability* (TM_S), *Timed Failures* (TM_F) and *Timed Failures-Stability* (TM_{FS}). Reed's original axioms for these models, which do not consider termination, can be found elsewhere [9,11].

If a termination axiom (TA) were added to a Timed CSP model, then it would be necessary to prove that each new model, for example ($TM_{FS} + TA, d$), is a well-defined metric space, where the processes satisfying the axioms $TM_{FS} + TA$ form the "set of points" of the metric space. The "metric", d, is a function that measures the "distance" between two points, i.e. processes. The distance metric used by Reed and Roscoe [9,11] is based on the length of time it takes to tell two processes apart, based on the observations made (e.g. timed traces and timed refusals). In essence, if two processes can be distinguished very quickly, then the "distance" between them is large; but if it takes a long time to distinguish between them, then the "distance" is small.

Processes are formally denoted by their observable behaviours, $\mathcal{S}$, where what is observed depends on the semantic model (e.g. $\mathcal{S}$ is a set of timed failures in the TM_F model). These observations, $\mathcal{S}$, are elements in the point set of the metric space, provided they satisfy the axioms of the model under consideration. So, a process P is denoted by a set of observations $\mathcal{S}$ that satisfy the axioms for the particular semantic model.

3.3.1. A Termination Axiom for TM_T

The Timed Trace model TM_T, models processes by denoting them by sets of timed traces s. The termination axiom that we believe is the most suitable candidate for this model is the following:

$$s \in \mathcal{S} \wedge \checkmark \in \Sigma(s) \;\Rightarrow\; \exists\, t \in [0, \infty) \bullet \tilde{s} = s' ^\frown \langle (t, \checkmark) \rangle \wedge \checkmark \notin \Sigma(s') \qquad (TA_T)$$

The TA_T axiom[8] captures the notion that if a process can perform the timed trace s and a ✓ has occurred in the trace then:

(i) only one ✓ can have occurred; and

(ii) it must have been the last event.

In other words, this axiom would then ensure that the ✓-requirement was enforced on the timed trace s.

3.3.2. A Termination Axiom for TM_S

The Timed Stability model TM_S models processes by denoting them by sets of pairs (s, α) consisting of a timed trace s and an associated stability value α. The termination axiom we believe is the most suitable candidate for this model is the following:

$$(s, \alpha) \in \mathcal{S} \wedge \checkmark \in \Sigma(s) \;\Rightarrow \tilde{s} = s' ^\frown \langle (\alpha, \checkmark) \rangle \wedge \checkmark \notin \Sigma(s') \qquad (TA_S)$$

The TA_S axiom captures the notion that if a process can perform the timed trace s with a stability value of α and a ✓ has occurred in the trace then:

(i) the ✓-requirement is satisfied; and

(ii) the ✓ occurred at time α, the time of stability.

This axiom and the trace prefix closure axiom together would ensure that the ✓-requirement was enforced and that stability on termination was maintained.

3.3.3. A Termination Axiom for TM_F

The Timed Failures model TM_F models processes by denoting them by sets of timed failures $(s, \aleph)$, consisting of a timed trace s and a timed refusal set $\aleph$. The termination axiom we believe is the most suitable candidate is the following:

$$\begin{array}{ll} & (s, \aleph) \in \mathcal{S} \wedge \checkmark \in \Sigma(s) \qquad (TA_F) \\ \Rightarrow & \exists\, t : [0, \infty) \bullet s = s' ^\frown \langle (t, \checkmark) \rangle \wedge \checkmark \notin \Sigma(s') \wedge (s, \aleph \cup \aleph_1) \in \mathcal{S} \end{array}$$

where $I(\aleph_1) \subseteq [t, \infty)$. The TA_F axiom captures the notion that if a process can perform the timed trace s while refusing $\aleph$ and a ✓ has occurred in the trace then:

(i) the ✓-requirement is satisfied; and

(ii) from the time t at which the ✓ occurred it can henceforth refuse all further events.

---

[8] See Section 2.2.1 for the reason why there is an $\tilde{s}$ in axioms TA_T and TA_S, but not TA_F and TA_{FS}.

3.3.4. A Termination Axiom for TM_{FS}

The Timed Failures-Stability model TM_{FS} models processes by denoting them by sets of timed failures paired with associated stabilities. For simplicity, we flatten these pairs into triples $(s, \alpha, \aleph)$, consisting of a timed trace s, a stability value α and a timed refusal set $\aleph$. We believe the most suitable termination axiom for this model is the following:

$$(s, \alpha, \aleph) \in \mathcal{S} \wedge \checkmark \in \Sigma(s) \qquad (TA_{FS})$$

$$\Rightarrow \quad s = s' \frown \langle(\alpha, \checkmark)\rangle \wedge \checkmark \notin \Sigma(s') \wedge (s, \alpha, \aleph \cup \aleph_1) \in \mathcal{S}$$

where $I(\aleph_1) \subseteq [\alpha, \infty)$.

The TA_{FS} axiom captures the notion that if a process can perform the timed trace s with a stability value of α while refusing $\aleph$ and a $\checkmark$ has occurred in the trace then:

(i) the $\checkmark$-requirement is satisfied;
(ii) the $\checkmark$ occurred at the time of stability α; and
(ii) that from time α it can henceforth refuse all further events.

4. Related Work

The authors are only aware of these particular issues regarding the modelling of successful termination within Timed CSP, having been addressed by the work of Schneider [27]. In [27] Schneider, presents an updated version of Reed and Roscoe's original Timed-Failures model [9], but does not consider any of the other timed models.

Schneider's Timed-Failures model has removed the use of the after action delay constant "δ"; all actions are instantaneous and there is no lower bound on the time between sequential actions, as there was in the original models of Reed and Roscoe [8,9,10]. In addition, an equivalent $\checkmark$-requirement has also been introduced, captured in the definition of the set of timed traces *TT* (see page 336 of [27]), i.e. that if a $\checkmark$ occurs in a timed trace then it must be the final event, even if other events occur at the same time. Together, these modifications result in an improved treatment of termination, within the Timed-Failures model.

However, all of the parallel operators Schneider defines in this version of Timed CSP: *alphabetised* $({}_A\|_B)$, *interleaving* $(|||)$ and *interface* (generalised) $(\|_A)$ use synchronous termination semantics. Therefore, this version of Timed CSP could be augmented by the introduction of timed versions of our asynchronous $(|||_\Theta)$ and race $(|_\Theta)$ parallel operators, and thus, provide a more flexible language for designing parallel systems.

Currently, the authors are not aware of any other modification of the models for Timed CSP developed by Reed and Roscoe [9,8,11], by either the addition of any termination axioms or by the modification to the existing axioms, in order to capture a more consistent use of $\checkmark$ in these models.

5. Conclusions and Further Work

We began this paper by outlining how the authors defined an improved model for untimed CSP, with respect to termination, called CSP_T. This was achieved by adding a termination axiom to the original failure-divergence model for CSP and replacing the original parallel operators with more flexible ones. Based on our experience of having successfully achieved this process for untimed CSP, we outlined the various issues that determined our choice of a termination axiom that best captured our notion of successful termination within the context of an untimed model.

Our goal in this paper was to achieve the same outcome for Timed CSP. Thus, we decided to adopt a similar approach for defining alternative models for Timed CSP. To this end, we have proposed a collection of termination axioms for several of the models for Timed CSP. As with untimed CSP, we have considered several factors that influence the choice and definition of a termination axiom within the family of Timed CSP models, the most significant of these is the notion of stability. In particular, when should a process become stable on termination: either at termination or at some time afterwards? In the axioms we have proposed, we have chosen to adopt the former, which we believe is the most appropriate, within the Timed CSP world.

As noted in Section 3.3, the next step in this work is formally to add each axiom to its respective model, by proving that each new model, for example $(TM_{FS} + TA, d)$, represents a complete, bounded, ultra-metric space and further, that the operators are all well-defined in the new model. The authors assume that these proofs would be similar to those presented by Reed and Roscoe [9,11] for the existing axioms of TM_{FS}.

In analysing the timed versions of CSP our intention is to lay the ground work for the definition of timed equivalents of our three untimed parallel operators (synchronous, asynchronous and race) that were introduced in CSP_T. To be able to achieve this goal, it will first be necessary to ensure that successful termination is well-defined within the models of Timed CSP, which we believe we have begun by the definition of these candidate termination axioms.

Acknowledgements

We are extremely indebted to the anonymous reviewers of this paper who provided very valuable feedback and helped us to improve the readability of the paper. Thank you. We would like to gratefully acknowledge the advice and assistance of the following individuals from whom at various stages we benefited a great deal over several very detailed discussions on our work: Jonathan Bowen, Michael Luck and Steve Schneider. We would also like to acknowledge both Springer and Elsevier for granting us permission to use extracts from our previous papers. Finally, we would like to acknowledge the help and assistance of the CPA 2013 Editors, especially Peter Welch, in preparing this paper.

References

[1] S. D. Brookes, C. A. R. Hoare, and A. W. Roscoe. A Theory of Communicating Sequential Processes. *Journal of the ACM*, 31(7), 1985.

[2] S. D. Brookes and A. W. Roscoe. An improved failures model for Communicating Sequential Processes. In *Proceedings of Pittsburgh Seminar on Concurrency*, LNCS 197, pages 281–305. Springer-Verlag, 1985.

[3] C. A. R. Hoare. *Communicating Sequential Processes*. Prentice-Hall International, 1985. ISBN: 0-131-53271-5.

[4] A. W. Roscoe. *The Theory and Practice of Concurrency*. Prentice-Hall International, 1997. ISBN: 0-13-674409-5.

[5] A.W. Roscoe. *Understanding Concurrent Systems*. Springer, 2010. ISBN: 978-1-84882-257-3.

[6] P. Howells and M. d'Inverno. A CSP model with flexible parallel termination semantics. *Formal Aspects of Computing*, 21(5):421–449, 2009.

[7] P. Howells and M. d'Inverno. Specifying Termination in CSP. *Theoretical Computer Science*, 2013. DOI: 10.1016/j.tcs.2013.05.008.

[8] G.M. Reed and A.W. Roscoe. A Timed Model for Communicating Sequential Processes. In *Proceedings of ICALP'86*, LNCS 226, pages 314–323. Springer-Verlag, 1986. (Also, Theoretical Computer Science, 58, pages 249–261, 1988.).

[9] G.M. Reed. *A Uniform Mathematical Theory of Distributed Computing*. PhD thesis, Oxford University, 1988.

[10] G.M. Reed and A.W. Roscoe. Analyzing TM_{FS}: A Study of Nondeterminism in Real-Time Concurrency. In *Proceedings of 2nd UK-Japan CS Workshop*, LNCS 491, pages 36–63. Springer-Verlag, 1991.

[11] G.M. Reed and A.W. Roscoe. The timed failures - Stability model for CSP. *Theoretical Computer Science*, 211(12):85 – 127, 1999.

[12] P. Howells and Mark d'Inverno. Specifying and Analysing Networks of Processes in CSP$_T$ (or In Search of Associativity). In *Proceedings of Communicating Process Architectures 2013 (CPA13)*, 2013.

[13] P. Howells. *Communicating Sequential Processes with Flexible Parallel Termination Semantics*. PhD thesis, University of Westminster, 2005.

[14] S. D. Brookes. *A Model for Communicating Sequential Processes*. PhD thesis, Oxford University, 1983.

[15] H. Tej and B. Wolff. A Corrected Failure-Divergence Model for CSP in Isabelle/HOL. In *Proceedings of the FME '97 – Industrial Applications and Strengthened Foundations of Formal Methods*, LNCS 1313. Springer-Verlag, 1997.

[16] C. A. R. Hoare and He Jifeng. *Unifying Theories of Programming*. Prentice-Hall, 1998.

[17] J. Davies and S. Schneider. An Introduction to Timed CSP. Technical Monograph PRG-75, Programming Research Group, Oxford University, 1989. Available from `https://www.cs.ox.ac.uk/files/3399/PRG75.pdf`.

[18] J. Davies. *Specification and Proof in Real-Time Systems*. Cambridge University Press, 1993.

[19] J. Davies, D. Jackson, and S. Schneider. Broadcast Communication for Real-time Processes. In *Proceedings of Symposium on Real-time and Fault-tolerant Systems*, LNCS 571, pages 149–169. Springer-Verlag, 1991.

[20] J. Davies and S. Schneider. Factorizing Proofs in Timed CSP. Technical Monograph PRG-75, Programming Research Group, Oxford University, 1989. (Also, LNCS 442, pages 129–159, Springer-Verlag, 1990).

[21] J. Davies and S. Schneider. Using CSP to Verify a Timed Protocol over a Fair Medium. In *Proceedings of CONCUR'92*, LNCS 630, pages 355–369. Springer-Verlag, 1992.

[22] S. Schneider. Correctness and Communication in Real-time Systems. Technical Monograph PRG-84, Programming Research Group, Oxford University, 1990. Available from `https://www.cs.ox.ac.uk/files/3408/PRG84.pdf`.

[23] S. Schneider. Timewise Refinement for Communicating Processes. In *Proceedings of 9th Workshop on Mathematical Foundations of Programming Language Semantics*, LNCS 802, pages 162–193. Springer-Verlag, 1993.

[24] S. Schneider. An Operational Semantics for Timed CSP. *Information and Computation*, 116(2):193–213, 1995.

[25] J. Davies, D.M. Jackson, G.M. Reed, J.N. Reed, A.W. Roscoe, and S. Schneider. Timed CSP: Theory and Practice. In *Proceedings of REX Workshop on Real-Time: Theory in Practice*, LNCS 600, pages 640–675. Springer-Verlag, 1992.

[26] J. Davies and S. Schneider. A Brief History of Timed CSP. *Theoretical Computer Science*, 138(2):243–271, 1995.

[27] S. Schneider. *Concurrent and Real-time Systems: The CSP Approach*. Wiley, 2000. ISBN: 0-471-62373-3.

[28] G.M. Reed. A Hierarchy of Domains for Real-time Distributed Computing. In *Proceedings of 5th Workshop on Mathematical Foundations of Programming Language Semantics*, LNCS 442, pages 80–128. Springer-Verlag, 1990.

[29] G.M. Reed and A.W. Roscoe. Metric-spaces as Models for Real-time Concurrency. In *Proceedings of 3rd Workshop on Mathematical Foundations of Programming Language Semantics*, LNCS 298, pages 331–343. Springer-Verlag, 1987.

[30] W.A. Sutherland. *Introduction to Metric and Topological Spaces*. Oxford University Press, 1981.

Appendix: CSP$_T$ Semantic Definitions

This appendix contains a summary of the notation and definitions used in the paper; for a more detailed description of CSP see [3,4,5] and for CSP$_T$ see [6,7].

The notation used in CSP and CSP$_T$ is the following. Σ is the set of all events, and is countable; denoted by a, b, c. $\mathbb{P}(\Sigma)$ is the power set of Σ; denoted by X, Y, Z. $\mathbb{F}(X)$ is the set of finite subsets of X. Σ^* is the set of finite sequences of events, i.e. *traces*; denoted by r, s, t, u. $\langle\rangle$ represents the empty sequence. $\langle a, b, c, d\rangle$ represents the sequence with members a, b, c, d. $s^\frown t$ represents the concatenation of the two sequences s and t. $s \leq t$ is the sequence *prefix* relation and $s < t$ is the proper *prefix* relation. $\#s$ is the length of the sequence s.

a in s is sequence membership. $\overline{X}$ represents the set complement of X, with respect to Σ, i.e. $\overline{X} \cap X = \varnothing$ and $\overline{X} \cup X = \Sigma$. $X - Y$ is the set difference of X and Y, i.e. $X - Y = X \cap \overline{Y}$. Processes are denoted by P, Q, R and A, B, C denote their alphabets. The set of all process identifiers (used to define recursive processes) is denoted by Ide, and we use $p, q, \ldots$ to range over Ide.

In the failure-divergence models for CSP and CSP_T a process P is denoted by an ordered pair $\langle F, D\rangle$, where F is the set of *failures* of P and D is the set of *divergence traces* of P. A *failure* is an ordered pair (s, X), where s is a trace and X is a set of events called a *refusal set*. If the failure (s, X) is in the set of failures for P then P can perform the trace s and then may refuse to participate in any of the events in X. A *divergence trace* s is one that P can perform but after it has done so, P may then be *diverging*, i.e. performing an unbounded sequence of internal events.

The formal definition of the traces and alphabet (set of events that it can perform) for a process $P = \langle F, D\rangle$ are defined as follows:

$$\mathsf{traces}(P) = \{\, s \mid \exists X \in \mathbb{P}(\Sigma) : (s, X) \in F \,\}$$
$$\alpha(P) = \{a \mid \exists\, t \in \mathsf{traces}(P) : a \; in \; t\,\}$$

The failure set F and divergence set D for a process are then any sets that satisfy the following conditions:

$$F \subseteq \Sigma^* \times \mathbb{P}(\Sigma) \qquad\qquad D \subseteq \Sigma^*$$

In addition, they must also satisfy the following standard set of *process axioms* for the original model of CSP, see [2], as well as our CSP_T *Termination* axiom (T1):

$$s \in D \Rightarrow s^\frown t \in D \tag{D1}$$
$$s \in D \Rightarrow (s^\frown t, X) \in F \tag{D2}$$
$$(\langle\rangle, \varnothing) \in F \tag{N1}$$
$$(s^\frown t, \varnothing) \in F \Rightarrow (s, \varnothing) \in F \tag{N2}$$
$$(s, X) \in F \wedge Y \subseteq X \Rightarrow (s, Y) \in F \tag{N3}$$
$$(s, X) \in F \wedge (\forall c \in Y : (s^\frown\langle c\rangle, \varnothing) \notin F) \Rightarrow (s, X \cup Y) \in F \tag{N4}$$
$$(\forall\, Y \in \mathbb{F}(X) : (s, Y) \in F) \Rightarrow (s, X) \in F \tag{N5}$$
$$t \neq \langle\rangle \wedge (s^\frown\langle\checkmark\rangle^\frown t, \varnothing) \in F \;\Rightarrow\; s \in D \tag{T1}$$

A natural language interpretation of these axioms is as follows:

(D1) states that the divergence set of a process is suffix closed. This captures the idea that once a process has started to diverge it does so for ever and that it is impossible for the process to recover, i.e. stop diverging, by perform some event, even $\checkmark$.

(D2) implies that if a process is diverging then it may also fail, i.e. it may refuse any set of events offered to it at any later stage. This captures the totally nondeterministic and chaotic nature of a diverging process in that it is seen as being *catastrophic*. This axiom enforces the consistency requirement between the divergence set and the failure set of a process.

(N1) and (N2) together imply that the traces of a process form a non-empty prefix closed set, i.e. the traces of a process form a *tree*.

(N3) if a process can refuse a set of events X then it can refuse all the subsets of X. If a process is unable to perform any of the events in X then it could not perform a subset of them.

(N4) if at some point a process can refuse the set of events X and there is another set of events Y that it also can can refuse at that point then clearly the process can refuse both of them together, i.e. $X \cup Y$. Basically this means that if it is impossible for a process to perform an event at some point then it can be added to the refusal set at that point.

(N5) means that if a process can refuse all of the finite subsets Y of a (possibly infinite) set X then it can also refuse the set X. This is a *closure* property for refusal sets which allows us to deduce that infinite sets are refusable if all of their finite subsets are refusable.

(T1) means that if a process indicates that it has terminated (by means of the $\checkmark$) but continues to perform events (t), then it must have started diverging before it performed the $\checkmark$ (i.e. $s \in D$). See Sections 1 and 3.1.

Finally, there is a *refinement ordering* ($\sqsubseteq$) defined on CSP processes in terms of the reverse subset ordering on the failure and divergence set components of processes. This ordering then induces an *equivalence* relation ($\equiv$) on processes where: two processes are equivalent in the model N_T when their failure and divergent set components are equal. This space of processes together with the ordering, i.e. ($PROC_T$, $\sqsubseteq$) is a *partially ordered set*, with least element $\perp$. The semantic functions for CSP$_T$ can be found in [6,7].

Communicating Process Architectures 2013
P.H. Welch et al. (Eds.)
Open Channel Publishing Ltd., 2013

Verifying the CPA Networking Stack using SPIN/Promela

Kevin CHALMERS and Jon KERRIDGE

Institute for Informatics & Digital Innovation, Edinburgh Napier University, UK

{K.Chalmers, J.Kerridge}@napier.ac.uk

Abstract. This paper presents a verification of the CPA Networking Stack, using the SPIN Model Checker. Our work shows that the system developed for general networking within CPA applications works under the conditions defined for it. The model itself focuses on ensuring deadlock freedom, and work still needs to be undertaken to verify expected behaviour of the architecture.

Keywords. distributed systems, model checking

Introduction

The net2 package of JCSP [1] introduced a general protocol and architecture that enables communication between CPA runtimes. For example, JCSP has been shown to interact with CSP .NET [2]. In this paper, a model of the general protocol and architecture is verified using the SPIN Model Checker [3]. By doing this work, it can be shown that the CPA Networking Stack is deadlock free, even under communication failure.

The rest of this paper is as follows. In Section 1 the motivation for this model is presented. In Section 2 a brief introduction to the SPIN Model Checker is given. Section 3 presents the architecture and operation of the CPA Networking Stack. Section 4 describes the model of the CPA Networking Stack built in SPIN and in Section 5 the results of checking this model is presented. Finally, in Section 6 conclusions and future work is discussed.

1. Motivation

One problem of the original networking implementation was poor error handling. Output net channels can fail due to link failure when exceptions are not passed to the application level. In the second networking implementation, output net channels are registered with the outgoing link. If the connection fails, the link signals each registered channel in turn. To determine if registering with the link avoids output channels hanging, a model of the architecture and protocol was developed with the SPIN Model Checker [3]. The model allows general verification of the architecture to check deadlock freedom and other properties of interest.

The CPA Networking stack itself was inspired in part by the T9000 virtual channel processor [4] which has been model checked in the past [5]. However, the CPA Networking Stack has fundamental differences (the use of channel mobility and a new protocol). Other developed CPA networking implementations, for example pony [6] and C++CSP Networked [7] have not been model checked at present.

2. SPIN/Promela

SPIN (Simple Promela INterpreter) [3] is a state space model checker allowing verification of a number of correctness requirements including assertion, deadlock, fairness and liveness checking of a model. The language of SPIN is Promela (PROcess MEta LAnguage) which has similar semantics to CSP (e.g. channels, processes, choice). The similarity enables composition of models from JCSP / .NET CSP source code into a SPIN model for verification.

To verify a model, SPIN converts Promela into C which is then compiled. The compiled application attempts to verify the model by exploring the state space which can involve the correctness requirements mentioned.

2.1. Using SPIN/Promela to Verify CPA Networking

Traditionally, CSP models are verified using FDR [8], or more recently PAT [9]. Although enabling refinement checking of the networking model would be advantageous, FDR does not currently support channel mobility.

Although it is possible to overcome this limitation [10], Promela permits channel mobility by passing channels as parameters in a message. The Promela channel is therefore similar to a channel in JCSP [11] / .NET CSP [12], although Promela does permit guarded operations on shared channels.

Because of the mobility of Promela channels the produced model is very similar to the implementation code. The processes in Section 3 that are also implemented in Section 4 are almost identical in JCSP and Promela code. Other processes described in Section 4 (e.g. InputNode, OutputNode, Network) are only necessary to verify the model.

3. CPA Networking Stack

In this section, an overview of the architecture and operation of the CPA Networking Stack will be provided. The information presented is important when the SPIN model is described in Section 4.

3.1. Architecture

Two architectural views of the networking stack are presented. The first provides a layered model allowing separation of functionality. The second examines the internal components of the layers, describing how they interact together to support the underlying distributed channel mechanism.

3.1.1. Layered Architecture

A layered view of CPA networking consists of four layers as represented in Figure 1:

Application Layer user level processes.

Event Layer networked synchronization primitives. Interfaces are provided to application layer processes, and the communication functionality of the components encapsulated.

Link Layer connections to other nodes, including receive (RX), transmit (TX), server, and manager processes.

Communication Layer the underlying communication mechanism that the CPA Networking stack sits upon.

On the left of Figure 1 the addressing mechanism is given, on the right the message types are given, and down the centre the interfaces between the layers is given. The interfaces are:

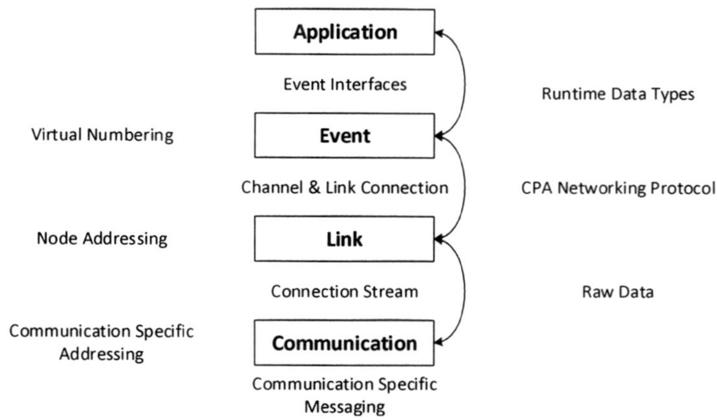

Figure 1. Layered Architecture of CPA Networking Stack.

Event Interfaces Interfaces with networked functionality added.

Channel & Link Connection Channels The crossbar allowing multiple links to communicate to multiple channel ends and multiple channel ends to communicate to multiple links. The crossbar is implemented using Any-2-One channels in both directions.

Connection Stream Links communicate with the communication mechanism using streams. These streams are communication specific.

Communication Specific Messaging The communication layer's messaging protocol (e.g. TCP/IP).

Addressing mechanisms between each layer are:

Virtual Numbering Number allocated for addressing and lookup purposes.

Node Addressing Each Node is uniquely identifiable to allow inter-Node connections. Link management relies on addressing to ensure that only one Link to a remote Node exists. An address takes the form $\langle \mathrm{Protocol} \rangle \backslash \langle \mathrm{Address} \rangle$. Protocol identifies the underlying communication mechanism (e.g. tcpip) and Address is the unique address of the Node based on the addressing mechanism of the communication mechanism.

Communication Specific Addressing Addressing mechanism enforced by the communication mechanism, for example $\langle \mathrm{IP\_Address} \rangle : \langle \mathrm{Port} \rangle$.

Each layer only understands certain message types. These are:

Runtime Data Types The Application Layer operates on data types dependent on the runtime used.

CPA Network Protocol The responsibility of the Event Layer is to convert outgoing messages into Network Protocol messages for communication via the Link Layer, and conversion of incoming Network Protocol messages from the Link Layer to communicate with the Application Layer. This protocol will be discussed further in Section 3.2

Raw Data Data leaving a Node is sent as bytes, which aids other platforms to interpret the incoming message. In particular protocol messages are transmitted as primitive data.

3.1.2. High Level Architecture

The individual components and how they are connected is presented in Figure 2.

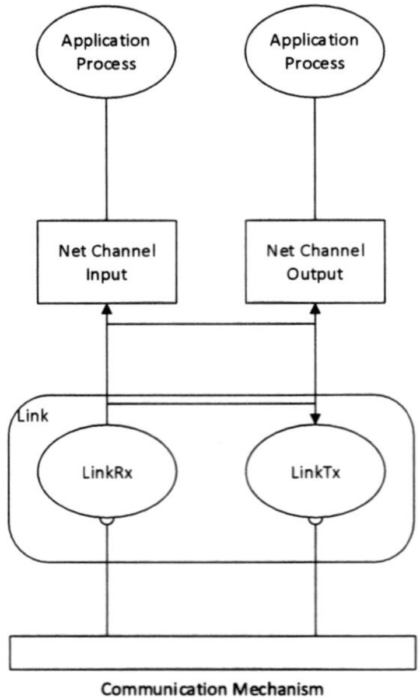

Figure 2. High-Level Architecture of CPA Networking Stack.

Link The Link component is responsible for connecting a node (a single running CPA application) to another node. The Link and its relevant subcomponents are designed to allow operation upon any communication mechanism if the necessary addressing and connection functionality is developed. The Link component has two sub-components which provide input and output operations between Nodes:

LinkTx The LinkTx process is responsible for transmitting messages to a remote node.

LinkRx The LinkRx process is responsible for receiving messages from a remote node. LinkRx interprets incoming messages and acts on the message type, accessing the destination channel if required.

Net Channel Output The networked output channel component provides the interface to the writing end of a networked channel, and hides the underlying interactions with the Link. The channel receiving messages from the Link is infinitely buffered.

Net Channel Input The networked input channel component provides the interface to the reading end of a networked channel.

3.2. Protocol

The primary goal of the CPA Networking protocol is that messages should be platform independent by using simple data primitives helps to achieve this goal.

3.2.1. Protocol Definition

The protocol is based on the original JCSP Networking implementation [13], the pony Framework [6] and C++CSP Networked [7]. The key feature is a virtual channel across a communication medium. Sent messages have a destination, and to permit synchronisation the message must be acknowledged, and therefore messages also have a source. Thus there are two attributes for a basic send message. The type of the message must also be included, providing a message triple. All required messages can be defined with a triple. There is also the optional data segment for data messages, providing the following message signature:

$$(\langle \text{type} \rangle, \langle \text{attr1} \rangle, \langle \text{attr2} \rangle, [\langle \text{data} \rangle])$$

Inclusion of data depends on $\langle \text{type} \rangle$, data size being sent as a header.

The basic channel message types are:

SEND Basic send message, requiring a source, destination, and data segment.
(`SEND`, $\langle \text{dest} \rangle, \langle \text{source} \rangle, \langle \text{data} \rangle$)

ACK `SEND` acknowledgement notifying the sender that a message has been read. Only the destination of the acknowledgement is required.
(`ACK`, $\langle \text{dest} \rangle$, null)

REJECT_CHANNEL Response to a `SEND` to a non-existent or destroyed channel. Only a destination is required.
(`REJECT_CHANNEL`, $\langle \text{dest} \rangle$, null)

POISON A `POISON` message requires a destination and a poison strength.
(`POISON`, $\langle \text{destination} \rangle, \langle \text{strength} \rangle$)

LINK_LOST When a Link fails, connected channels must be informed. `LINK_LOST` messages will never be transmitted between Nodes, but by a Link to local components. No extra information is required within the message.
(`LINK_LOST`, null, null)

ASYNC_SEND Unacknowledged `SEND` message. Asynchronous messages are used by server processes to allow typical distributed application request-respond interactions without blocking. `ASYNC_SEND` takes the same form as `SEND`.
(`ASYNC_SEND`, $\langle \text{dest} \rangle, \langle \text{source} \rangle, \langle \text{data} \rangle$)

3.3. Operation

In this section, the basic operations of CPA Networking are outlined. The methods to capture Link failure and data conversion are also covered. First a brief description of the new virtual channel is presented.

3.3.1. Virtual Channel

Figure 3 illustrates how components interact to form a networked channel. Messages between components illustrate the data that is being communicated.

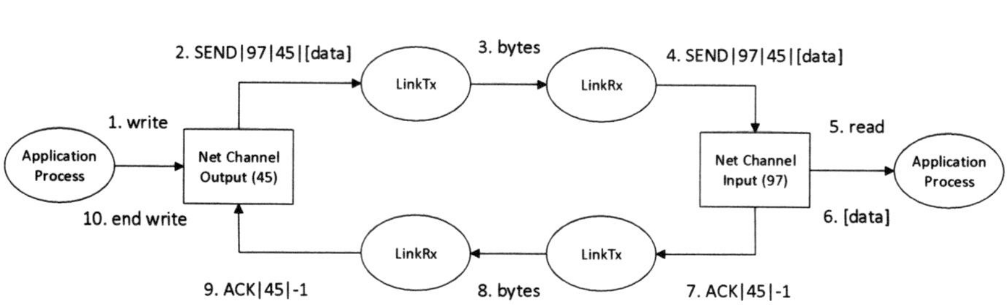

Figure 3. Virtual Networked Channel.

3.3.2. Basic SEND / ACK Operation

Figure 3 illustrates a standard read-write operation. A description of the steps involved is:

1. Application process writes to a networked output channel.
2. The networked output channel constructs a network message, setting the type as `SEND`, attr1 as the destination value (97) and attr2 as the source value (45). The sent data is attached to the network message and sent to the LinkTx. The networked output channel does not complete the write until the acknowledgement is received.
3. The LinkTx reads in the network message and writes the type (1) and two attributes (97 and 45) to the stream. The LinkTx examines the type of message, and as it is `SEND` there is a data portion. The LinkTx writes the size of the data and the bytes to the stream.
4. The receiving Node's LinkRx reads in the type attributes. The LinkRx process examines the message type, which is `SEND` and thus contains data. The size is read from the stream and used to read the required number of bytes. The LinkRx then retrieves the destination channel end from the channel index and checks its state. If the channel is in an `OK_INPUT` state the channel connecting to the partner LinkTx is added to the message, and the message sent to the networked input channel.
5. Application process reads the incoming message.
6. The networked input channel reads in the network message and checks the message type. As the type is `SEND` the message is to be delivered. The data is returned to the application process.
7. During step 6, a network message is created with the type `ACK`. attr1 is set to attr2 of the incoming message (the original source) and attr2 is set to -1. This message is written on the channel contained in the original message; the channel to the LinkTx process connected to the sending Node.
8. The LinkTx process reads the network message and writes the type (2) and the two attributes (45 and -1) to the stream. The LinkTx examines the type of the message, and as the type is `ACK` there is no data.
9. The original sending Node's LinkRx reads in the type and two attributes creating a network message from them. The LinkRx then examines the message type, and as it is a type that contains no data there is no need to read data from the stream. The LinkRx retrieves the channel from the channel manager and checks its state. If the channel is in an `OK_OUTPUT` state the network message is written to the networked output channel.
10. The networked output channel reads in the network message and checks the message type. As the type is `ACK` the write operation completes normally, freeing the application process.

The steps provided describe the operation under normal conditions. If the networked output channel is connected locally to a networked input channel, the same operations occur although at step 2 the message is sent directly to the local networked input channel component with the acknowledge channel of the networked output channel attached for direct acknowledgement.

As the architecture utilises I/O there is the possibility that erroneous behaviour can occur. The following sub-sections illustrate how this is handled.

3.3.3. SEND / REJECT operation

It is possible that erroneous message delivery can occur due to channel destruction or I/O operations. Message rejection is therefore implemented within the Link Layer. Figure 4 illustrates the component interactions that occur. The sequence of operations is:

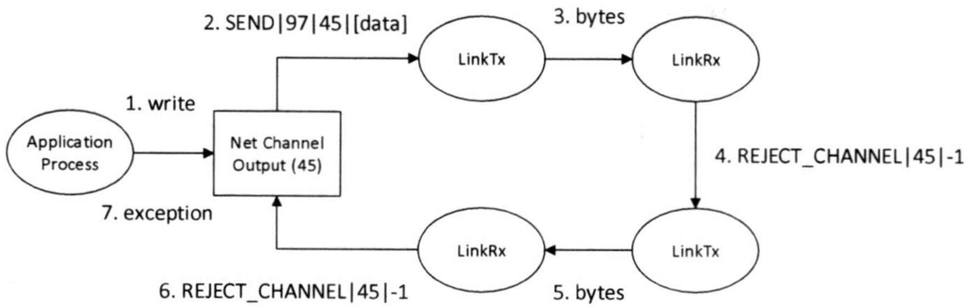

Figure 4. Reject Operation.

1. As normal operation
2. As normal operation
3. As normal operation
4. Initially this operation occurs as before. When the LinkRx attempts to retrieve channel 97 from the channel manager, the channel may not exist or its state may not be `OK_INPUT`. In either case, the LinkRx generates a network message and assigns the type `REJECT_CHANNEL`. attr1 is set to attr2 of the original message (45), and attr2 is not required. The network message is sent to the partner LinkTx.
5. As normal operation step 8, the LinkTx writes the message to the stream. There is no data segment.
6. The LinkRx reads in the type and two attributes. As the type contains no data segment, no data is read from the stream. The LinkRx then retrieves the necessary channel from the channel manager and checks the channel's state. If the channel is `OK_OUTPUT` the message is sent to the networked output channel.
7. The networked output channel reads in the message and checks the message type. As the message type is `REJECT_CHANNEL`, it is determined that the previous send was rejected. The networked output channel changes its state to `BROKEN` and removes itself from the channel manager. An exception is raised and causes the application process to continue but with an exception.

3.3.4. SEND / LINK_LOST

Another form of erroneous behaviour occurs when the connection to the node where the networked input channel resides fails. To overcome this, a networked output channel registers itself with a Link when it is created. As a networked output channel will only connect to one networked input channel, a Link can retain a set of all connected output channels. If the connection to the remote node is lost, the Link can inform all its registered channels by sending them a `LINK_LOST` message. Link failure may occur at any stage and therefore cannot easily be mapped into operational steps. There are two possibilities:

- Prior to a write operation, the Link to the remote node hosting the networked input channel fails, causing a `LINK\_LOST` message to be sent to the networked output channel on its acknowledgement channel. When write is called on the networked output channel, the acknowledgement channel is first checked for pending messages. As `LINK_LOST` will be present, the networked output channel can behave as if a message was rejected.
- After performing a write, but prior to receiving the `ACK`, the Link to the networked input channel fails. The Link informs all registered channel ends with a `LINK_LOST` message on their acknowledgement channels. The networked output channel will read

in this message, discover it is a LINK_LOST message and act as if the message was rejected.

By having all networked output channels register with Links, link failure can be transmitted as required. Networked input channels do not have this requirement as they may service multiple incoming connections. To avoid deadlock, the LinkTx remains active to black hole any outgoing messages. This is a current limitation of the CPA Networking Stack implementation - LinkTx processes do not die and remain active.

4. Building a Model of CPA Networking in Spin/Promela

4.1. Protocol Definition

SPIN uses the mtype keyword to define message types. From the discussion presented in Section 5.2, six message types within the protocol are relevant to channels. The ASYNC_SEND operation cannot be modelled as it can occur at any point during execution and requires no synchronisation between communicating components. This would increase the state space of the model beyond the capabilities of the model checker. An argument on its verification shall be presented in Section 5.

Discounting the ASYNC_SEND message, mtype is defined as follows:

```
mtype = {SEND, ACK, REJECT_CHANNEL, POISON, LINK_LOST};
```

4.2. Channel

A networked channel has a number of required definitions: the possible channel states, the data structure representing a channel, and the processes that represent networked input and output channels.

4.2.1. Channel States

Previous operational descriptions have mentioned channel states to determine how the Link Layer and Event Layer should behave. These state objects are shared between separate processes. The individual channel states are as follows:

INACTIVE initial channel state.

OK_INPUT a networked input channel willing to receive incoming messages.

OK_OUTPUT a networked output channel willing to send outgoing messages.

DESTROYED the channel end has been destroyed by an Application Layer process.

BROKEN a networked channel output end that has become broken due to some form of erroneous behaviour.

POISONED a channel end that has become poisoned, either by receiving a POISON message or by an Application Layer process invoking poison.

Figure 5 illustrates the transitions that occur between states within the channel.

4.2.2. Channel Data Structure

Each channel is provided with a data structure that contains the Virtual Channel Number (VCN), the state and the channel that the Link uses to communicate with the channel object. This is defined as follows:

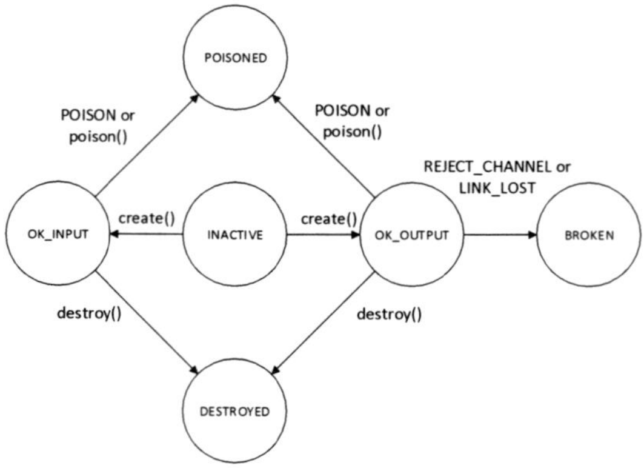

Figure 5. Channel States.

```
typedef CHANNEL_DATA
{
    byte vcn; byte state = INACTIVE; chan toChannel;
};
```

4.2.3. Channel Process

SPIN uses processes to represent components. A net channel process is given a `CHANNEL_DATA` structure to represent the channel, and an interface of channels that represent the possible calls that can be made on the channel. There are two channel types, NetChannelInput and NetChannelOutput. Figure 6 presents the NetChannelOutput process.

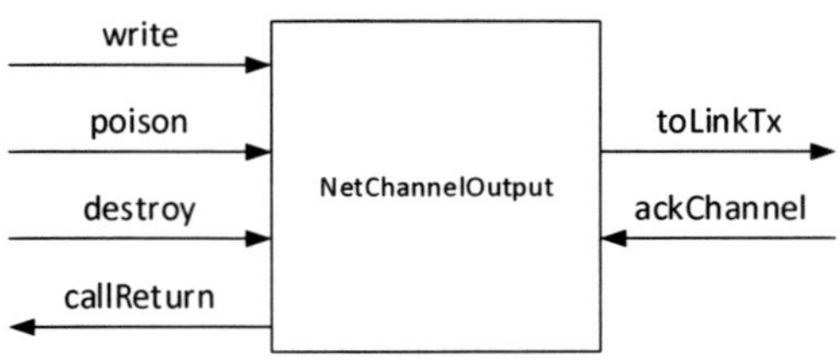

Figure 6. NetChannelOutput.

On the left of Figure 6 the interface channels are provided. Each channel represents the calling of a method on NetChannelOutput, except callReturn which is read to simulate the end of a call on the process.

On the right of Figure 6 are the channels connecting the channel process to the Link process. toLinkTx is a fixed channel that connects to the LinkTx where the input end of the virtual channel is connected. ackChannel is the channel coming from the Link, and is the channel defined in the `CHANNEL_DATA` type.

The Promela definition of the NetChannelOutput interface is as follows:

```
typedef OUTPUT_CHANNEL_INTERFACE
{
    chan write = [0] of { bool };
    chan poison = [0] of { bool };
    chan destroy = [0] of { bool };
    chan callReturn = [0] of { bit };
};
```

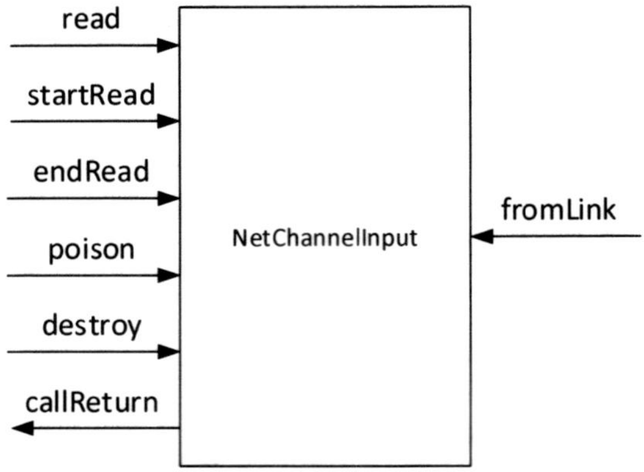

Figure 7. NetChannelInput.

Figure 7 presents the NetChannelInput process. The method interface is on the left, and includes extended rendezvous and poison operations. For completeness these operations are added to the SPIN model.

The NetChannelInput process has only one connection to the Link processes, the fromLink channel. This channel is as declared in the `CHANNEL_DATA` type. When a Link sends the NetChannelInput a message, it also sends the channel to send the response back to the Link. This is where channel mobility is required.

The Promela definition of the NetChannelInput interface is as follows:

```
typedef INPUT_CHANNEL_INTERFACE
{
    chan read = [0] of { bool };
    chan startRead = [0] of { bool };
    chan endRead = [0] of { bool };
    chan poison = [0] of { bool };
    chan destroy = [0] of { bool };
    chan callReturn = [0] of { bit };
};
```

4.3. Link Process

Link contains two processes: LinkTx and LinkRx. LinkTx receives messages from the channel processes and sends them to the remote LinkRx. It has two channels (one in and one out). input receives messages from the channel processes. txStream represents the connection to the remote LinkRx process.

LinkRx receives messages from a remote LinkTx and sends them to the correct channel. It also has two channels. rxStream represents the incoming stream from the remote LinkTx. toLinkTx connects to the complement LinkTx, and is used to send messages directly to the LinkTx and to attach to incoming messages to allow a subsequent acknowledgement to be sent directly to the LinkTx.

A Link represents a connection to another node, and is composed of a LinkTx and a LinkRx. Figure 8 represents the Link process.

As the LinkRx contains the protocol behaviour, the full Promela code listing is provided in Appendix A. The majority of the networking stack behaviour is contained in this process.

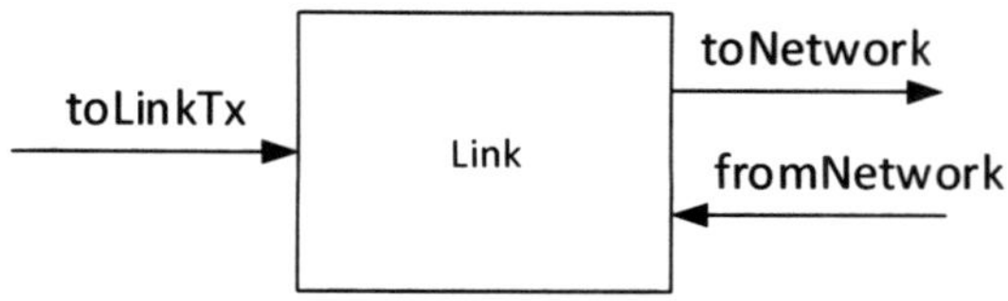

Figure 8. Link Process.

4.4. Application Processes

There are two types of application process: an outputting application and an inputting application. These processes operate on the complement end interface channels that connect to a NetChannelInput or NetChannelOutput. The application process chooses non-deterministically to write to one of the method call channels and then reads from the callReturn channel, thus waiting for the operation to complete. callReturn returns either 0 or 1 to represent either an `EXCEPTION` or an OK return message. If an `EXCEPTION` is returned, then the application process terminates.

4.5. Node Process

Within the model, two node types are defined: InputNode and OutputNode. An InputNode starts a number of Receiver processes with relevant NetChannelInput processes. An OutputNode starts a number of Sender processes and relevant NetChannelOutput processes. Figure 9 presents the InputNode process. The connection between the Link process(es) and the NetChannelInput process(es) is shown, although this is dynamic.

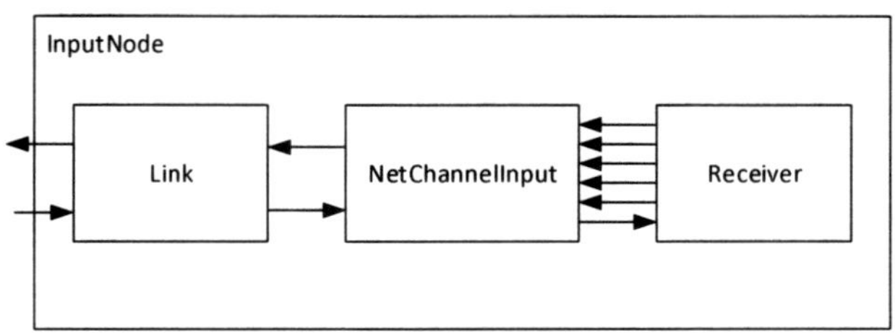

Figure 9. InputNode.

Figure 10 illustrates the OutputNode process. In this circumstance, the connection between the NetChannelOutput and Link is static.

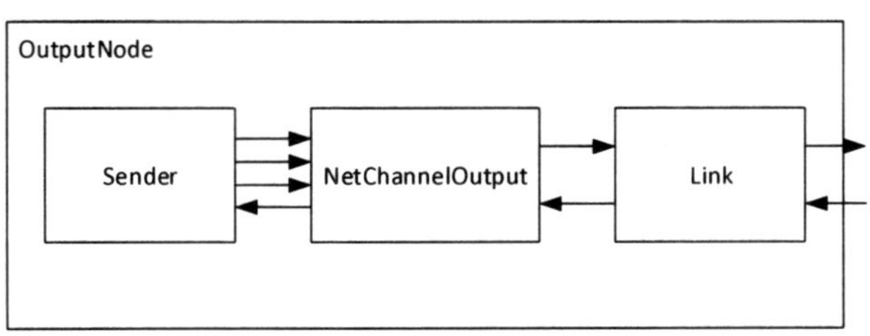

Figure 10. OutputNode.

For both Nodes, the toNetwork and fromNetwork channels represent the txStream and rxStream across the network connecting the two remote Nodes. A process is also added that allows simulation of the network connection itself.

4.6. Network Process

To simulate network failure, a simple process to represent the network is added to the model. The process non-deterministically chooses to either send a message from the OutputNode to the InputNode, from the InputNode to the OutputNode, or fail and break the connection. In the latter case, a `LINK_LOST` signal is sent to the two corresponding LinkRx processes, and a flag is set which the LinkTx processes check to determine if they should fail. In reality the latter occurrence is detected when the LinkTx process tries to write to a closed stream. The setting of a flag achieves the same outcome.

Although I/O can fail in different manners, the CPA Networking Stack expects the underlying I/O mechanism to handle this and simply inform the Link processes of a closed connection. As such, the model does not need to check behaviour of a partial communication being recieved - either the whole message is available or it is not.

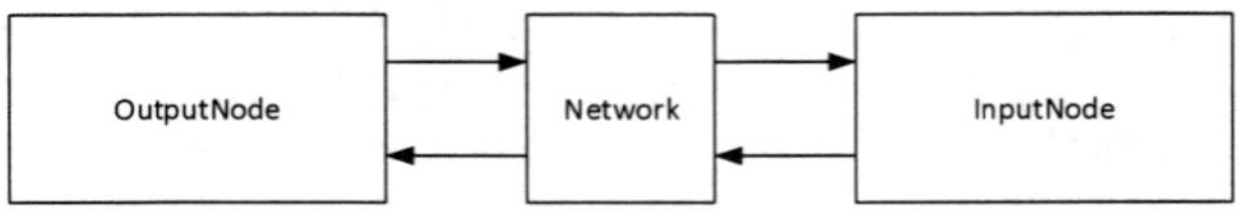

Figure 11. SPIN Model of CPA Networking.

Figure 11 presents the overall SPIN model developed for CPA Networking. The two nodes are connected via the Network process.

The Promela definition of the Network process is as follows:

```
proctype Network(chan in0; chan in1; chan out0; chan out1)
{
    mtype type;
    byte attr1;
    byte attr2;
end_network:
    do
    :: atomic
       {
            in0 ? type, attr1, attr2 -> out1 ! type, attr1, attr2
       }
    :: atomic
       {
            in1 ? type, attr1, attr2 -> out0 ! type, attr1, attr2
       }
    /* Non determistically choose to break link */
    :: atomic
       {
            linkLost = true ->
            out0 ! LINK_LOST(-1, -1);
            out1 ! LINK_LOST(-1, -1);
       }
       break;
    od
}
```

4.7. Global Values

There are a number of global values within the model, and these are summarised below:

NUMBER_INPUTS the total number of input channels within the model.

NUMBER_OUTPUTS the number of output channels connected to a single input channel

TOTAL_OUTPUTS the total number of output channels NUMBER_OUTPUTS * NUMBER_INPUTS

BUFFER_SIZE the size of the buffer to the channel processes. This is used to simulate the infinite buffer within the actual application. For the application to operate, BUFFER_SIZE should equal NUMBER_OUTPUTS. This value is manipulated to verify this assumption

CHANNEL_ARRAY a type declaration for the array of channel ends on a particular node CHANNEL_DATA channels[TOTAL_OUTPUTS]. SPIN does not permit arrays to be passed as parameters into processes, therefore this must be declared globally. For channels above NUMBER_INPUTS on the InputNode, the channel state is set to INACTIVE.

chans all the channels within the model. As CHANNEL_ARRAY cannot be passed as a parameter to an individual process, this is declared globally.

linkLost the flag used to indicate Link failure. This is initially set to false.

```
byte linkLost = false;

typedef CHANNEL_ARRAY
{
    CHANNEL_DATA channels[TOTAL_OUTPUTS];
};

CHANNEL_ARRAY chans[2];
```

5. Results

5.1. Basic Verification

Simple verification can be carried on a model comprising of a single NetChannelOutput connected to a NetChannelInput with BUFFER_SIZE = 1. This is the default assumption that for every connected output channel end to an input channel end, there is required a single place in the buffer to avoid deadlock. This assumption is required for the worst case scenario - the input end does not read while all output channels attempt to write, and therefore require a buffer space for their message to be stored. When checked with SPIN, the model is verified with no deadlock errors. This is enough to reasonably assume that having the Link processes inform the relevant output channels of connection failure overcomes the deadlock problem in the original JCSP Networking implementation.

The model also allows some indication of other problems in the original JCSP Networking implementation. In the new implementation, the LinkRx process retrieves a channel from the channel manager and locks the state object of the channel before checking said state. Thereby, LinkRx is the only process acting on the channel state at any one time. This allows various behaviours to occur based on the state of the channel object. This feature was added to the implementation of JCSP Networking when the model originally pointed out deadlock due to this occurrence not being taken into consideration. As the channel object can change state based on certain calls (poison, destroy), this would have caused inconsistent behaviour within the implementation. The original implementation used no such state variable, and LinkRx would send a message to a channel object based purely on availability within the channel manager. The channel manager would only return the connecting output channel to the networked channel object. When a channel object was destroyed, it was removed from the channel manager prior to any clean up operations (rejection of pending messages). Therefore, a channel either existed within the channel manager or it did not. There were no other possible states as no common protected state value was exposed. This meant that much of the behaviour required for more advanced functionality (poison, mobility, barriers) was not possible as there was no method to expose these states without reimplementation of the underlying mechanisms of JCSP Networking. As the new implementation exposes these properties, this problem has been overcome.

5.2. Advanced Verification

The simple verified model does not allow analysis of the assumption that the a networked input channel NetChannelInput requires exactly one buffer space for each connected NetChannelOutput. With manipulation of the BUFFER_SIZE value, this can be analysed to provide a stronger insight into this assumption. Table 1 presents results from different verification scenarios. To enable verification of the model, the option within SPIN to use minimal automata to search is activated.

Table 1. SPIN Verification Results.

NUMBER_OUTPUTS	1	2	3	4
BUFFER_SIZE				
0	FAIL	FAIL	FAIL	FAIL
1	PASS	FAIL	FAIL	FAIL
2	PASS	PASS	FAIL	FAIL
3	PASS	PASS	PASS	FAIL
4	PASS	PASS	PASS	PASS

Table 1 illustrates that a NetChannelInput requires one buffer space for each connected NetChannelOutput for connected NetChannelOutputs less than four. The number of states does not increase when the buffer size is increased beyond the required buffer size, except when a single NetChannelOutput to NetChannelInput has the buffer increased from 1 to 2, although search depth does not increase. The reason for the reduction in state space could be the usage of the minimal automata search option within SPIN, or that the NetChannelOutput requires less state space in conjunction with the Link processes at BUFFER_SIZE = 2. The NetChannelOutput also utilises the same size buffer as the NetChannelInput in the model, and this could have an effect in total required states.

6. Conclusions and Future Work

This paper has shown that the CPA Networking Stack is deadlock free even under network failure, a major issue in the original JCSP Networking implementation. The work has provided an overview of the model built using SPIN, and shown the relation between connected networked output channels and the size of buffer required to service them.

Future work required in the formal verification of the CPA Networking Stack involves refinement checking to determine if the general behaviour of the CPA Networking Stack is correct. The model will have to show that a virtual networked communication behaves similarly to a standard channel communication.

References

[1] Kevin Chalmers, Jon Kerridge, and Imed Romdhani. A Critique of JCSP Networking. In Peter H. Welch, S. Stepney, F.A.C Polack, Frederick R. M. Barnes, Alistair A. McEwan, G. S. Stiles, Jan F. Broenink, and Adam T. Sampson, editors, *Communicating Process Architectures 2008*, pages 271–291, sep 2008.

[2] Kevin Chalmers. Performance of the Distributed CPA Protocol and Architecture on Traditional Networks. In Peter H. Welch, Adam T. Sampson, Jan Baekgaard Pedersen, Jon Kerridge, Jan F. Broenink, and Frederick R. M. Barnes, editors, *Communicating Process Architectures 2011*, pages 227–242, jun 2011.

[3] G.J. Holzmann. The model checker SPIN. *Software Engineering, IEEE Transactions on*, 23(5):279–295, 1997.

[4] D. May, R. Shepherd, and P. Thompson. The T9000 transputer. In *Computer Design: VLSI in Computers and Processors, 1992. ICCD '92. Proceedings, IEEE 1992 International Conference on*, pages 209–212, 1992.

[5] Geoff Barrett. Model checking in practice: the T9000 virtual channel processor. *Software Engineering, IEEE Transactions on*, 21(2):69–78, 1995.

[6] Mario Schweigler and Adam T. Sampson. pony - The occam-pi Network Environment. In Peter H. Welch, Jon Kerridge, and Frederick R. M. Barnes, editors, *Communicating Process Architectures 2006*, pages 77–108, sep 2006.

[7] Neil C.C. Brown. C++CSP Networked. In Ian R. East, David Duce, Mark Green, Jeremy M. R. Martin, and Peter H. Welch, editors, *Communicating Process Architectures 2004*, pages 185–200, sep 2004.

[8] Formal Systems (Europe) Ltd. *FDR: User Manuel and Tutorial, version 2.82*, 2005.

[9] Jun Sun, Yang Liu, Jin Song Dong, and Jun Pang. PAT: Towards flexible verification under fairness. volume 5643 of *Lecture Notes in Computer Science*, pages 709–714. Springer, 2009.

[10] Peter H. Welch and Frederick R. M. Barnes. A CSP Model for Mobile Channels. In Peter H. Welch, S. Stepney, F.A.C Polack, Frederick R. M. Barnes, Alistair A. McEwan, G. S. Stiles, Jan F. Broenink, and Adam T. Sampson, editors, *Communicating Process Architectures 2008*, pages 17–33, sep 2008.

[11] P. H. Welch. Process oriented design for Java: concurrency for all. pages 51–57. CSREA Press, 2000.

[12] Kevin Chalmers and Sarah Clayton. CSP for .NET Based on JCSP. In Peter H. Welch, Jon Kerridge, and Frederick R. M. Barnes, editors, *Communicating Process Architectures 2006*, pages 59–76, sep 2006.

[13] Peter H. Welch, Jo R. Aldous, and Jon Foster. CSP networking for Java (JCSP.net). In *Proceedings of the International Conference on Computational Science-Part II*, ICCS '02, pages 695–708, London, UK, UK, 2002. Springer-Verlag.

A. Promela Code for LinkRx

```
proctype LinkRx(chan toTxProcess; chan rxStream; bit nodeNumber)
{
    /* Attributes read in with incoming message */
    byte attr1;
    byte attr2;

    /* This process reads an incoming message from the message and
       processes it. Generally the message
       is forwarded onto the correct destination, although erroneous
          behaviour must be dealt with */
    /* Valid end state. Waiting for input from the network */
end_lrx:
    do
    /* SEND received. */
    :: atomic
      {
            rxStream ? SEND(attr1, attr2) ->
            /* First check if the message is going to a valid channel
               */
            if
            /* Destination channel is outside range. Reject message
               */
            :: (attr1 > TOTAL_OUTPUTS) -> toTxProcess ! REJECT_CHANNEL
               (attr2, -1)
            :: else ->
                /* Message is for a valid channel. Check channel
                   state and deal with accordingly */
```

```
                if
                :: (chans[nodeNumber].channels[attr1].state ==
                   OK_INPUT) ->
                    /* Channel is OK to receive messages.  Forward the
                         message onto the channel process */
                    chans[nodeNumber].channels[attr1].toChannel ! SEND
                        (attr2, toTxProcess)
                :: (chans[nodeNumber].channels[attr1].state ==
                   POISONED) ->
                    /* Channel has been poisoned.  Propogate the
                        poison back to the writer */
                    toTxProcess ! POISON(attr2, 0)
                :: (chans[nodeNumber].channels[attr1].state ==
                   DESTROYED) ->
                    /* Channel has been destroyed.  Reject the message
                         */
                    toTxProcess ! REJECT_CHANNEL(attr2, 0)
                :: (chans[nodeNumber].channels[attr1].state == BROKEN)
                    ->
                    /* Channel is broken.  This should only happen
                        during Link failure, but just in case reject */
                    toTxProcess ! REJECT_CHANNEL(attr2, 0)
                :: else ->
                    /* Channel is in some other state.  This could be
                        a channel trying to send to an output
                        or some other problem.  We reject the message
                            in this instance and continue */
                    toTxProcess ! REJECT_CHANNEL(attr2, 0)
                fi
            fi
    }
  /* Acknowledgement operation */
  :: atomic
    {
      rxStream ? ACK(attr1, attr2) ->
            /* First check if the message is going to a valid channel
                */
            if
            /* Destination channel is outside range.  Ignore message
                */
            :: (attr1 > TOTAL_OUTPUTS) -> skip
            :: else ->
                /* Message is for a valid channel.  Check channel
                   state and deal with accordingly */
                if
                :: (chans[nodeNumber].channels[attr1].state ==
                   OK_OUTPUT) ->
                    /* ACK being sent to an output channel.  Forward
                        the message onto the channel process */
                    chans[nodeNumber].channels[attr1].toChannel ! ACK
                /* In all other cases, we drop the message.  The
                   message has been sent to a channel that was
                   not in a state to accept it. */
                :: else -> skip
                fi
            fi
    }
  /* Reject channel message received */
  :: atomic
```

```
  {
    rxStream ? REJECT_CHANNEL(attr1, attr2) ->
        /* First check if the message is going to a valid channel
           */
        if
        /* Destination channel is outside range.  No point in
           rejecting (we could end up with a continuous
           cycle of rejects). Simply ignore the message. */
        :: (attr1 > TOTAL_OUTPUTS) -> skip
        :: else ->
            /* Message is for a valid Channel.  Check channel
               state and deal with accordingly */
            if
            :: (chans[nodeNumber].channels[attr1].state ==
               OK_OUTPUT) ->
                /* Channel can accept the reject message.  Pass
                   onto the channel process */
                   chans[nodeNumber].channels[attr1].toChannel !
                       REJECT_CHANNEL
            /* In all other cases ignore the message.  The channel
                 is in no state to receive it */
            :: else -> skip
            fi
        fi
  }
/* Poison message received */
:: atomic
  {
    rxStream ? POISON(attr1, attr2) ->
        /* First check if the message is going to a valid channel
           */
        if
        /* Destination channel is outside range.  No point in
           rejecting (we could end up with a continuous
           cycle of rejects). Simply ignore the message. */
        :: (attr1 > TOTAL_OUTPUTS) -> skip
        :: else ->
            /* Message is for a valid Channel.  Check channel
               state and deal with accordingly */
            if
            :: (chans[nodeNumber].channels[attr1].state ==
               OK_OUTPUT) ->
                /* Channel is an output.  Simply send POISON to it
                   . */
                chans[nodeNumber].channels[attr1].toChannel !
                   POISON
            :: (chans[nodeNumber].channels[attr1].state ==
               OK_INPUT) ->
                /* Channel is an input.  Simply send POISON to it
                   */
                chans[nodeNumber].channels[attr1].toChannel !
                   POISON(attr1, attr2)
            /* In all other cases we ignore the poison.  Either
               the channel is poisoned, and in the model
               nothing else needs to be done (in the
                  implementation we increase the poison strength
                  if
               necessary), or it is destroyed or broken, which is
                  considered to be greater than poison */
```

```
                :: else -> skip
                fi
            fi
      }
    /* Link lost received */
    :: rxStream ? LINK_LOST(attr1, attr2) ->
        atomic
        {
            /* Inform all output ends */
            byte idx = 0;
            do
            :: (idx < TOTAL_OUTPUTS) ->
                if
                :: (chans[nodeNumber].channels[idx].state == OK_OUTPUT
                   ) ->
                    chans[nodeNumber].channels[idx].toChannel !
                        LINK_LOST
                :: else -> skip
                fi;
                idx = idx + 1;
            :: else -> break
            od;
        }
        break;
    od;
}
```

Communicating Process Architectures 2013
P.H. Welch et al. (Eds.)
Open Channel Publishing Ltd., 2013

Improving the Performance of Periodic Real-time Processes: a Graph Theoretical Approach

Antoon H. BOODE [a,1,2], Hajo BROERSMA [b] and Jan F. BROENINK [a]
[a] *Robotics and Mechatronics,*
[b] *Formal Methods and Tools,*
Faculty of Electrical Engineering, Mathematics and Computer Science,
University of Twente, The Netherlands
{A.H.Boode, H.J.Broersma, J.F.Broenink}@utwente.nl

Abstract. In this paper the performance gain obtained by combining parallel periodic real-time processes is elaborated. In certain single-core mono-processor configurations, for example embedded control systems in robotics comprising many short processes, process context switches may consume a considerable amount of the available processing power. For this reason it can be advantageous to combine processes, to reduce the number of context switches and thereby increase the performance of the application. As we consider robotic applications only, often consisting of processes with identical periods, release times and deadlines, we restrict these configurations to periodic real-time processes executing on a single-core mono-processor. By graph theoretical concepts and means, we provide necessary and sufficient conditions so that the number of context switches can be reduced by combining synchronising processes.

Keywords. CSP, graph transformations, acyclic multi-graphs, real-time periodic processes, synchronised products

Introduction

In certain single-core mono-processor configurations, for example embedded control systems in robotics comprising many short processes, process context switches may consume a considerable amount of the available processing power.

Li et al. [1] show, that the average cost of a context switch varies from $3.8\,\mu$s to over 1 ms. Veldhuizen [2] shows that the cost of a context switch is on average $7.7\,\mu s$. Clearly these figures depend on the hardware and software being used[3]. To what extent a system is suffering from context switches depends roughly on the ratio between the context switch and the process action; the higher the time consumption of an action, the less relevant the time consumption of the context switch.

As we are considering systems with many short processes, it can be advantageous to combine processes, in order to reduce the number of context switches, thereby increasing the performance of the application. In this paper we restrict these configurations to robotic appli-

[1]Corresponding Author: *Ton Boode, Robotics and Mechatronics, Faculty EEMCS, University of Twente, P.O. Box 217 7500 AE Enschede, The Netherlands*. E-mail: A.H.Boode@utwente.nl.

[2]funded by InHolland University of Applied Sciences, Alkmaar, The Netherlands

[3]occam-π context switch overheads, under the KRoC CCSP multicore scheduler [3], are of the order of 100 nanoseconds.

cations. We consider periodic real-time processes executing on a single-core mono-processor, because robotic applications (like embedded control systems) often consist of processes with identical periods, release times and deadlines. The processes typically have a period of 1 ms. This observation makes it reasonable to assume that the release time, the periods and the deadlines for the constituent processes of the application are the same. As we consider periodic real-time processes, for every process activity (i.e. action), there must be an upper bound for which the action has finished executing; otherwise one cannot guarantee the timeliness of the process. As an example, consider 100 very short processes, containing in average 3 actions, running at 1 KHz, so a period of 1 ms. Using the minimum context switch time consumption given by Li [1], the 300 context switches will need more than the available processing time in one period.

When looking at programs, we distinguish between the specification level and the execution level. On the one hand, there is the specification of a set of parallel processes (for example, in CSP [4]); on the other hand, there is the execution of processes representing the specification, on a computer system, running under an operating system.

On specification level, a process defines a series of actions. Processes sharing the same action can only perform this action if all processes sharing this action are ready to perform this action; this is atomic and performed as one action.

On execution level, as soon as a process has to synchronise with another process[4], a context switch has to be executed, to let the execution be continued by that other process. Such a context switch consumes time. One can reduce the number of these synchronisation related context switches by combining communicating processes.

At specification level, a set of parallel real-time processes can be represented by a graph consisting of several components. A single process is represented by one component, which is a finite labelled weighted directed multi-graph[5], consisting of vertices, arcs between pairs of vertices and labels associated with the arcs. A label is a string representing the (name of an) action and a number representing the worst-case execution time of the action. With each name exactly one worst-case execution time value is associated. Our interpretation of a component, representing a process, is that the vertices represent states and the arcs together with their labels represent the actions that are necessary to move from one state to another. Components have different arc sets, but some of their arcs may have the same labels, meaning that they represent the same action.

The execution of a process is, from a graph-theoretical point of view, represented by a series of arcs: a path through the graph. In process terms this is called a *trace*. Such a path has a length, which is the summation over the worst-case execution time values of the labels associated with the arcs in the path. Our goal is to reduce the worst-case execution time of the set of parallel processes, which is represented by the summation over the maximum path length of each graph, by combining synchronising processes. In graph-theoretical terms this leads to combining graphs, using notions like the Cartesian product of graphs and the synchronised product of graphs.

Via a design methodology, a process specification has to be transformed into a program. We insert into this transformation three steps, of which this paper describes the second one. Firstly, we transform the process specification into a set of graphs, secondly, where possible and meaningful (in terms of performance gain), we take synchronised products of subsets of the set of graphs, and thirdly, this set of synchronised products is transformed into a process specification.

---

[4]To synchronise actions, both processes have to do extra work and at least one of them will have to yield the processor (assuming single-core execution), causing a context switch.

[5]These graphs are (slightly) more general than *labelled transition systems* in that they may have more than one starting and finishing points (used in intermediate stages of the graph transformations described later).

The paper is organised as follows. Before we specify the model that we use to analyse the performance of periodic real-time processes, we introduce the necessary graph-theoretic and process algebra terminology in Section 1. In Section 2, we introduce periodic real-time processes as finite labelled weighted directed acyclic multi-graphs, and we present an overview of synchronised products. In Section 3, we discuss the transformation of a set of graphs to its Cartesian product, where we show that the longest path length for a set of graphs is identical to the longest path length of the Cartesian product of this set of graphs. In Section 4 the synchronisation constraints (disregarded by the Cartesian product) are met by means of the weak synchronised product of a set of parallel processes. In Section 5 the reduced weak synchronised product of a set of parallel processes is introduced, where not-specified behaviour represented by the Cartesian product is removed. In Section 6 the synchronised product is introduced and necessary and sufficient conditions are proved for the longest path length of that product to be less than the longest path length from the original set of graphs (representing parallel processes). We finish with Section 7, where we give our conclusions followed by a discussion and ideas for future work.

1. Terminology

We use [5] and [6] for terminology and notation on graphs and processes not defined here and consider finite labelled weighted directed acyclic multi-graphs only.

So, if we use H to denote a graph, we will always mean a finite labelled weighted directed acyclic multi-graph. Thus H consists of a set of vertices V, a multi-set of arcs A, and a mapping $\lambda : A \to L$, where L is a set of label pairs[6]. An arc $a \in A$ which is directed from a vertex $u \in V$ (the tail) to a vertex $v \in V$ (the head) will usually be denoted as $a = uv$; the reverse arc will be denoted as vu. Note that we allow multiple arcs from u to v, but that we do not allow uv and vu to be present in the same graph. For each arc $a \in A$, $\lambda(a) \in L$ consists of a pair $(l(a), t(a))$, where $l(a)$ is a string representing an action and $t(a)$ is a positive real number representing the worst-case execution time of the action represented by $l(a)$. If an arc has multiplicity $k > 1$, then all copies have different labels, otherwise we could replace two copies of an arc with identical labels by one arc, because they represent exactly the same action at the same stage of the process. If two arcs $a, b \in A$ have labels $\lambda(a) = (l(a), t(a))$ and $\lambda(b) = (l(b), t(b))$ such that $l(a) = l(b)$, then this implies that $t(a) = t(b)$; this follows since $l(a) = l(b)$ means that the arcs a and b represent the same action at different stages of a process.

A directed path in H is a sequence of *distinct* vertices $v_1 v_2 \ldots v_k$ of H such that $v_j v_{j+1} \in A$ for $j = 1, \ldots, k-1$. The length of a path $v_1 v_2 \ldots v_m$ is defined as $\sum_{i=1}^{m-1} t(v_i v_{i+1})$. A directed path defines a *total ordering* on its arcs: $v_1 v_2 < v_2 v_3 < \ldots < v_{k-1} v_k$.

A directed cycle is a directed path $v_1 v_2 \ldots v_k$ together with an additional arc $v_k v_1$, and is denoted by $v_1 v_2 \ldots v_k v_1$. An acyclic graph does not contain any directed cycles.

We consider finite acyclic graphs, H, only. In general, such a graph consists of several components, where each component, H_i, is *weakly connected* (i.e. all vertices are connected by sequences of arcs, ignoring arc directions) and corresponds to one sequential process. For such components, $\ell(H_i)$ is defined as the maximum length taken over all directed paths in H_i. For the whole graph, which corresponds to a parallel set of sequential processes that must each run to completion, the maximum path length, $\ell(H)$, is the sum of all the individual $\ell(H_i)$. A *partial ordering* on the arcs of a weakly connected graph is induced from the total orderings of its directed paths: $a < b$ if and only if a and b are so ordered in some directed path within the graph.

[6] We shall also use the notation $V(H)$ and $A(H)$ to denote the vertices and arcs for any graph H.

Time and processes are thoroughly described in CSP (for example, by Schneider [6]). Our view of time in a process is that each action takes some time to execute and this time is directly linked to the label of the action. For every process P_i, the actions of the process constitute the process alphabet set A_{P_i}, which consists of labels. A label in a process is identical to a label in a graph: both are identical strings of characters with identical associated values.

For components H_i and H_j, an arc a_i with label $\lambda(a_i)$ in component H_i is a synchronising arc with respect to components H_i and H_j, if and only if there exists an arc a_j with label $\lambda(a_j)$ in component H_j and $\lambda(a_i) = \lambda(a_j)$. If it is clear from the context, we omit the '*with respect to components H_i and H_j*' part of the definition.

Viewed as CSP processes, components are combined in parallel using the CSP *alphabetised* parallel operator with alphabet sets defined by the labels on their respective arcs. For an arc of a component H_i whose label does not occur on an arc of another component H_j, the corresponding action is not blocked from execution.

For a set of parallel processes that contains a deadlock, the graph H representing this set is said to be ill-defined or inconsistent. An example of such a set of parallel processes is called a pathological case.

Components $H_i = (V_i, A_i, \{\lambda(a) | a \in A_i\})$ and $H_j = (V_j, A_j, \{\lambda(a) | a \in A_j\})$ are said to be independent if and only if $\{\lambda(a) | a \in A_i\} \bigcap \{\lambda(a) | a \in A_j\} = \emptyset$.

The in-degree (out-degree) of a vertex v in a graph H is defined as the number of arcs with head v (tail v) and denoted by $d_H^-(v)$ ($d_H^+(v)$).

The Cartesian product $H_1 \times H_2$ of H_1 and H_2 is defined as the graph on vertex set $V_{1,2} = V_1 \times V_2$ (the Cartesian product of the vertex sets) with two types of arcs. Arcs of type 1 (type 2) are between pairs $(v_1, v_2) \in A_{1,2}$ and $(w_1, w_2) \in A_{1,2}$ with $(v_1, w_1) \in A_1$ and $v_2 = w_2$ (with $v_1 = w_1$ and $(v_2, w_2) \in A_2$), so arcs of type 1 and 2 correspond to arcs of H_1 and H_2, respectively. For $k \geqslant 3$, the Cartesian product $H_1 \times H_2 \times \ldots \times H_k$ is defined recursively as $((H_1 \times H_2) \times \ldots) \times H_k$.

Since we only consider labelled weighted directed acyclic graphs, paths, etc., for convenience we skip the adjective finite labelled weighted directed acyclic where possible in the sequel.

2. Periodic Real-time Processes as Labelled Weighted Directed Acyclic Graphs

The rationale behind modelling processes by graphs is, that a process is always in a certain state, where via performing an action another state is reached. Similarly, from a specific vertex in a graph another vertex can be reached by passing through the arc between them. A process can be defined as a labelled weighted directed acyclic graph, therefore a periodic real-time process also can be defined as a weakly connected labelled directed graph [6]. If the process specification contains cycles, due to the real-time constraints (timeliness, leading to an upper bound for the number of cycles), one can unfold the cycles leading to an acyclic path, which is a directed acyclic graph.

Hence, a set of parallel real-time periodic processes can be modelled as a graph H with components H_i. For our purpose of improving the performance of real-time periodic applications, we are going to show how execution time is reduced by combining components of graphs. A set of parallel processes and its combination into one process has to have identical behaviour (i.e. the *traces* and *failures*[7] of the set of parallel processes must be the same as those of the combined process).

Several products have been defined in the literature, like strong products, synchronised products, etc. None of these products are sufficient. The strong product as defined by [5] is a

[7]There are no *divergences* as the processes being combined are finite (repeated *periodically* by the real-time application).

labelled acyclic directed graph that is order preserving, but the arcs that produce a synchronised arc are not removed from the graph. In other words, behaviour is added to the original process set. Synchronised products have been defined by authors like Aiguier et al. [7], Caucal and Hassen [8], Hammal [9] and Wöhrle and Thomas [10]. The definition by [9] does not take into account that for a certain graph a certain label may occur more than once in a path, so the definition does not preserve the order. The definition of [8] is used to synchronise languages where the synchronised product of languages G and H is the disjunction of these languages and is also not order preserving.

For these reasons, these definitions do not meet our requirements, as the definition of the synchronised product has to be order preserving and our graphs have to reflect the behaviour of the processes on which the graphs are based. The definition by Aiguier et al. [7] stems from Input Output Symbolic Transition Systems and is almost similar to our product, although the terminology is different. Also we allow the source of the graph to be a set of vertices and Aiguier et al. require a single start state. Even the definition by Wöhrle and Thomas [10] does not fit our needs, although this product preserves the order as shown in Figure 1 (the dashed arc is the synchronising arc). In their approach it is possible for the synchronised product of two weakly connected graphs to contain again two weakly connected graphs (where the diamond-shaped component represents states and transitions unreachable according to the synchronisation rules).

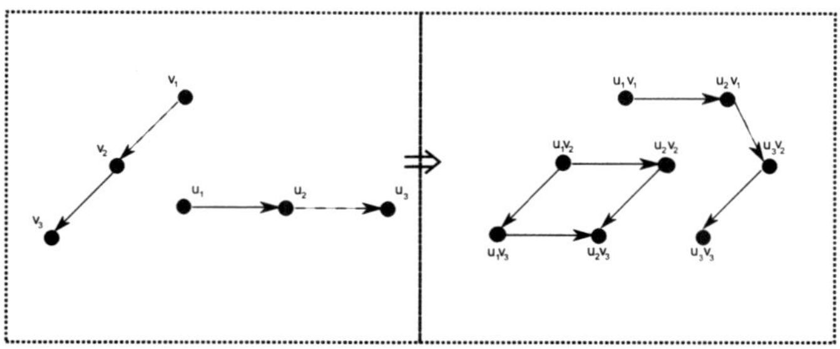

Figure 1. Synchronising product according to Wöhrle and Thomas.

At first sight, the Cartesian product seems to be a good way to express the combination of parallel processes. However, this product does not take care of necessary synchronisations. Therefore, we propose a modification of the product by [10] that will be developed in a number of steps. Figure 2 shows an example, where dashed arcs are synchronising actions. It shows five steps, where a synchronised product is built from a set of graphs. These five steps are elaborated in Sections 3 through 6.

The first diagram ($H1 + H2$) shows state-action transition graphs for two parallel processes, synchronising on one common action.

The second diagram ($H_1 \Box H_2$) shows the Cartesian product of those two graphs. The vertices are the cross-product of the original vertices and the transitions between them are in one of two dimensions (one for each of the original graphs). This corresponds to the CSP interleaving of the two processes (i.e. where each is free to engage in actions, regardless of whether they are held in common). Clearly, this is not a suitable serialisation of the original parallel system. This is presented in Section 3.

The third diagram ($H_1 \boxminus H_2$) is called the Weak Synchronised product of the (original) two graphs. It is derived from the Cartesian product by removing arcs representing common actions, if those arcs proceed from a vertex in only one of the dimensions (i.e. the action was engaged in by only one of the original processes). Common arcs remain always in the form

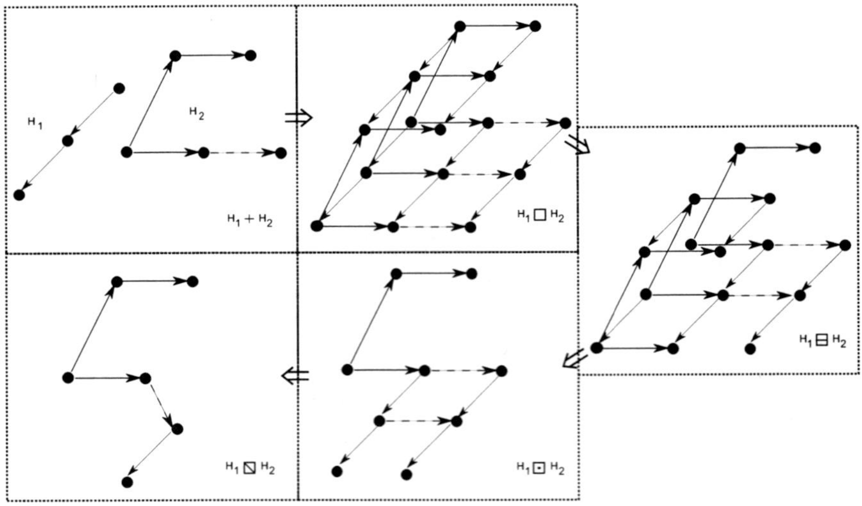

Figure 2. Transformations from parallel (+) through Cartesian (□), weak synchronised (⊟), reduced weak synchronised (⊡) to synchronised (⧅) product.

of two dimensional parallelograms (one dimension for each original process engaging in the action). If there is deadlock in the system, this will appear as vertices with no out-flowing arcs. This is in Section 4.

The fourth diagram ($H_1 \boxdot H_2$) is called the Reduced Weak Synchronised product. It is derived from the third by (iteratively) removing all vertices that have been left with no in-flowing arcs (other than those in the Cartesian product that had none – i.e. the starting points), together with the out-flowing arcs from those removed vertices. This is Section 5.

Finally, the fifth diagram ($H_1 \boxbslash H_2$) is the Synchronised product. This collapses the common action parallelograms into single action arcs across the diagonal, leaving the two isolated vertices. The same iterative process from the fourth stage (for removing vertices with no in-flowing arcs and their out-flowing arcs) cleans up. This is Section 6.

3. The Cartesian Product of a Set of Parallel Processes

To visualise all our transformations we use a simplified version of an untimed example by [11] given in Listing 1, shown in Figure 3. The example contains three serial processes running in parallel, synchronising on their common actions respectively. Clearly they can be serialised simply by concatenating them , removing the middle `SKIP`s and merging the common actions to a single occurrence. The example is chosen to illustrate the stages of transformation and kept simple for this purpose. In the Appendix, Listing 2 and the related graph transformation in Figure 12 give a (slightly) more complex, and interesting example.

The process `SEQUENCE_CONTROL` is tail recursive, each element of the recursion being one period of the control logic. We use the constituent processes of this period for our transformations (starting in Figure 4). We assume that the actions have a given upper bound time value. We abbreviate the actions and their related upper bound time values in the several product figures: for example, (`read_distance_sensors`, 120 μs) will become *rds*. As before, a dashed arc represents a synchronising arc.

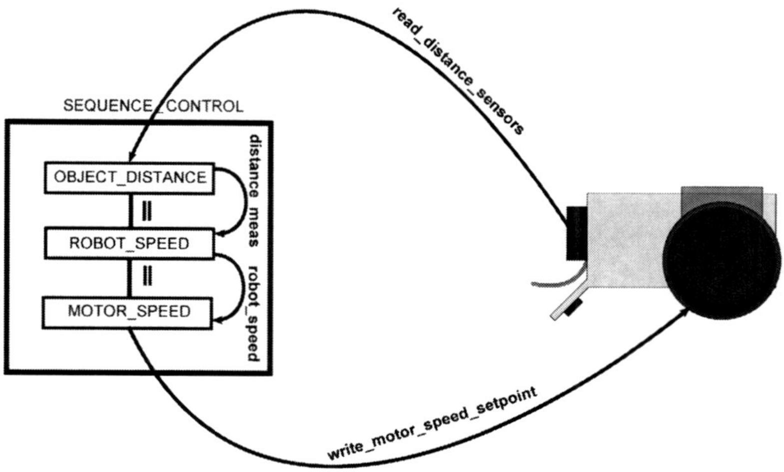

Figure 3. Untimed sequence control processes of a mobile robot.

```
OBJECT_DISTANCE   = read_distance_sensors →
                    compute_object_distance → distance_meas → SKIP

ROBOT_SPEED = distance_meas → compute_robot_speed → robot_speed → SKIP

MOTOR_SPEED       = robot_speed →
                    compute_motor_speed → write_motor_speed_setpoint → SKIP

SEQUENCE_CONTROL = (OBJECT_DISTANCE || ROBOT_SPEED || MOTOR_SPEED);
                    SEQUENCE_CONTROL;
```

Listing 1. Description of the SEQUENCE_CONTROL process.

The graph $SQ = MS + RS + OD$ representing the processes MS = MOTOR_SPEED, RS = ROBOT_SPEED and OD = OBJECT_DISTANCE and using abbreviated actions is given by:

$$\begin{aligned} MS = &\ (V(H_1), A(H_1), \{\lambda(a) | a \in A(H_1)\}) \\ = &\ (\{v_1, v_2, v_3, v_4\}, \{v_1v_2, v_2v_3, v_3v_4\}, \{(v_1v_2, rs), (v_2v_3, cms), (v_3v_4, wmss)\}) \end{aligned}$$

$$\begin{aligned} RS = &\ (V(H_2), A(H_2), \{\lambda(a) | a \in A(H_2)\}) \\ = &\ (\{v_5, v_6, v_7, v_8\}, \{v_5v_6, v_6v_7, v_7v_8\}, \{(v_5v_6, dm), (v_6v_7, crs), (v_7v_8, rs)\}) \end{aligned}$$

$$\begin{aligned} OD = &\ (V(H_3), A(H_3), \{\lambda(a) | a \in A(H_3)\}) \\ = &\ (\{v_9, v_{10}, v_{11}, v_{12}\}, \{v_9v_{10}, v_{10}v_{11}, v_{11}v_{12}, \}, \{(v_9v_{10}, rds), (v_{10}v_{11}, cod), (v_{11}v_{12}, dm)\}) \end{aligned}$$

The Cartesian product of the graph SQ, $MS \square RS \square OD$, contains 64 states; therefore, we do not show the formal definition of the graph. From Figure 4, it may be checked that $\ell(MS + RS + OD)$, which is $\ell(MS) + \ell(RS) + \ell(OD)$, is equal to $\ell(MS \square RS \square OD)$. Next, we will show that this holds in the general case for finite labelled weighted directed acyclic multi-graphs.

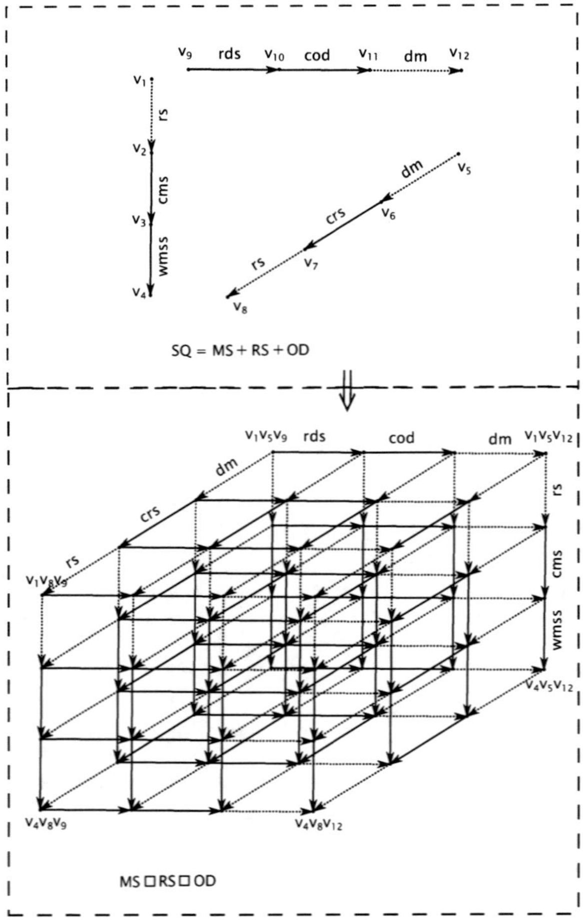

Figure 4. Sequence control processes of a mobile robot, from + to $\Box$.

In the Cartesian product $H_1 \Box H_2$ of H_1 and H_2, we distinguish between two types of arcs. Arcs of type H_1 (type H_2) are between pairs $(v_1, w_1) \in V(H_1 \Box H_2)$ and $(v_2, w_2) \in V(H_1 \Box H_2)$ with $v_1 v_2 \in A(H_1)$ and $w_1 = w_2$ (with $v_1 = v_2$ and $w_1 w_2 \in A(H_2)$), so arcs of type H_1 and type H_2 correspond to (are in fact copies of) arcs of H_1 and H_2, respectively.

For $k \geqslant 3$, the Cartesian product $\Box_{i=1}^{k} H_i = H_1 \Box H_2 \Box \ldots \Box H_k$ is defined recursively as $((H_1 \Box H_2) \Box \ldots) \Box H_k$. If no ambiguity can arise, we write $\Box H_i$ for $\Box_{i=1}^{k} H_i$. In this product of k directed graphs, we distinguish between arcs of type H_i for $i = 1, \ldots, k$, analogously as for the case $k = 2$. Note that in case the H_i are labelled, the labels of the arcs of type H_i in $\Box H_i$ correspond to the labels of the arcs of H_i: each copy of arc $a \in A(H_i)$ in $\Box H_i$ has label $\lambda(a)$. If H_i is a multi-graph, an arc $a \in A(H_i)$ can appear more than once in H_i, but in that case the copies of a in H_i have distinct labels, so each of the copies can be identified by its label. In $\Box H_i$ similarly, we can distinguish the copies of a by their labels ($\lambda_1(a)$, $\lambda_2(a)$, ...).

For the sequel, we need a number of useful properties of acyclic directed graphs. Most of these properties are straightforward and easy to prove – see [12].

Let G be an acyclic directed (multi-)graph. Then G has at least one vertex v_1 with in-degree 0. If we delete v_1 and all the arcs with tail v_1 from G, we obtain a new acyclic directed (multi-)graph, so we can again find a vertex v_2 with in-degree 0, etc. We can repeat this procedure as long as there are vertices, and we obtain a so-called acyclic ordering $v_1, v_2, \ldots$ of the vertex set of G. It is important to observe that this ordering implies that arcs of G can only exist from v_i to v_j with $i < j$. We will use a slightly different (partial) ordering for our purposes, as follows.

We assume throughout that all our graphs H_i are acyclic directed multi-graphs. For the moment, we disregard the labels and weights, so in the following paragraphs the length of a directed path is just the number of arcs.

For each H_i we define S_0^i to denote the set of vertices with in-degree 0 in H_i, S_1^i the set of vertices with in-degree 0 in the graph obtained from H_i by deleting the vertices of S_0^i and all arcs with tails in S_0^i, and so on, until the final set $S_{t_i}^i$ contains the remaining vertices with in-degree 0 and there are no arcs in the remaining graph. As in the acyclic ordering, this ordering implies that arcs of H_i can only exist from a vertex in $S_{j_1}^i$ to a vertex in $S_{j_2}^i$ if $j_1 < j_2$. This also implies that the vertices of $S_{t_i}^i$ have out-degree 0 in H_i, and that t_i is the length of a longest directed path in H_i, so $t_i = \ell(H_i)$. In fact, all longest directed paths of H_i have their starting vertex in S_0^i and their terminating vertex in $S_{t_i}^i$. If a vertex $v \in V(H_i)$ is in the set S_j^i in the above ordering, we also say that v is at level j in H_i. Note that a vertex v of level $j > 0$ can only be reached from a vertex of level smaller than j, and that there always exists at least one vertex u of level $j-1$ with $uv \in A(H_i)$. Similarly, there exists a directed path of length p between some (not any) vertex at level j and some (not any) vertex at level $j+p$, but no longer directed paths (but possibly shorter directed paths). So, in particular, if there is a directed path of length p from a vertex u to a vertex v, and u is at level j, then v is at level at least $j+p$.

Apart from the inheritance of (copies of) the arcs and labels, the Cartesian product preserves some other important properties for our analysis. First of all, we show that the Cartesian product of a series of acyclic graphs $H_1, H_2, \ldots, H_k$ is again an acyclic graph, and that the length of a longest path in the Cartesian product is the sum of the lengths of longest paths in H_i, $i = 1, 2, \ldots, k$. In fact, we prove the stronger statement that each longest path P in $\Box H_i$ corresponds to longest paths in all H_i, in the sense that P contains exactly one copy of each of the arcs of a longest path Q_i in H_i, $i = 1, 2, \ldots, k$. We say that P is the interleaved concatenation of these Q_i.

Lemma 1. *Let H_i be an acyclic graph for $i = 1, 2, \ldots, k$, where $k \geqslant 2$. Then $\Box H_i$ is acyclic and every longest path in $\Box H_i$ is the interleaved concatenation of longest paths Q_i in H_i, $i = 1, 2, \ldots, k$. In particular, $\ell(\Box H_i) = \ell(H_1) + \ell(H_2) + \ldots + \ell(H_k)$.*

Proof. First note that it suffices to prove the statements for $k = 2$, since for integers $k \geqslant 3$, $H_1 \Box H_2 \Box \ldots \Box H_k$ is $((H_1 \Box H_2) \Box \ldots) \Box H_k$, hence $H_1' \Box H_2'$, and the result follows by induction. So we want to prove that $H_1 \Box H_2$ is acyclic and that every longest path in $H_1 \Box H_2$ is the interleaved concatenation of longest paths Q_1 and Q_2 in H_1 and H_2, respectively.

It is easy to show that there exists a path in $H_1 \Box H_2$ that is the interleaved concatenation of two longest paths Q_1 and Q_2 in H_1 and H_2, respectively. In fact, if $P = p_1 p_2 \ldots p_{k_1}$ and $Q = q_1 q_2 \ldots q_{k_2}$ are two vertex-disjoint (longest) paths, then clearly $P \Box Q$ contains the path $(p_1, q_1)(p_1, q_2) \ldots (p_1, q_{k_2})(p_2, q_{k_2}) \ldots (p_{k_1}, q_{k_2})$ with a length that is the sum of the lengths of P and Q.

The keys to the remaining part of the proof are the following observations on paths in $H_1 \Box H_2$. Consider a (longest) path P in $H_1 \Box H_2$ that starts with a subpath Q_1 of type H_1 arcs only, followed by a subpath R_1 with a first arc of type H_2 (and with R_1 possibly containing arcs of type H_1 as well). Then Q_1 corresponds directly to a path Q_1' in H_1 (with the first coordinates of the vertex pairs corresponding to the vertices in Q_1 as the vertices of Q'_1; all the second coordinates are identical and equal to one particular vertex of $V(H_2)$), while the vertex pairs corresponding to the two vertices of the arc connecting the end of Q_1 to the beginning of R_1 have the same first coordinate $x \in V(H_1)$. The vertex pairs corresponding to the vertices of R_1 keep this first coordinate x as long as the arcs are of type H_2. In case these arcs are followed by an arc of type H_1, this arc of type H_1 corresponds to an arc in H_1 starting from x. So, all the subsequent subpaths of P with only arcs of type H_1 correspond directly to paths $Q'_1, Q'_2, \ldots$ in H_1, and similarly all the subsequent subpaths of P with only arcs of type H_2 correspond directly to paths $R'_1, R'_2, \ldots$ in H_2. Moreover, there is an arc in H_1 between the end vertex of Q'_1 and the first vertex of Q'_2 (if any), and so on, and similarly for R'_1 and R'_2, and so on (if any) in H_2. By symmetry, the same observations can be made if the path P starts with an arc of type H_2, and contains arcs of both types.

To prove that $H_1 \square H_2$ is acyclic, suppose that it is not and contains a cycle C. Then the first and last vertices of C are identical, say equal to (p_1, q_1). It is clear that C contains arcs of both types; otherwise C corresponds directly to a cycle in H_1 or H_2, contradicting our assumption that H_1 and H_2 are both acyclic. Assuming, without loss of generality, that the first $k \geqslant 1$ arcs of C are of type H_1, p_1 is the first vertex of a path $Q'_1 = p_1 p_2 \ldots p_{k+1}$ in H_1, with the corresponding subpath of C in $H_1 \square H_2$ consisting of vertex pairs (p_i, q_1), $i = 1, \ldots, k+1$. Then the first arc of type H_2 in C we encounter is from (p_{k+1}, q_1) to (p_{k+1}, q_2) for some $q_1 q_2 \in A(H_2)$, and so on, so q_1 is the first vertex of a path R'_1 in H_2, as in the above argumentation. Since (p_1, q_1) also appears as the last vertex of C, by similar arguments p_1 and q_1 both appear as the last vertex of two paths Q'_t and R'_s in H_1 and H_2, respectively. Since by the above argumentation all the subpaths of type Q'_i are connected, this implies that H_1 contains a cycle: a *contradiction*. This proves that $H_1 \square H_2$ is acyclic.

Suppose now that P is a longest path in $H_1 \square H_2$. Assume that P has length $\ell(H_1 \square H_2) > \ell(H_1) + \ell(H_2)$. Using the above argumentation and the fact that $H_1 \square H_2$ is acyclic, the two paths Q and R formed by the Q'_i in H_1 and the R'_i in H_2, respectively, together have length $\ell(H_1 \square H_2) > \ell(H_1) + \ell(H_2)$, but this contradicts that the length of Q is at most $\ell(H_1)$ and the length of R is at most $\ell(H_2)$. Together with the above arguments, this shows that P has length exactly $\ell(H_1) + \ell(H_2)$ and that P is the interleaved concatenation of the two longest paths Q and R in H_1 and H_2, respectively. This completes the proof of Lemma 1. □

Remark 3.1

The expression in Lemma 1 on the length of longest paths in the Cartesian product is not valid if we drop the condition that each of the H_i is acyclic. It is easy to present counterexamples. For instance, consider $H_1 \square H_2$, where H_1 consists of two arcs connecting 3 vertices and H_2 has two arcs between two vertices, but in opposite directions (i.e. a cycle). As can be observed in Figure 5, the longest directed path lengths of H_1 and H_2 are 2 and 1, respectively[8], making their sum 3. However, $H_1 \square H_2$ contains a directed path of length 5.

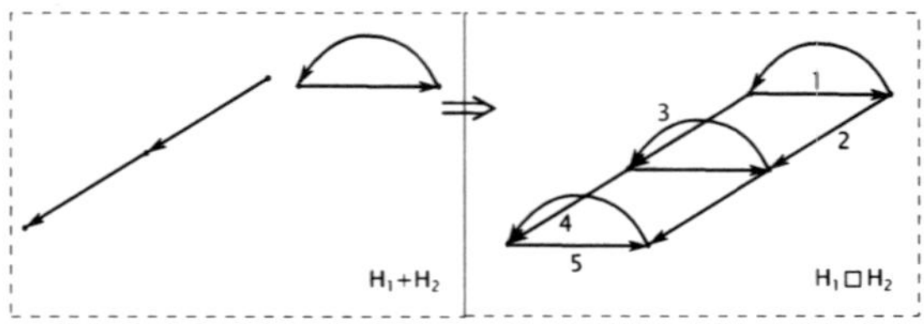

Figure 5. Cyclic graph counterexample.

[8]The vertices in a directed path must be distinct – see Section 1.

Remark 3.2

The notion of the level of a vertex in an acyclic directed graph, that we introduced before, has a natural extension to the Cartesian product, in the following sense. For a vertex $(v_1, v_2, \ldots, v_k) \in V(\Box H_i)$ we define the level vector $(f_1, f_2, \ldots, f_k)$, in which f_i denotes the level of vertex v_i in H_i. Then the vertices with in-degree 0 in $\Box H_i$ are precisely all vertices with level vector $(0, 0, \ldots 0)$, whereas level vector $(t_1, t_2, \ldots, t_k)$ with $t_i = \ell(H_i)$ corresponds to all vertices that are terminals of some longest path in $\Box H_i$. For each integer vector $(x_1, x_2, \ldots, x_k)$ with $0 \leqslant x_i \leqslant f_i$, there exists a vertex in $\Box H_i$ with this level vector, and if $x_i < f_i$, there also exists an arc of type H_i between a vertex with level vector $(x_1, x_2, \ldots, x_i, \ldots, x_k)$ and $(x_1, x_2, \ldots, x_i + 1, \ldots, x_k)$. This implies there are several longest paths in $\Box H_i$, each represented by adding one of the total of $t_1 + \ldots t_k$ units to one of the coordinates in the level vector between subsequent vertices on the path. On the other hand, there cannot be any arcs in $\Box H_i$ between a vertex with level vector $(x_1, x_2, \ldots, x_i, \ldots, x_k)$ and a vertex with level vector $(y_1, y_2, \ldots, x_i - 1, \ldots, y_k)$; all arcs imply an increase (by 1 or more) in precisely one entry of the level vector, while the other entries remain the same.

This shows how the partial ordering on the vertices of an acyclic directed graph has a natural extension to the Cartesian product. The same holds for the partial ordering on the arcs. Since the Cartesian product is again acyclic, we can define the same ordering there. So we define that for $a, b \in A(\Box H_i)$, $a < b$ if and only if a precedes b on some directed path in $\Box H_i$. From the structure it then follows that the ordering of the arcs in the individual H_i is preserved in the Cartesian product, in the following sense. If $a < b$ for two arcs $a, b \in A(\Box H_i)$ of the same type H_i, then for the corresponding arcs a' of a in H_i and b' of b in H_i, it holds that $a' < b'$ in H_i. The simplest way to see this is by the level vectors: if $b' < a'$, then the level of the head of b' in H_i is smaller than the level of the tail of a' in H_i, but then the corresponding coordinate in the level vector of the head (and thus of the tail) of b is also smaller than that of the tail (and thus of the head) of a in $\Box H_i$, contradicting that there is a directed path in $\Box H_i$ in which a precedes b.

Remark 3.3

In the above proof, we did not specifically consider the possibility of having multiple arcs, and we did not use the labels and weights in our arguments, for convenience. It is obvious that the proof is the same if we allow multiple arcs, because any directed path can contain at most one of these arcs, and we have already indicated how we can identify the corresponding arc in H_i from the inherited labels. It is also rather straightforward how the proof should be adapted if we consider weighted arcs and the length of a path is the sum of the weights of its arcs. The crucial observation is that every arc with a specific weight in $H_1 \Box H_2$ corresponds to either an arc in H_1 or an arc in H_2 with exactly the same weight, so instead of a contribution of 1 (which can be interpreted as weight 1) of an arc to the length of a path, we then have to use this specific weight. The weights have no influence on the level vectors, since the levels of the vertices are determined by the (non)existence of arcs, not by their weights. Of course, longest paths in terms of the highest total weight do not necessarily coincide with longest paths in terms of the largest number of arcs, so longest paths in the weighted sense may jump more than one unit in one of the coordinates of the level vector. Note, however, that these paths still start in a vertex with level vector $(0, 0, \ldots, 0)$ and terminate in a vertex with level vector $(t_1, t_2, \ldots, t_k)$ (where the t_i refer to the unweighted case of Remark 3.2 above).

Remark 3.4

An (acyclic) directed graph can have an exponentially high number of longest paths in terms of its number of vertices n. Consider for instance such a graph G with a square number of vertices, with $\sqrt{n}$ vertices of level i for all $i = 1,2,\ldots,\sqrt{n}$, and arcs between any two vertices from a lower level to a higher level, all with weight 1. Then the number of longest paths in G is $\sqrt{n}^{\sqrt{n}}$, so clearly exponential in n.

4. The Weak Synchronised Product of a Set of Parallel Processes

The Cartesian product of graphs is an adequate model for the interleaved execution of processes as long as the graphs represent independent processes. The model fails if the processes are not independent, for instance in case the processes must synchronise over certain actions. The different paths in the Cartesian product represent all possible (interleaved) traces of the constituent processes, thereby also representing behaviour that is simply impossible, due to synchronisation. For this reason we need a more restrictive notion than the Cartesian product of graphs. As for the Cartesian product, this has to be order-preserving. This product we are going to introduce next is based on the synchronised product by [10]. Figure 6 gives for our example the transformation of *SQ* consisting of the Cartesian product of the graphs $MS\Box RS\Box OD$ to the weak synchronised product $MS\boxminus RS\boxminus OD$.

The weak synchronised product $H_1\boxminus H_2$ of H_1 and H_2 is defined as the graph on vertex set $V(H_1)\times V(H_2)$ (the Cartesian product of the vertex sets) and arc set $A_{1,2}$ with four types of arcs.

The first two types correspond to arcs in H_1 and H_2 that have labels that only appear in one of H_1 and H_2. We call this set of arcs the *asynchronous arc set* and denote it by $A^a_{1,2}$. Therefore, $A^a_{1,2}$ is the set of all pairs $(v_1,x)(v_2,x)$ with $x\in V(H_2)$ and the associated label $\lambda(v_1v_2)$ (*a-type* H_1 arcs) or $(y,w_1)(y,w_2)$ with $y\in V(H_1)$ and the associated label $\lambda(w_1w_2)$ (*a-type* H_2 arcs), where for arcs $v_1v_2\in A(H_1)$ label $\lambda(v_1v_2)$ does not appear in H_2 and for arcs $w_1w_2\in A(H_2)$ label $\lambda(w_1w_2)$ does not appear in H_1.

The other types correspond to arcs in H_1 and H_2 with the same label. We call this set of arcs the *synchronous arc set* and denote it by $A^s_{1,2}$. Therefore, $A^s_{1,2}$ is the set of all arcs $(v_1,w_1)(v_2,w_1)$, $(v_1,w_1)(v_1,w_2)$, $(v_1,w_2)(v_2,w_2)$, $(v_2,w_1)(v_2,w_2)$, with the associated label $\lambda(v_1v_2)$, where for arcs $v_1v_2\in A(H_1)$ and $w_1w_2\in A(H_2)$ label $\lambda(v_1v_2)=\lambda(w_1w_2)$. The first and third are *s-type* H_1-arcs and the others are *s-type* H_2-arcs.

For $k\geqslant 3$, the weak synchronised product $H_1\boxminus H_2\boxminus\ldots\boxminus H_k$ is defined recursively as $((H_1\boxminus H_2)\boxminus\ldots)\boxminus H_k$. If no confusion can arise, we denote it as $\boxminus H_i$.

Although this weak synchronised product, like the Cartesian product, might represent behaviour that is not allowed by the original process specification, we will use it as an intermediate result. For example, in Figures 2 and 6, it is possible in both the Cartesian product and the weak synchronised product to reach a vertex that represents a non-reachable state in the process specification, by using only one of the synchronous arcs of a parallelogram.

We will first show that longest paths cannot be longer than in the Cartesian product, and that this new product also preserves acyclicity and the order on the arcs.

Lemma 2. *Let H_i be an acyclic graph for $i=1,2,\ldots,k$, where $k\geqslant 2$. Then $\boxminus H_i$ is acyclic and $\ell(\boxminus H_i)\leqslant\ell(\Box H_i)$.*

Proof. As in the proof of Lemma 1, it suffices to prove the statements for $k=2$, since for integers $k\geqslant 3$, the weak synchronised product $H_1\boxminus H_2\boxminus\ldots\boxminus H_k$ is defined recursively as $((H_1\boxminus H_2)\boxminus\ldots)\boxminus H_k$, hence $H_1'\boxminus H_2'$, and the result follows by induction.

It is obvious from the definitions that $H_1\boxminus H_2$ is a spanning subgraph of $H_1\Box H_2$ (i.e. that the vertex set of $H_1\Box H_2$ equals the vertex set of $H_1\boxminus H_2$, and that the arc set of $H_1\boxminus H_2$ is a subset of the arc set of $H_1\Box H_2$). From this observation, it follows by Lemma 1 that $H_1\boxminus H_2$ is acyclic and that $\ell(H_1\boxminus H_2)\leqslant\ell(H_1\Box H_2)$. □

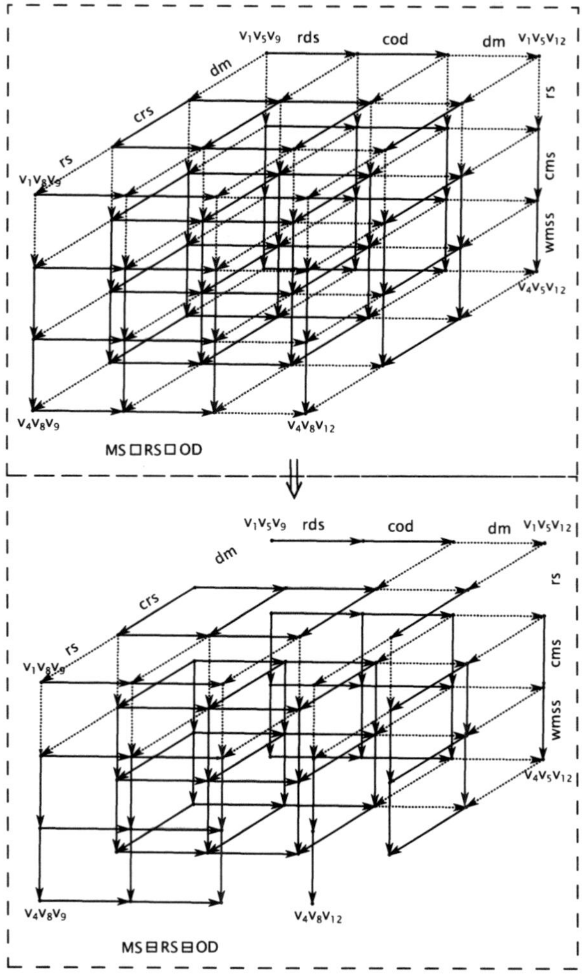

Figure 6. Sequence control processes of a mobile robot, from □ to ⊟.

Remark 4.1

It is not difficult to give examples of labelled directed acyclic graphs H_1 and H_2 with $\ell(H_1 \boxminus H_2) < \ell(H_1 \Box H_2)$. For example, Figure 7 shows H_1 consisting of one directed path $v_1v_2v_3$ with labels $\lambda(v_1v_2) = a$ and $\lambda(v_2v_3) = b$, and H_2 also consisting of one directed path $w_1w_2w_3$ with $\lambda(w_1w_2) = b$ and $\lambda(w_2w_3) = a$. In the context of processes, this example is ill-defined in the sense that it is immediately clear that the two processes are deadlocked from the start. The graph representing this pathological example is inconsistent. In Figure 8, we give a three dimensional example where we show that such a pathological case can occur distributed over several graphs, although only the final step reduces the out-degree of the source to zero. The source vertex in Figure 8 is marked by a circle around the vertex. From three dimensions it is not difficult to extend to n-dimensions. A set of graphs H is said to be consistent if it does not contain a (possibly n-dimensional) pathological case.

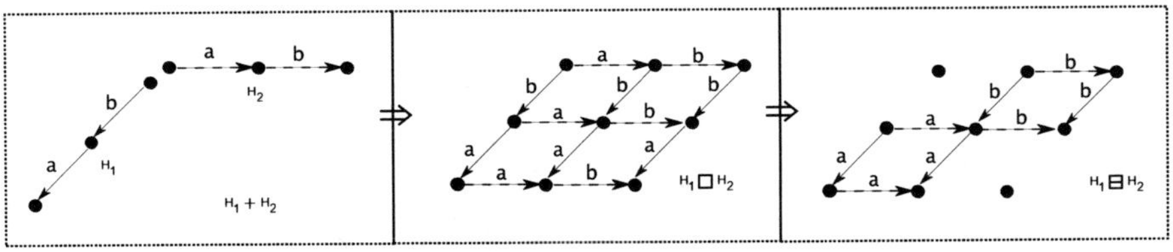

Figure 7. Two-dimensional pathological case.

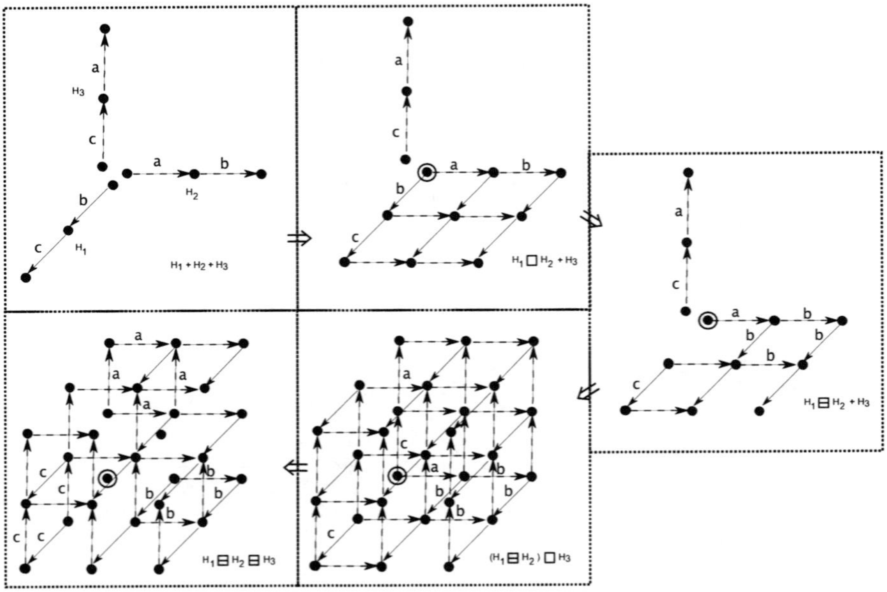

Figure 8. Three-dimensional pathological case.

We are now going to show that inconsistency can always be concluded if $\ell(\boxminus H_i) < \ell(\square H_i)$. By induction, we can again restrict our attention to the case that $k = 2$.

Let P be a longest path in $H_1 \square H_2$. Then P is the interleaved concatenation of two longest paths R and Q in H_1 and H_2, respectively. Let $R = r_1 r_2 \ldots r_{k_1}$ and $Q = q_1 q_2 \ldots q_{k_2}$. If P is not a longest path in $H_1 \boxminus H_2$, there is at least one label λ that appears on arcs in both R and Q. Consider the first label on R (starting from r_1) that also appears in Q – say this is label λ_1 that appears on r_j. If λ_1 is also the first label on Q that appears in both Q and R, say on q_t, then $R \boxminus Q$ contains a path of length $j+t$ corresponding to the subpath of P of length $j+t$ from the starting vertex.

Continuing this way, if all labels that appear in both Q and R also appear in the same order in Q and R (with possible repetitions, also in the same order), then $H_1 \boxminus H_2$ contains a path with the same length as P. So, if $\ell(H_1 \boxminus H_2) < \ell(H_1 \square H_2)$, we may assume there is an ith instance of a label λ_r on R that is also the ith instance of that label on Q, and a jth instance of a label λ_q on Q that is also the jth instance of that label on R, and such that λ_r is after λ_q on R but before λ_q on Q. But then the process is ill-defined in a similar way as in the pathological example, a situation that can and should be avoided.

For the sequel, we are going to assume that the processes are defined and specified in such a way, that the above unwanted situation does not occur and that the related graphs H_i are therefore consistent. This then automatically implies that for consistent graphs H_i, $\ell(\boxminus H_i) = \ell(\Box H_i)$.

Remark 4.2

From the fact that $\boxminus H_i$ is a spanning subgraph of $\Box H_i$, it follows that the ordering of the arcs in the individual H_i is preserved in the weak synchronised product, in the following sense. If $a < b$ for two arcs $a, b \in A(\boxminus H_i)$ of a-type or s-type for the same H_i, then for the corresponding arcs a' of a in H_i and b' of b in H_i, it holds that $a' < b'$ in H_i.

This can be seen by using the level vectors. Suppose $b' < a'$. Then, the level of the head of b' in H_i is smaller than the level of the tail of a' in H_i. This means that the corresponding coordinate in the level vector of the head (and thus of the tail) of b is also smaller than that of the tail (and thus of the head) of a in $\boxminus H_i$, contradicting that there is a directed path in $\boxminus H_i$ in which a precedes b. Therefore, the supposition is false.

Remark 4.3

The weak synchronised product, like the Cartesian product, may still represent behaviour that is not possible by the specification of the corresponding set of processes, as can be seen in the examples in Figures 2 and 6 ($\Box \Rightarrow \boxminus$). One obvious thing we can do about this is, that we iteratively remove vertices (and the related arcs) that have an in-degree $\neq 0$ in the Cartesian product, but have an in-degree $= 0$ in the weak synchronised product.

5. The Reduced Weak Synchronised Product of a Set of Parallel Processes

The reduced weak synchronised product $H_1 \boxdot H_2$ of H_1 and H_2 is defined as the graph obtained from the synchronised product $H_1 \boxminus H_2$ by first removing all vertices with level 0 in $H_1 \boxminus H_2$ that have level > 0 in $H_1 \Box H_2$, together with all the arcs that have one of these vertices as a tail. This is then repeated in the newly obtained graph, and so on, until there are no more vertices with level 0 in the current graph that have level > 0 in $H_1 \Box H_2$. The resulting graph for our standard example is shown in Figure 9.

For $k \geqslant 3$, the reduced weak synchronised product $H_1 \boxdot H_2 \boxdot \ldots \boxdot H_k$ is defined recursively as $((H_1 \boxdot H_2) \boxdot \ldots) \boxdot H_k$, and denoted as $\boxdot H_i$ if no confusion can arise.

Lemma 3. *Let H_i be an acyclic graph and let $\boxdot H_i$ be the reduced weak synchronised product of H_i for $i = 1, 2, \ldots, k$, where $k \geqslant 2$. Then $\ell(\boxdot H_i) = \ell(\boxminus H_i)$.*

Proof. One direction is clear: since $\boxdot H_i$ is a subgraph of $\boxminus H_i$, we have that $\ell(\boxdot H_i) \leqslant \ell(\boxminus H_i)$. For the other direction, consider a longest path P in $\boxminus H_i$. By previous arguments, we know that P starts in a vertex v with level vector $(0, 0, \ldots, 0)$ and terminates in a vertex w with level vector $(t_1, t_2, \ldots, t_k)$. For each vertex $x \neq v, w$ on P, $d^-(x) \geqslant 1$ and $d^+(x) \geqslant 1$. This implies that none of the vertices of P is removed from $\boxminus H_i$, so P is a path and therefore a longest path in $\boxdot H_i$. This completes the proof of Lemma 3. $\Box$

Remark 5.1

Note that the paths that represent the behaviour of the specified processes, all start in the source of the graph and end in the sink of the graph. Because we only remove vertices that are not in the source of the graph and have an in-degree of zero, behaviour not specified by the original set of processes is removed.

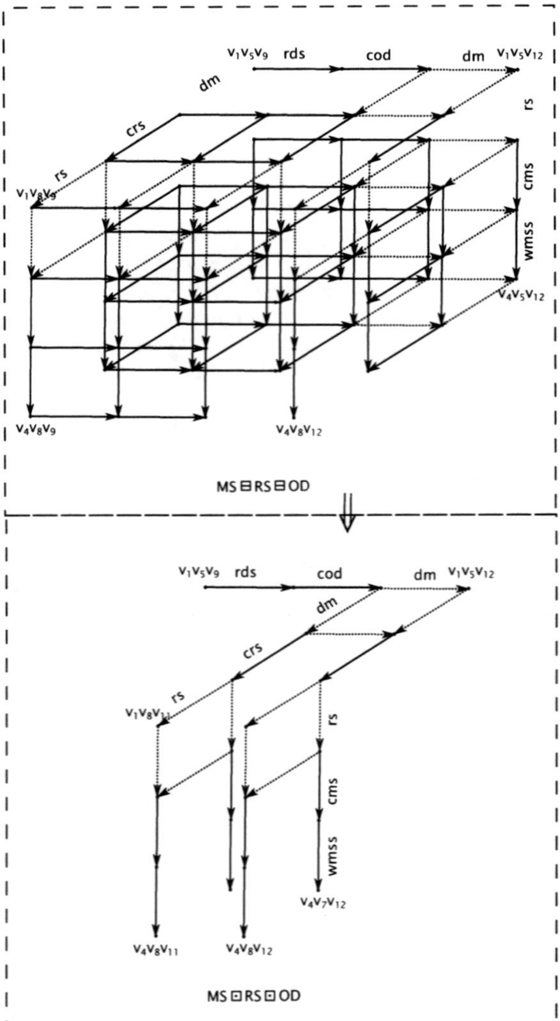

Figure 9. Sequence control processes of a mobile robot, from ⊟ to ⊡.

Remark 5.2

Also note that, although this newly introduced product may filter out vertices and arcs representing unwanted process behaviour, it does not filter out all unwanted behaviour, see Figure 9. In Section 6, we translate additional restrictions into our product.

6. The Synchronised Product of a Set of Parallel Processes

The synchronised product $H_1 \boxbslash H_2$ of H_1 and H_2 is defined from the reduced weak synchronised product, by first replacing quadruples of arcs that represent synchronised arcs as follows.

Replace each parallelogram of arcs $(v_1,w_1)(v_2,w_1)$, $(v_1,w_1)(v_1,w_2)$, $(v_1,w_2)(v_2,w_2)$ and $(v_2,w_1)(v_2,w_2)$, with $\lambda((v_1,w_1)(v_2,w_1)) = \lambda((v_1,w_1)(v_1,w_2)) = \lambda((v_1,w_2)(v_2,w_2)) = \lambda((v_2,w_1)(v_2,w_2))$, by one diagonal arc $(v_1,w_1)(v_2,w_2)$ with label $\lambda((v_1,w_1)(v_2,w_2)) = \lambda((v_1,w_1)(v_2,w_1))$. These new arcs of $H_1 \boxbslash H_2$ are called synchronous arcs, and the set of these arcs is denoted as $A^s_{1,2}$. This intermediate stage is shown in Figure 10.

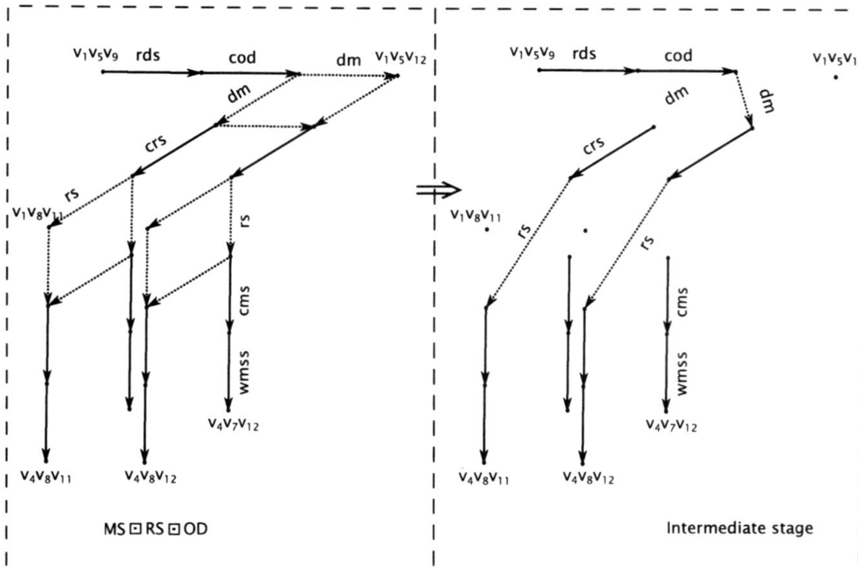

Figure 10. Sequence control processes of a mobile robot, from $\boxdot$ to intermediate stage.

Secondly, all vertices with level 0 in the resulting graph that have level > 0 in $H_1 \Box H_2$ are removed, together with all the arcs that have one of these vertices as a tail. This is then repeated in the newly obtained graph, and so on, until there are no more vertices with level 0 in the current graph that have level > 0 in $H_1 \Box H_2$. The resulting graph is called the synchronised product and denoted as $H_1 \boxbslash H_2$. The set of arcs consisting of the other remaining (asynchronous) arcs of $H_1 \boxbslash H_2$ is denoted as $A^a_{1,2}$. The resulting graph for our standard example is shown in Figure 11.

For $k \geqslant 3$, the synchronised product $H_1 \boxbslash H_2 \boxbslash \ldots \boxbslash H_k$ is defined recursively as $((H_1 \boxbslash H_2) \boxbslash \ldots) \boxbslash H_k$, and denoted as $\boxbslash H_i$ if no confusion can arise.

Lemma 4. *Let H_i be an acyclic graph and let $\boxbslash H_i$ be the synchronised product of H_i for $i = 1,2,\ldots,k$, where $k \geqslant 2$. Then $\ell(\boxbslash H_i) \leqslant \ell(\boxdot H_i)$.*

Proof. As in the proof of Lemma 1, it suffices to prove the statement for $k = 2$, since for integers $k \geqslant 3$, the synchronised product $H_1 \boxbslash H_2 \boxbslash \ldots \boxbslash H_k$ is defined recursively as $((H_1 \boxbslash H_2) \boxbslash \ldots) \boxbslash H_k$, hence $H'_1 \boxbslash H'_2$, and the result follows by induction.

From the definitions of reduced weak synchronised product and synchronised product, it follows that the vertex set of $H_1 \boxbslash H_2$ is a subset of the vertex set of $H_1 \boxdot H_2$, and the asynchronous arc set $A^a_{1,2}$ of $H_1 \boxbslash H_2$ is a subset of the asynchronous arc set of $H_1 \boxdot H_2$. For the synchronous arc set $A^s_{1,2}$ of $H_1 \boxbslash H_2$ every arc replaces a quadruple of arcs in $H_1 \boxdot H_2$, as follows: $\{(v_1,w_1)(v_2,w_1), (v_1,w_1)(v_1,w_2), (v_1,w_2)(v_2,w_2), (v_2,w_1)(v_2,w_2)\}$ with an associated label $\lambda((v_1,w_1)(v_2,w_1)) = \lambda((v_1,w_1)(v_1,w_2)) = \lambda((v_1,w_2)(v_2,w_2)) = \lambda((v_2,w_1)(v_2,w_2))$

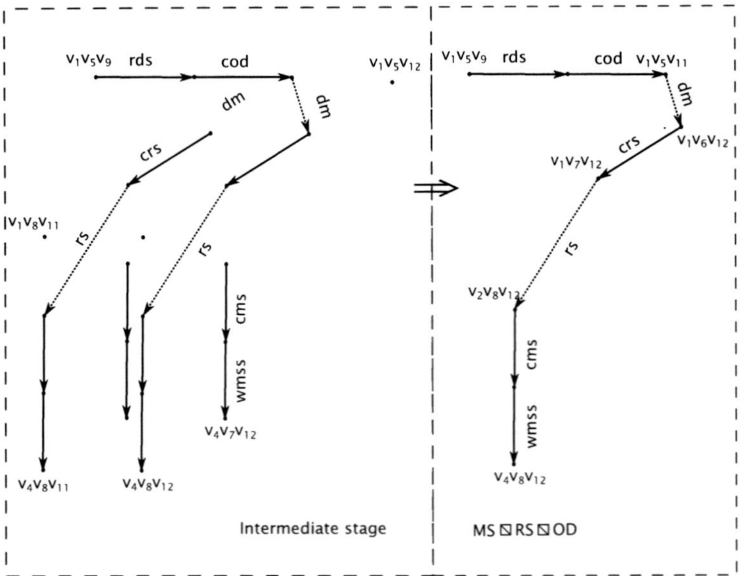

Figure 11. Sequence control processes of a mobile robot, from intermediate stage to ⧅.

in $H_1 \boxdot H_2$ is replaced by $(v_1,w_1)(v_2,w_2)$ with the associated label $\lambda((v_1,w_1)(v_2,w_2)) = \lambda((v_1,w_1)(v_2,w_1))$. Clearly, the length of (a longest path in) the graph with vertex set $\{(v_1,w_1), (v_1,w_2), (v_2,w_1), (v_2,w_2)\}$ and arc set $\{(v_1,w_1)(v_2,w_1), (v_1,w_1)(v_1,w_2), (v_1,w_2)(v_2,w_2), (v_2,w_1)(v_2,w_2)\}$ is twice the length of the arc $(v_1,w_1)(v_2,w_2)$. This shows that the length of a longest path in the synchronised product is not greater than the length of a longest path in the reduced synchronised product (but it will be smaller if synchronisation occurs between the constituent paths). From these observations it follows that, because $H_1 \boxdot H_2$ is acyclic, $H_1 ⧅ H_2$ is acyclic and $\ell(H_1 ⧅ H_2) \leqslant \ell(H_1 \boxdot H_2)$. □

Remark 6.1

The proof of Lemma 4 shows that combining processes *may* lead to a performance gain, where the gain $\mathcal{G}$ is defined by $\mathcal{G} = \sum_{i=1}^{k} \ell(H_i) - \ell(⧅_{i=1}^{k} H_i)$. It is clear from the above that a gain is only guaranteed if $\ell(⧅H_i) < \ell(\boxdot H_i)$. Logically, this means that we can only be sure of a gain if there exist distinct indices i and j such that for every longest path P in H_i and for every longest path Q in H_j, the paths P and Q contain at least one synchronising arc, so there are arcs $a \in A(P)$ and $b \in A(Q)$ with $\lambda(a) = \lambda(b)$. To get a performance gain we need necessary and sufficient conditions that will reduce the length of the synchronised product with respect to the length of its constituent graphs. It is obvious (follows from Lemma 4) that a reduction can only be achieved by synchronising arcs. As the length of a graph is defined as the size of its longest paths, we only have to consider the synchronisation of synchronising arcs in longest paths.

Lemma 5. *Let H_i be an acyclic graph for $i = 1,2,\ldots,k$, where $k \geqslant 2$. Then $\ell(⧅H_i) = \ell(H_1) + \ell(H_2) + \ldots + \ell(H_k)$ if and only if every H_i has at least one longest path without synchronising arcs.*

Proof. First note that it suffices to prove the statement for $k = 2$, since for integers $k \geqslant 3$, $H_1 \boxbslash H_2 \boxbslash \ldots \boxbslash H_k$ is $((H_1 \boxbslash H_2) \boxbslash \ldots) \boxbslash H_k$, hence $H_1' \boxbslash H_2'$, and the result follows by induction.

$\Leftarrow$ By Lemma 1, $\ell(H_1 \Box H_2) = \ell(H_1) + \ell(H_2)$. If $P = p_1 p_2 \ldots p_{k_1}$ and $Q = q_1 q_2 \ldots q_{k_2}$ are two vertex-disjoint longest paths without synchronising arcs of H_1, H_2 respectively, then clearly $P \Box Q$ contains the path PQ, where PQ denotes the path $PQ = (p_1, q_1)(p_1, q_2) \ldots (p_1, q_{k_2})(p_2, q_{k_2}) \ldots (p_{k_1}, q_{k_2})$. By the definition of $H_1 \boxminus H_2$, it follows that $H_1 \boxminus H_2$ contains the path PQ, even so by definition $H_1 \boxdot H_2$ and $H_1 \boxbslash H_2$ contain the path PQ. As $\ell(PQ) = \ell(H_1) + \ell(H_2)$ it follows that $\ell(H_1) + \ell(H_2) = \ell(H_1 \boxbslash H_2)$.

$\Rightarrow$ The proof is by contra-position. Suppose that all longest paths $P = p_1 p_2 \ldots p_{k_1}, Q = q_1 q_2 \ldots q_{k_2}$ of H_1 and H_2, without loss of generality, contain one synchronising arc a with label $\lambda(a)$, say from p_i to p_{i+1} and q_j to q_{j+1}. The synchronised product of paths P and Q is $P' \Box Q' \bigcup (p_i p_{i+1} \boxbslash q_j q_{j+1}) \bigcup P'' \Box Q''$, with $P' = p_1 p_2 \ldots p_i$, $P'' = p_{i+1} \ldots p_{k_1}$, $Q' = q_1 q_2 \ldots q_j$, $Q'' = q_{j+1} \ldots q_{k_2}$. Therefore it follows that $\ell(P \boxbslash Q) = \ell(P' \Box Q') + \ell((p_i p_{i+1} \boxbslash q_j q_{j+1})) + \ell(P'' \Box Q'')$.

Note that $p_i p_{i+1}$ and $q_j q_{j+1}$ have the same label and therefore the same weight t. Therefore $\ell(p_i p_{i+1} \Box q_j q_{j+1}) = 2 \times t = 2 \times \ell(p_i p_{i+1} \boxbslash q_j q_{j+1})$ (due to the synchronisation constraint) and it follows that $\ell(P \boxbslash Q) = \ell(P' \Box Q') + \ell((p_i p_{i+1} \boxbslash q_j q_{j+1})) + \ell(P'' \Box Q'') = \ell(P' \Box Q') + t + \ell(P'' \Box Q'') < \ell(P' \Box Q') + 2 \times t + \ell(P'' \Box Q'') = \ell(P' \Box Q') + \ell((p_i p_{i+1} \Box q_j q_{j+1})) + \ell(P'' \Box Q'')$, $\ell(P \Box Q)$ so the synchronised product will reduce the length of the longest paths in H_1 and H_2. This leads to $\ell(H_1 \Box H_2) > \ell(H_1 \boxbslash H_2)$. □

We need necessary and sufficient conditions to get to $\ell(\boxbslash H_i) < \ell(\boxdot H_i)$.

Lemma 6. *Let H_i be an acyclic graph for $i = 1, 2, \ldots, k$, where $k \geqslant 2$. Then $\ell(\boxbslash H_i) < \ell(\boxdot H_i)$ if there exists H_n, H_m, $n \neq m$, $1 \leqslant n, m \leqslant k$, such that each longest path in H_n, H_m, contains at least one same labelled synchronising arc.*

Proof. Again it suffices to prove the statements for $k = 2$, since for integers $k \geqslant 3$, $H_1 \boxbslash H_2 \boxbslash \ldots \boxbslash H_k$ is $((H_1 \boxbslash H_2) \boxbslash \ldots) \boxbslash H_k$, hence $H_1' \boxbslash H_2'$, and the result follows by induction.

From Lemma 5, we have that every H_i has at least one longest path without synchronising arcs if and only if $\ell(H_1 \boxbslash H_2 \boxbslash \ldots \boxbslash H_k) = \ell(H_1) + \ell(H_2) + \ldots + \ell(H_k)$, therefore as both H_1 and H_2 contain only longest paths with at least one synchronisation arc, both H_1 and H_2 do not contain a longest path without synchronising arcs. From this observation it follows that $\ell(H_1 \boxbslash H_2) \neq \ell(H_1) + \ell(H_2)$. By Lemma 4, $\ell(H_1 \boxbslash H_2) \leqslant \ell(H_1 \boxdot H_2)$, it follows that $\ell(H_1 \boxbslash H_2) < \ell(H_1) + \ell(H_2)$. Together with the observation that $\ell(H_1 \boxdot H_2) = \ell(H_1) + \ell(H_2)$ this gives $\ell(H_1 \boxbslash H_2) < \ell(H_1 \boxdot H_2)$. □

Lemma 6 is rather restrictive. We can loosen the requirements on two graphs containing only longest paths with synchronisation arcs, to one graph containing only longest paths with synchronisation arcs and another graph containing at least one longest path containing a synchronisation arc. The rationale behind it is that a longest path P_1 without a synchronisation arc, and a longest path P_2 with a synchronisation arc, both in H_1, combined with a longest path Q with a synchronisation arc in H_2, will lead to graphs consisting of the reduced weak synchronised products of P_1 and the part of Q up to the synchronisation arc, and the reduced weak synchronised products of P_2 and Q. It is obvious that $\ell(P_1 \boxdot Q)$ is smaller than $\ell(P_2 \boxdot Q)$.

Theorem 1. *Let H_i be an acyclic graph for $i = 1, 2, \ldots, k$, where $k \geqslant 2$. Then $\ell(\boxbslash H_i) < \ell(\boxdot H_i)$ if there exists H_n, H_m, $n \neq m$, $1 \leqslant n, m \leqslant k$, such that each longest path in H_n, contains at least one synchronising arc and there is at least one longest path with a same labelled synchronisation arc in H_m.*

Proof. Again it suffices to prove the statements for $k = 2$, since for integers $k \geqslant 3$, $H_1 \boxslash H_2 \boxslash \ldots \boxslash H_k$ is $((H_1 \boxslash H_2) \boxslash \ldots) \boxslash H_k$, hence $H_1' \boxslash H_2'$, and the result follows by induction.

Let all longest paths of H_1 be of the structure $P = p_1 p_2 \ldots p_{k_1}$, without loss of generality, containing one synchronising arc a with label $\lambda(a)$, say from p_i to p_{i+1}. Let there be at least one longest path $Q = q_1 q_2 \ldots q_{k_2}$ of H_2 containing one synchronising arc a with label $\lambda(a)$, say from q_j to q_{j+1}. Note that $p_i p_{i_1}$ and $q_j q_{j+1}$ have the same label and therefore the same weight t. Let there be at least one longest path $R = r_1 r_2 \ldots r_{k_3}$ of H_2 containing no synchronising arc.

Let $P' = p_1 p_2 \ldots p_i, P'' = p_{i+1} p_{i+2} \ldots p_{k_1}, Q' = q_1 q_2 \ldots q_j, Q'' = q_{j+1} q_{j+2} \ldots q_{k_2}$. Then $\ell(P \boxslash Q) = \ell(P' \boxslash Q') + \ell(p_i p_{i+1} \boxslash q_j q_{j+1}) + \ell(P'' \boxslash Q'') = \ell(P' \square Q') + 1 \times t + \ell(P'' \boxslash Q'') < \ell(P' \square Q') + 2 \times t + \ell(P'' \boxslash Q'') = \ell(P' \boxdot Q') + \ell(p_i p_{i+1} \boxdot q_j q_{j+1}) + \ell(P'' \boxdot Q'') = \ell(P \boxdot Q)$.

Because both Q and R are longest paths of H_2, $\ell(Q) = \ell(R)$. Due to the synchronisation constraints, $\ell(P \boxdot R) = \ell(P' \boxdot R) = \ell(P') + \ell(R) < \ell(P) + \ell(R) = \ell(P) + \ell(Q) = \ell(P \boxdot Q)$.

These two results, $\ell(P \boxslash Q) < \ell(P \boxdot Q)$ and $\ell(P \boxdot R) < \ell(P \boxdot Q)$, complete the proof of Theorem 1. □

Remark 6.2

We may have the case that there are no more H_n, H_m that can be combined in the manner of Theorem 1. Still further synchronisation is possible, if there exists $H_{m_i}, m_i \neq n$, where for each longest path in H_n, there is a longest path containing a synchronising arc in $\mathop{\square}\limits_{i=1}^{l} H_{m_i}, l < k$.

7. Conclusions

With Theorem 1, we have proved that if one wants to reduce the worst-case performance of periodic real-time parallel processes, one can combine processes, where all longest traces for at least one process must contain synchronising actions and at least one other process must contain at least one longest trace with a synchronising action. To reach this point we have introduced graph products that can help us to analyse and combine a number of parallel processes. We were able to identify the pathological case in a natural manner by introducing the weak synchronised product. This made it visible that a set of parallel processes may contain unwanted behaviour, for example a deadlocked state. We have shown in the proof of Lemma 4 and Remark 4.1, that we can filter out this unwanted or ill-defined behaviour.

We informally introduced the notion of a consistent and an inconsistent set of graphs (representing real-time periodic processes). The latter represents behaviour of processes that is unwanted, but might appear in a non-trivial process specification. From our proof, it follows that one can detect whether such a situation occurs in a process specification: one just has to find paths that shrink when the weak synchronised product is taken.

Finally, we have shown how to get to the synchronised product, which can be used to improve the worst case performance of parallel processes, and how processes might be combined on synchronising actions in order to obtain a performance gain.

7.1. Discussion

The performance gain is significant if the set of parallel processes will miss deadlines if not synchronised, but will meet its deadlines if synchronised. Whether such a significant performance gain is achieved by combining processes is not clear. Firstly, a tool that will produce a synchronised product of parallel processes based on the transformations described in our paper does not exist yet. Secondly, whether or not a significant performance gain is achieved by combining processes depends on the ratio of the context switch time and the calculation time of the processes itself; clearly this depends on the type of hardware and operating system used.

For context switches, Li [1] distinguishes between direct and indirect costs with respect to the processing power. The direct costs consist of issues like saving and restoring registers, translation table look aside buffer entries that need to be reloaded, flushing of the processor pipe-line, but also kernel code that has to execute. Indirect costs include cache misses caused when context switches to a process whose cache lines have been reused. Such costs may degrade performance in a significant way. Li [1] also shows that the average direct cost is 3.8 μs. The indirect cost varies from a few microseconds to more than one millisecond, all on a 2.0 GHz Intel Pentium Xeon CPU, with 512 kB L2 cache. The operating system is a Linux 2.6.17 kernel with Redhat 9.

Veldhuijzen [2] shows that the cost is on average 7.7μs on a 560 MHz Pentium IV processor, running under the QNX operating system. A typical control loop as used in [2] takes 70 μs. Together with the motion profile and the many context switches it takes up to 650 μs. This is well within the boundary of 1 *ms*, the period of a control loop. Veldhuijzen [2] gives a gain of 100 - 140 μs. This would mean a gain up to 15 to 20 %.

As a contrast, Ritson et al. [3] show context switch overheads for occam-π, under the KRoC CCSP multicore scheduler, of the order of 100 nanoseconds - often much less (around 30 nanoseconds). For such systems, the value of the transformations described occurs when the granularity of concurrency becomes too fine even for that language and scheduler – and our ambitions for ever-more complex behaviour from systems drive us in that direction.

7.2. Future Work

The graph products we introduced will form the basis for further research. One of the main aims of further study is to develop exact algorithms and heuristics for optimising the performance gain by combining processes. To get the set of all longest paths in a graph is exponential in n, as shown in Remark 3.4 and therefore not tractable. It is essential that *all* longest paths are found. It is obvious that, for example, a breadth-first search will give an answer to the length of the longest path; but this is not sufficient. If there is a longest path in graph H_1 that does not have a synchronising arc in common with a longest path in graph H_2, the synchronised product will have the same length as the Cartesian product and no gain is achieved. If such a situation exists in a hard real-time system, then, if the original parallel specification has a deadline-miss due to these two (interleaved) traces, also the specification represented by the synchronised product may have the same deadline-miss.

In our case, robotic applications, although the number of states and the number of actions in a process is limited, this may lead to the need for a heuristic giving a reasonable performance gain.

This research is restricted to periodic real-time applications, where the periods, release time and deadlines are the same. This can be extended to applications where this is not the case. An issue that will arise is the scheduling of the synchronised product, because this product will have internal deadlines.

Furthermore the aspect of memory usage is not taken into account. Combining graphs leads to a state space explosion, although synchronisation may reduce the magnitude of the explosion. To reduce such an explosion, it might be necessary to combine only a subset of the set of graphs representing the parallel process specification. Decomposition of the original graph into its prime factors, will give the optimal set of graphs from which synchronised products can be taken. This leads to a set of graphs that still fits in the available memory and has a maximal performance gain. Another application of the decomposition into prime products is in case the original specification does not fit in the available memory. It might not always be possible to just extend the available memory. As an example, imagine a robot, running around on Mars, needs a software update, but the model just does not fit in the available memory. It would be difficult to get a memory extension in place. The same is applicable in robots operating in, for example, nuclear fusion reactors. Maintenance under these (hot) circumstances of the robot is also difficult. Clearly these situations apply to much more complex applications than we are considering. For the above mentioned reasons we have to show the associativity and commutativity of our synchronised product. Furthermore we have to define the constraints for the prime factors of our synchronised product and give an algorithm that calculates such factors.

A valid question is: what is the maximum gain that can be achieved by combining processes within a certain amount of available memory? With that knowledge, we can improve the performance of the systems produced with tools like LUNA [13] and TERRA [14]. These issues are for future research.

Acknowledgements

We would like to express our gratitude to the anonymous reviewers for the very useful suggestions and comments.

References

[1] Chuanpeng Li, Chen Ding, and Kai Shen. Quantifying the Cost of Context Switch. In *Proceedings of the 2007 Workshop on Experimental Computer Science*, ExpCS '07, New York, NY, USA, 2007. ACM.

[2] B. Veldhuijzen. Redesign of the CSP Execution Engine. MSc thesis 036CE2008, Control Engineering, University of Twente, February 2009.

[3] Carl G. Ritson, Adam T. Sampson, and Frederick R. M. Barnes. Multicore Scheduling for Lightweight Communicating Processes. In John Field and Vasco Thudichum Vasconcelos, editors, *Coordination Models and Languages, COORDINATION 2009, Lisboa, Portugal, June 9-12, 2009*, volume 5521 of *Lecture Notes in Computer Science*, pages 163–183. Springer, June 2009. `http://www.cs.kent.ac.uk/pubs/2009/2928/`.

[4] Charles A. R. Hoare. *Communicating Sequential Processes*. Prentice-Hall, 1985.

[5] J.A. Bondy and U.S.R. Murty. *Graph Theory*. Springer, Berlin, 2008.

[6] S. Schneider. *Concurrent and Real Time Systems: the CSP Approach*. John Wiley Sons, Inc., New York, NY, USA, 1st edition, 1999.

[7] M. Aiguier, C. Gaston, P. Le Gall, D. Longuet, and A. Touil. A Temporal Logic for Input Output Symbolic Transition Systems. In *Software Engineering Conference, 2005. APSEC '05. 12th Asia-Pacific*, pages 8 pp.–, 2005.

[8] Caucal, D and Hassen, S. Synchronization of Grammars. In *Proceedings of the 3rd International Conference on Computer Science: Theory and Applications*, CSR'08, pages 110–121, Berlin, Heidelberg, 2008. Springer-Verlag.

[9] Youcef Hammal. A Component-based Approach for Consistency Checking of UML Dynamic Diagrams. In *Proceedings of the 11th IASTED International Conference on Software Engineering and Applications*, SEA '07, pages 192–197, Anaheim, CA, USA, 2007. ACTA Press.

[10] S. Wöhrle and W. Thomas. Model Checking Synchronized Products of Infinite Transition Systems. In *Proc. 19th LICS, IEEE Comp. Soc*, pages 2–11. IEEE Computer Society Press, 2004.

[11] O. Oguz, J.F. Broenink, and A. H. Mader. Schedulability Analysis of Timed CSP Models Using the PAT Model Checker. In P. H. Welch, F. R. M. Barnes, K. Chalmers, J. B. Pedersen, and A. T. Sampson, editors, *Communicating Process Architectures 2012, Dundee, Scotland*, pages 65–88. Open Channel Publishing, August 2012. WoTUG-34.

[12] J Bang-Jensen and G.Z. Gutin. *Digraphs: Theory, Algorithms and Applications*. Springer Publishing Company, Incorporated, 2nd edition, 2008.

[13] M. M. Bezemer, R. J. W. Wilterdink, and J. F. Broenink. LUNA: Hard Real-Time, Multi-Threaded, CSP-Capable Execution Framework. In P. H. Welch, A. T. Sampson, J. B. Pedersen, J. M. Kerridge, J. F. Broenink, and F. R. M. Barnes, editors, *Communicating Process Architectures 2011, Limerick, Ireland*, volume 68 of *Concurrent System Engineering Series*, pages 157–175, Amsterdam, November 2011. IOS Press BV. WoTUG-33.

[14] M. M. Bezemer, R. J. W. Wilterdink, and J. F. Broenink. Design and Use of CSP Meta-Model for Embedded Control Software Development. In P. H. Welch, F. R. M. Barnes, K. Chalmers, J. B. Pedersen, and A. T. Sampson, editors, *Communicating Process Architectures 2012, Dundee, Scotland*, pages 185–199, England, August 2012. Open Channel Publishing Ltd. WoTUG-34.

Appendix

In Listing 2, we give an example for the serialisation of two processes containing choice. Two processes synchronise over the actions a, c, and e. According the process specification of Listing 2, two traces can occur, $d \rightarrow c \rightarrow e$ and $a \rightarrow b \rightarrow e$. The last stage of Figure 12 shows the graph representing these two traces.

```
H1  =  (a → b → H1')
        □
        (d → c → H1')
H1' =  e → SKIP

H2  =  (a → H2')
        □
        (c → H2')
H2' =  e → SKIP
H   =  H1 {a,b,c,d,e}||{a,c,e} H2
```

Listing 2. Description of the CHOICE in two parallel processes.

Assuming that the weight of all arcs is 1, the graph consisting of the two components H_1 and H_2 has 8 vertices and a length $\ell(H_1) + \ell(H_2) = 5$, whereas the synchronized product $\ell(H_1 \boxbslash H_2)$ has 5 vertices and a length $\ell(H_1 \boxbslash H_2) = 3$. This example shows a gain for both the memory occupancy, as the performance of the application.

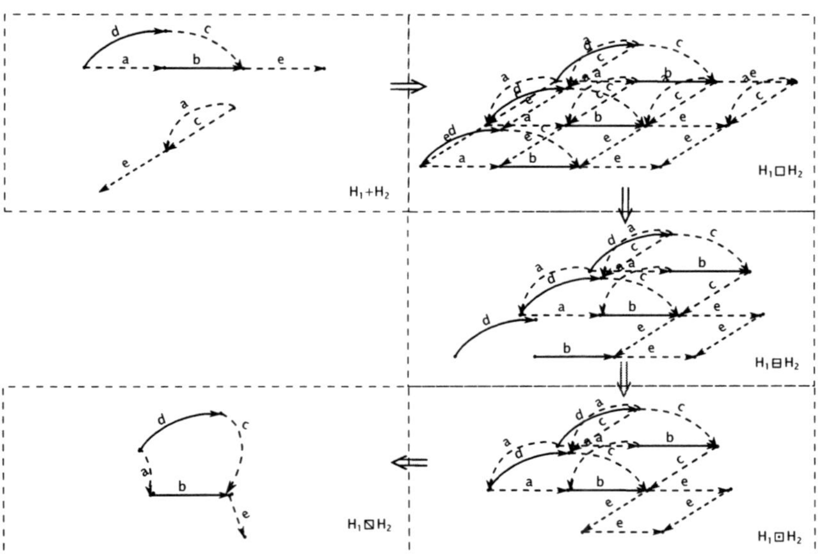

Figure 12. Choice in two parallel processes, from $+$ through $\boxbslash$.

Communicating Process Architectures 2013
P.H. Welch et al. (Eds.)
Open Channel Publishing Ltd., 2013

Scaling PyCSP

Rune Møllegaard FRIBORG [a,1], John Markus BJØRNDALEN [b] and Brian VINTER [c]
[a] *BiRC, University of Aarhus, Denmark*
[b] *Department of Computer Science, University of Tromsø, Norway*
[c] *Niels Bohr Institute, University of Copenhagen, Denmark*

Abstract. PyCSP is intended to help scientists develop correct, maintainable and portable code for emerging architectures. The library uses concepts from Communicating Sequential Processes (CSP) and is implemented in the Python programming language. This paper introduces a new channel implementation and new process types for PyCSP that are intended to simplify writing programs for clusters. The new processes and channels are investigated by running three benchmarks on a cluster, using up to 512 CPU cores. The results show that PyCSP scales well, and that the introduction of a mobile channel home improves the performance of all three benchmarks.

Keywords. PyCSP, CSP, clusters, channels, processes

Introduction

Maintaining scientific code is a well-known challenge; many applications are written by scientists without any formal computer science or software engineering qualifications, and are usually grown "organically" from a small kernel to hundreds of thousands of lines of code.

These applications have traditionally targeted simple single core systems and have still grown to a complexity where the cost of maintaining the codes is a limiting factor, and where the continued correctness of the code is often questionable.

PyCSP [1] is intended to help scientists develop correct, maintainable and portable code for emerging architectures. Python is highly suited for scientific applications. While it is interpreted and thus very slow, scientific libraries efficiently utilize the underlying hardware. CSP provides a formal and verifiable approach to controlling concurrency [2], fits directly into scientific workflows, and maps directly onto many graphical tools that present scientific workflows such as Taverna [3], Knime [4] and LabView [5].

The intended users for PyCSP are not computer scientists, but scientists in general. General scientists can not be expected to learn CSP as formulated by Hoare, thus the approach to controlling concurrency is based on CSP, but does not require any knowledge in CSP. The key elements of controlling concurrency using PyCSP are presented in Section 3.

CPU's are produced with multiple cores today and every announced future CPU generation [6] seems to feature an ever increasing number of cores. As single core performance increases very slowly, researchers are required to exploit this parallel hardware for increased performance. To this end a number of parallel libraries like BLAS and programming tools like Intel Parallel Studio [7] are appearing. Unfortunately parallel libraries are often not enough to achieve acceptable speed and even with advanced tools parallel programming remains a source of added complexity and new bugs in software development.

This paper investigates using PyCSP for running scientific applications on a cluster with up to 512 GPU cores. The cluster is comprised of several computers, each with multi-core

---

[1]Corresponding Author: *Rune Møllegaard Friborg*, E-mail: `runef@birc.au.dk`.

CPU's. Thus, there are several layers of concurrency to control and coordinate. For this reason, PyCSP has been updated with a channel supporting distributed communication and a new process type supporting remote execution.

1. Related Work

During the last decade we have seen numerous new libraries and compilers for CSP. Several implementations are optimized for multi-core CPU's that are becoming the de-facto standard when buying even small desktop computers. occam-π [8], C++CSP [9] and JCSP [10] are three robust CSP implementations of CSP. C++CSP and JCSP are libraries for C++ and Java, while occam-π uses CSP inherently in the programming language.

Programming clusters of multi-cores is at best a nontrivial exercise, clusters may be programmed using MPI [11] or modern PGAS languages like UPC [12] for a C-style language, distNumPy [13] for Python style programming or even Fortress, X10, or Chapel [14] if the programmer is willing to learn a completely new programming language. These languages all suffer from the fact that they are fundamentally parallel programming languages, not concurrent languages. Google's language [15] Go. is concurrent but channels are not supported directly across the network so it does not scale seamlessly.

Two other CSP libraries have previously introduced cluster computing capabilities. occam-π through PONY [16] and JCSP [17]. The presented PyCSP aims for complete transparency using a single channel model, while both occam-π and JCSP requires the explicit creation of networked channels or channel ends. If all channels in occam-π is created with network-enabled shared channel ends, then it may be comparable to the presented PyCSP.

2. CSP

The Communicating Sequential Processes algebra, CSP [2], was introduced more than 25 years ago and while it was highly popular and thoroughly investigated in its first years, interest dropped off in the late 1980s because the algebra appeared to be a solution in search of a problem, namely modeling massively concurrent processes and providing tools to solve many of the common problems associated with writing parallel and concurrent applications.

CSP provides many attractive features with respect to the next generation processors; it is a formal algebra with automated tools to help prove correctness, it works with entirely isolated process spaces, thus the inherent coherence problem is eliminated by design, and it lends itself to being modeled through both programming languages and graphical design tools.

3. PyCSP

PyCSP provides an API that can be used to write concurrent applications based on CSP. PyCSP was introduced in 2007 [1] and has since been revised in 2009 [18] and in 2012 based on a distributed channel design [19]. PyCSP is now available in two separate versions: Parallel and greenlets. Both are packaged in single modules to motivate the developer to switch between them as needed. A common API is used for both implementations making it trivial to switch between them: compositional elements can be written once and used for both implementations. When switching from parallel to greenlets, the PyCSP application may execute differently as processes may be scheduled in another order and less fairly. Hidden latencies may also become more apparent when all other processes are waiting to be scheduled. Having several implementations sharing one PyCSP API was presented in [20]. The pycsp.greenlets implementation is for fine-grained CSP networks running on a single core and pycsp.parallel is for coarse-grained CSP networks running on multi-core, many-core and distributed architectures.

The two implementations:

- `pycsp.greenlets` - This implementation uses co-routines instead of threads. Greenlets [21] is a simple co-routine implementation available as a Python module. It provides the possibility of creating millions of CSP processes in a single CSP network. This version is optimal for single-core architectures since it provides the fastest communication, but with no parallelism.
- `pycsp.parallel` - All channels are network enabled and support communication with remote and local processes. This is the default implementation for PyCSP programs. Currently four different process types exists for this version.
  - `@process` - A CSP process implemented as an OS thread. The internal synchronization is handled by thread-locking mechanisms. Because of the Python Global Interpreter Lock[1], this is best suited for applications that spend most of their time in external routines that release the GIL.
  - `@multiprocess` - A CSP process implemented as an OS process. It is built on top of the multiprocessing module available in Python 2.6 and newer. This implementation is not affected by the Global Interpreter Lock, but has some limitations on a Windows OS, primarily caused by the absence of a Windows fork() call.
  - `@sshprocess` - A CSP process implemented as an OS process running on a remote host. The execution of the process is handled using the SSH protocol, thus this process requires a set of extra parameters to initiate the execution.
  - `@clusterprocess` - A CSP process implemented as an OS process running on a remote host listed in a node file. The remote host is selected based on a distribution scheme provided to the clusterprocess. The process implementation is suited for distributing large scale CSP networks on clusters and is explained in more detail in Section 3.2.

In this paper we focus exclusively on the pycsp.parallel implementation and particularly on the aspects facilitating the distribution of a single CSP network across cores and hardware nodes. These next sections explain the different aspects of the pycsp.parallel implementation.

3.1. Channels

All channels are network-enabled, untyped, any-2-any, can be buffered, and support input and output guards. The primary motivation behind the design of this channel is that scientific users of the PyCSP library should not have to think about whether they might be sending a channel end to a process that might be running in a remote location, or how they avoid invoking an output guard on a channel that does not support it. One of the powerful characteristics of CSP is that every process is isolated, which means that we can move it anywhere and as long as the channels still work, the process will execute the same.

Every Python interpreter importing pycsp.parallel will create one dispatch thread listening on one port and that port receives all the communication to channels and processes started in this Python interpreter.

When a new channel is created, it registers the dispatch thread at the local Python interpreter as the channel home. Processes can request channel ends for registered channels at the

[1]Python uses a Global Interpreter Lock, the GIL, to protect the interpreter when multiple threads execute Python code. The GIL limits concurrency when executing Python code, but libraries commonly mitigate the problem by releasing the GIL when executing external code.

```
# Python interpreter 1
A = pycsp.Channel("A") # Hosting channel A
print(A.address)       # Outputs ('192.168.1.16', 56562)

@pycsp.multiprocess
def Hello(cout, id)
  cout("Hello from Python interpreter 2")

# The channel location is passed to the other Python interpreter
# inside the channel end returned by A.writer()
pycsp.Spawn(Hello(A.writer()))
cin = A.reader()
print(cin())
```

Listing 1. Connecting channels implicitly.

```
# Python interpreter 1
A = pycsp.Channel("A") # Hosting channel A
print(A.address)       # Outputs ('192.168.1.16', 56562)
cin = A.reader()       # Request reading end
print(cin())

# Python interpreter 2
A = pycsp.Channel("A", connect=('192.168.1.16', 56562))
cout = A.writer()      # Request writing end
cout("Hello from Python interpreter 2")
```

Listing 2. Connecting channels explicitly.

dispatch thread. For a process to perform any action on a channel, a request must be posted to the channel home through the dispatch thread.

Each process registers a remote locking object at its local dispatch thread. Whenever the channel home tries to match a read request (process A) with a write request (process B), it does this in a critical region by acquiring locks at the remote locking objects of the involved processes A and B. This blocks all other channel homes from accepting requests owned by either process A or B. See [19] for a thorough description of the distributed channel model.

To connect to a hosted channel the location must be known, as there is no name server available for channels. The location of a channel can be fetched directly from the channel object and used implicitly by new processes (Listing 1), or channel locations can be specified explicitly (Listing 2). A channel end contains the address of the channel host and will automatically reconnect to the channel host if it is passed to a new Python interpreter.

The channel home is hosted where it is created, but our goal is that a simple application, such as the example in Listing 3, can be distributed on a set of nodes, simply by changing the process types. Unfortunately, when all channels are created on the same node and sent to processes spawned at other nodes, that node will be a bottleneck as all the channels will have that node as the home node. For this reason, we have made channel homes mobile: a process can request a channel to move its home to be hosted at another Python interpreter. A process may now acquire the home of a channel by invoking `ChannelEnd.become_home()` (Listing 4). This results in a more distributed communication pattern, which is able to scale with the number of processes (shown in Section 5).

```
@pycsp.process
def elementP(this_read, next_write):
  # Loop until poisoned
  while True:
    token = this_read()
    next_write(token + 1)

@pycsp.process
def connectionP(value, chan_out, chan_result):
  chan_out(value)
  print(chan_result())

  # Poison element processes
  pycsp.poison(chan_out)

if __name__ == "__main__":
  head = pycsp.Channel()
  current = head
  L = []
  # Create ring processes
  for i in xrange(N):
    next = pycsp.Channel()
    L.append(elementP(current.reader(), next.writer()))
    current = next

  # Start ring processes
  pycsp.Spawn(connectionP(0, head.writer(), current.reader()))
  pycsp.Parallel(L)

  # Wait for spawned processes and hosted channels to quit
  pycsp.shutdown()
```

Listing 3. Ring example.

When a process invokes `ChannelEnd.become_home()` on a channel that is hosted elsewhere, a new channel home is created locally and a MOVE request is posted to the active channel home. The old home temporarily suspends the channel and sends any buffered messages the channel may contain to the new location. It then notifies all the posted requests that the channel has been moved to the new address. After this, every process with a posted request updates the new address of the channel object and repost the request at the new location. The old home retains the MOVE notice to forward any requests until the PyCSP application quits.

The order of posted requests may easily be re-ordered during a call to `become_home`, thus priority and fairness can not be guaranteed during this operation. Also, if a process invokes `ChannelEnd.become_home()` on a channel that has already moved, the process is only updated with the location of the new channel and does not become the home of the channel.

3.2. Cluster Processes

To simplify cluster programming with PyCSP, a new process type has been introduced (Listing 5). The new type enables executing processes transparently on cluster nodes using a `@clusterprocess` decorator. The `@clusterprocess` decorator provides a similar environment to the decorated function as the `@process` (thread) and the `@multiprocess` decorators. Thus, to make a process with a computational workload run on a set of remote hosts, the only

```
@pycsp.clusterprocess
def elementP(this_read, next_write):

  # Move channel home to this Python interpreter
  this_read.become_home()

  # Loop until poisoned
  while True:
    token = this_read()
    next_write(token + 1)
```

Listing 4. Distributed element process from ring example in Listing 3.

```
@pycsp.clusterprocess(
  cluster_nodefile="$PBS_NODEFILE",       # Var or path to node file
  cluster_pin=None,                       # Index of host in node file
  cluster_hint='blocked', or 'strided'    # Distribution scheme
  cluster_ssh_port=22,                    # Port to ssh server
  cluster_python='python'                 # Python executable
)
def Remote_process(channel ends and function arguments):
  # process content
```

Listing 5. Specifying node distribution for cluster processes.

necessary change is to swap the `@process` decorator with the `@clusterprocess` decorator and provide a list of host names.

To start a Python interpreter on a remote host, we use the Python paramiko module[2], which implements the SSH protocol [22]. SSH was selected, since it is commonly used for logins on Linux clusters. The Python paramiko module and SSH servers are cross-platform and are also available for many other operating systems, such as Microsoft Windows and Mac OS X.

When the first clusterprocess is started, a node runner thread is initiated which handles each SSH connection. Only one SSH connection is made per cluster node and each executes the `pycsp.parallel.server` module in Listing 6. For the clusterprocess to work, it is assumed that all files are identical on all cluster nodes. This is a practical requirement, as Python functions and modules can not be transferred properly using serialisation, but must be loaded directly from source in lines 26-33 (Listing 6). Communication to `pycsp.parallel.server` is handled using PyCSP channels and not SSH transport protocols. Using PyCSP channels allows for a more concurrent design, where it is the task of the `RunFunc` process (lines 9-13) to send the return value of the cluster process directly to the process initiating `Parallel`, `Spawn` or `Sequence`. The node file information is passed along to new clusterprocesses together with the current state of distribution. This ensures, that a self-spawning chain will use the desired distribution scheme and not put all the new clusterprocesses on the first entry in the node file. For a self-spawning process tree, the distribution scheme will not be correct, as the state of distribution at parallel processes are not known. In such a case, the user must alternate between using 'strided' and 'blocked' distribution schemes.

[2]https://pypi.python.org/pypi/paramiko

```
cwd, ip, port, input_name = sys.argv[1:] # Parameters
os.chdir(cwd)                              # Change to working dir

# Connect to channel at parent and request input end
input_chan = Channel(input_name, connect=(ip, int(port)))
input_chan_end = input_chan.reader()

# (Spawn) PyCSP MultiProcess
def RunFunc(output_chan_name, fn, args, kwargs):
    output_chan = Channel(output_chan_name, connect=(ip, int(port)))
    send_return_value = output_chan.writer()
    val = Parallel(Process(fn, *args, **kwargs))
    send_return_value(val[0])

try:
    while True:
        # Get next process
        val = input_chan_end()
        h, p, output_chan_name, scriptPath, funcName,
          args, kwargs, available_nodes = val

        # Update cluster node information
        nodefile, group_state = available_nodes
        NodePlacement().set_nodegroup(nodefile, group_state)

        # Load script
        sys.path.insert(0, os.path.dirname(scriptPath))
        moduleName = os.path.basename(scriptPath)
        if moduleName[-3:] == '.py':
            moduleName = moduleName[:-3]
        m = __import__(moduleName)
        # Retrieve function object from imported script
        fn = getattr(m, funcName)().fn

        # Spawn a runner
        pycsp.Spawn(
          MultiProcess(RunFunc, output_chan_name, fn, args, kwargs),
          pycsp_host=h, pycsp_port=p)

except ChannelPoisonException:
    pass # shutdown nicely
```

Listing 6. python -m pycsp.parallel.server.

4. Experimental Platform

The genomedk cluster in Aarhus is comprised of 96 nodes. Each node has two Intel Xeon E5-2670 CPU's @ 2.67 GHz (total of 16 cores per node) and 64GB memory @ 1600 MHz. The nodes are interconnected by 10 GigE and 1 GigE NIC's. For the benchmarks in this paper, a maximum of 32 nodes were used. The cluster runs Centos 6.3 and is primarily used for bioinformatics tasks.

```
@clusterprocess
def find_kNN(body_in, body_out, result_out, locals, k):
  best = {}
  for local_id, _ in locals:
    best[local_id] = []

  if USING_BECOME_HOME:
    body_in.become_home()

  body_out(locals)
  while True:
    bodies = body_in()

    if locals[0][0] == bodies[0][0]:
      result_out(best)
      return

    for local_id, local_params in locals:
      B = best[local_id]
      for body_id, body_params in bodies:

        dist = numpy.sum((body_params-local_params)**2)
        dist = math.sqrt(dist)
        B.append((dist,body_id))

      # Sorts on first element.
      B.sort()
      best[local_id] = B[:k]

    body_out(bodies)
```

Listing 7. find_kNN process.

5. Experiments and Results

To investigate the performance of a distributed PyCSP network we have executed three different benchmarks: a kNN experiment to perform a k nearest neighbour search, a token ring experiment to show the scaling of communication on independent channels and finally a throughput experiment to test the maximum obtainable bandwidth on a 10GigE link for different message sizes. All runs were executed two times and the measured performance results did not deviate by more than 2% between the two runs. The average of the two runs was used for the benchmark plots.

Also, to demonstrate the effect of the new mobile channel home, all three experiments compare two executions with and without the use of the mobile channel home.

In the kNN experiment we run one `find_kNN` process per core (Listing 7). The input data set is split into smaller sets and divided among the `find_kNN` worker processes all connected in a ring. Every worker process is given a local set which it will pass around the ring. In each pass the worker execute a kNN algorithm on the local set and the received set, until finally all results are collected through a result channel and joined to produce the end result. Two different data input sizes have been selected, to test different computation to communication ratios. The experiment took 867 seconds for 204800 data elements on 512 cores and 266 seconds for 102400 data elements on 512 cores.

Figure 1 shows the results from running the kNN experiment on the cluster. To reduce the run time of the entire experiment, we did not run the experiment with fewer than 32

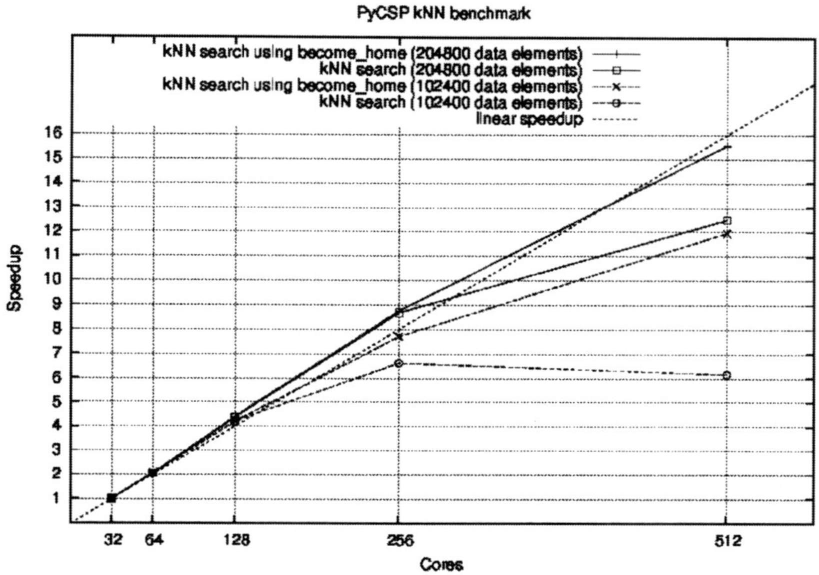

Figure 1. Running the kNN experiment. Speedup from 32 to 512 cores on the cluster. The runs using `become_home` moves the channel homes to the individual `find_kNN` processes.

cores. The results show that moving the channel homes to the individual `find_kNN` processes improves to performance considerably. The result for 204800 elements using `become_home()` show close to linear speedup for up to 512 cores. At 128 and 256 nodes, we observe a higher than linear speedup (super-linear speedup). The data elements fit better into the cache of the CPU's when split into smaller slices, which is the case for the input data size of 204800 elements split between 128 or 256 `find_kNN` processes.

Figure 2 shows the number of messages per second that are transmitted in a ring of PyCSP processes, similar to the one created in listing 3. The number of concurrent tokens were increased from 1 to 256. The linear scaling tells us that the channel homes have been successfully moved to the individually processes, thus eliminating a central bottleneck in the channel communication.

The throughput benchmark in Figure 3 shows the bytes/s transmitted from one 16-core node to another 16-core node on the cluster. The CSP network is a ring design, but instead of putting the first 16 processes on one node and the next 16 processes on another (blocked distribution scheme), the processes are distributed in a round-robin fashion (strided distribution scheme). This design makes every channel communication a communication from one node to another. The throughput on the 10GigE link has then been computed by running a ring network of size 32 with 16 concurrent tokens. For the local communication we ran the same experiment with multiprocessing instead of clusterprocesses. In both cases, the message size was varied from 1 byte to 100Mbyte, as this is the expected usage scenario. PyCSP supports larger messages than 100Mbyte, though the performance is lower for large messages.

The results show a throughput reaching 9.3Gbit on the 10GigE link when the message size is at 1Mbyte. Below 1Mbyte the latency of a channel communication results in a drop in throughput. Above 1Mbyte there is a overhead in handling the larger sizes for both local and 10GigE. This issue has been investigated through profiling of PyCSP and have shown that the time spent serialising Python data objects does not scale linearly with the size of data. Also,

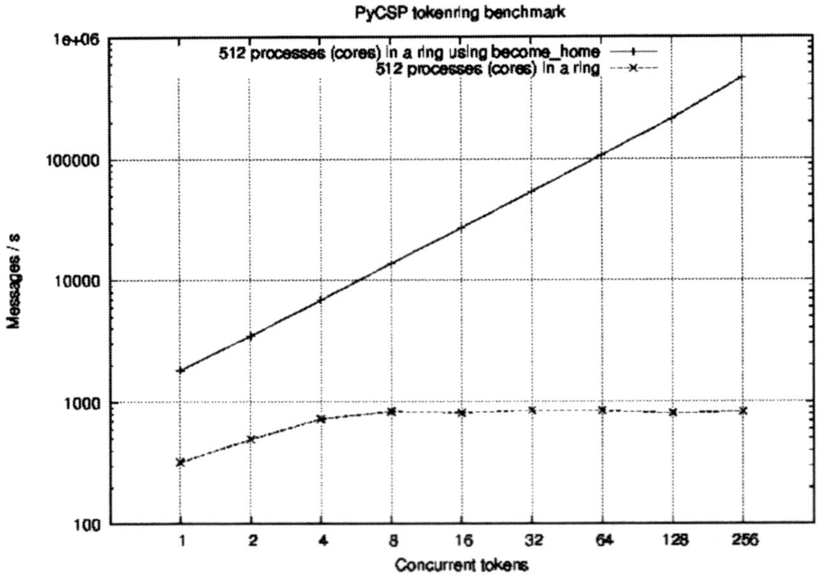

Figure 2. PyCSP token ring benchmark on the cluster.

there is a considerable drop in performance when the size of the data objects is above the available L3 cache on the CPU. Serialisation of Python data objects is a necessary step and is currently performed using the standard Python module pickle. During this serialisation, the data is written to memory and if the data can not fit in the cache of the CPU, the data must be sent to main memory before the operation can complete. A possible solution to this would be to make the serialisation step stream data directly to a socket stream. We are confident that this can be solved in the future.

From the results in Figures 2 and 3 we can also learn the following about the performance of a single dispatch thread hosting channels. It can communicate up to 850 messages per second when handling 512 channels for 512 cores in a distributed environment. It can reach a throughput of up to 550Mbit/s on a network link and up to 3.3Gbit/s on a local socket.

6. Conclusions

This paper introduces a new channel implementation and new process types for our Python CSP library, PyCSP, that are intended to simplify writing programs for clusters.

The new channel implementation adds the challenge of how to create a distributed channel network, without requiring the knowledge of each hostname and port number. This challenge is solved by making the channel home mobile, such that a process may request to become the home of a channel. Experiments in the paper shows that without the distribution of channel homes, PyCSP can not scale.

The new processes and channels are investigated by running three benchmarks on a cluster. The experiments show that PyCSP scales well for the application benchmark, with close to linear scaling for up to 512 CPU cores. The tokenring benchmark shows that the use of mobile channel homes have completely eliminated a communication bottleneck. For the throughput experiment, we are able to use nearly all available bandwidth with 1Mbyte messages, but the throughput quickly drops for larger messages. The serialisation of Python

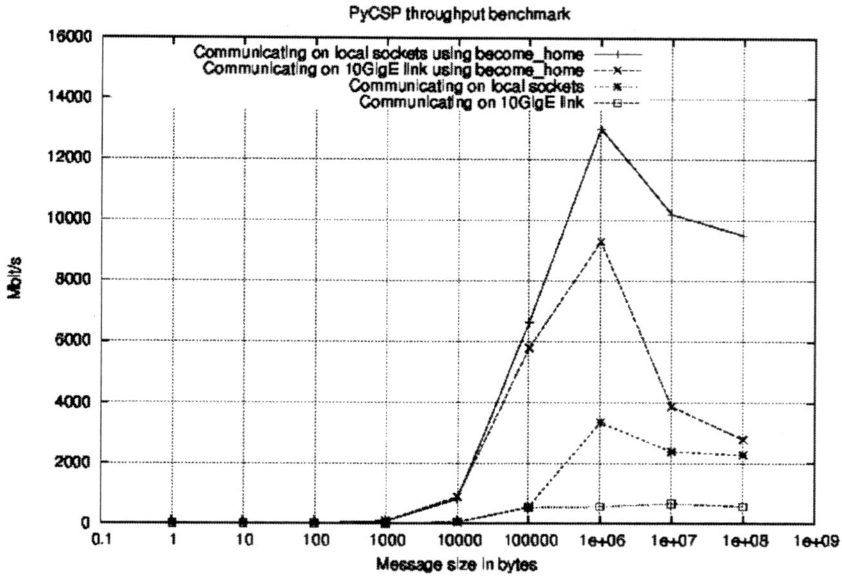

Figure 3. PyCSP throughput benchmark on the cluster.

objects is found to be the bottleneck and we expect that using streams between serialisation and sockets will solve the issue.

With these results PyCSP is now able to support scalable scientific applications on clusters. Scientific applications using PyCSP composed by other PyCSP networks. The kNN benchmark could be such a composed CSP network performing numerous massive kNN searches on-the-fly for other processes in a larger CSP network.

References

[1] John Markus Bjørndalen, Brian Vinter, and Otto Anshus. PyCSP - Communicating Sequential Processes for Python. In A.A.McEwan, S.Schneider, W.Ifill, and P.Welch, editors, *Communicating Process Architectures 2007*, pages 229–248. IOS Press, Netherlands, Jul 2007.

[2] C. A. R. Hoare. Communicating sequential processes. *Communications of the ACM, 21(8):666-677*, pages 666–677, August 1978.

[3] Taverna Project. http://taverna.sourceforge.net.

[4] Michael R. Berthold, Nicolas Cebron, Fabian Dill, Thomas R. Gabriel, Tobias Kötter, Thorsten Meinl, Peter Ohl, Christoph Sieb, Kilian Thiel, and Bernd Wiswedel. Knime: The konstanz information miner. In *Studies in Classification, Data Analysis, and Knowledge Organization (GfKL 2007)*. Springer, 2007.

[5] LabView. http://www.ni.com/labview/.

[6] Brian Vinter. Next Generation Processes. In B.H.V Topping and P. Ivanyi, editors, *Parallel, Distributed and Grid Computing for Engineering*, Computational Science, Engineering and Technology, pages 21–33. Saxo-Coburg Publications, 2009.

[7] Intel Parallel Studio. http://software.intel.com/en-us/intel-parallel-studio-home/.

[8] Peter H. Welch and Frederick R. M. Barnes. Communicating Mobile Processes - Introducing occam-π. In *Communicating Sequential Processes*, Lecture Notes in Computer Science, pages 175–210. Springer-Verlag, 2005.

[9] Neil C. Brown. C++CSP2: a Many-to-Many Threading. In Alistair A. McEwan, Wilson Ifill, and Peter H. Welch, editors, *Communicating Process Architectures 2007*, pages 183–206. IOS Press, Netherlands, Jul 2007.

[10] Peter H. Welch, Neil C. Brown, James Moores, Kevin Chalmers, and Bernhard Sputh. Integrating and Extending JCSP. In Alistair A. McEwan, Wilson Ifill, and Peter H. Welch, editors, *Communicating Process Architectures 2007*, pages 349–369. IOS Press, Netherlands, Jul 2007.

[11] MPI: A Message-Passing Interface Standard. *Message Passing Interface Forum*, March 1994.

[12] UPC Consortium. UPC Language Specifications, v1.2. Tech Report LBNL-59208, Lawrence Berkeley National Lab, 2005.

[13] Mads Ruben Burgdorff Kristensen, Yili Zheng, and Brian Vinter. PGAS for Distributed Numerical Python Targeting Multi-core Clusters. In *Parallel & Distributed Processing Symposium (IPDPS), 2012 IEEE 26th International*, pages 680–690. IEEE, 2012.

[14] Ewing Lusk and Katherine Yelick. Languages for high-productivity computing: the DARPA HPCS language project. *Parallel Processing Letters*, 17(01):89–102, 2007.

[15] Robert Griesemer, Rob Pike, Ken Thompson, et al. The Go programming language, 2009.

[16] Mario Schweigler and Adam T. Sampson. PoNY – the occam-π Network Environment. In Peter H. Welch, Jon Kerridge, and Frederick R. M. Barnes, editors, *Communicating Process Architectures 2006*, pages 77–108. IOS Press, Netherlands, Sep 2006.

[17] Peter H. Welch and Brian Vinter. Cluster Computing and JCSP Networking. In James S. Pascoe, Roger J. Loader, and Vaidy S. Sunderam, editors, *Communicating Process Architectures 2002*, pages 203–222. IOS Press, Netherlands, Sep 2002.

[18] Brian Vinter, John Markus Bjørndalen, and Rune M. Friborg. PyCSP Revisited. In *Communicating Process Architectures 2009*. IOS Press, Netherlands, Nov 2009.

[19] Rune M. Friborg and Brian Vinter. Verification of a Dynamic Channel Model using the SPIN Model Checker. In *Proceedings of WoTUG 33 / CPA 2011*. IOS Press, Netherlands, June 2011. ISBN 978-1-60750-773-4.

[20] Rune M. Friborg, John Markus Bjørndalen, and Brian Vinter. Three Unique Implementations of Processes for PyCSP. In *Communicating Process Architectures 2009*. IOS Press, Netherlands, Nov 2009.

[21] Greenlet: Lightweight in-process concurrent programming. http://pypi.python.org/pypi/greenlet.

[22] Tatu Ylonen and Chris Lonvick. The secure shell (SSH) protocol architecture. 2006.

Communicating Process Architectures 2013
P.H. Welch et al. (Eds.)
Open Channel Publishing Ltd., 2013

Service-Oriented Programming in MPI

Sarwar ALAM [1], Humaira KAMAL and Alan WAGNER
Department of Computer Science, University of British Columbia, Canada

Abstract. In this paper we introduce a service-oriented approach to the design of distributed data structures for MPI. Using this approach we present the design of an ordered linked-list structure. The implementation relies on Fine-Grain MPI (FG-MPI) and its support for exposing fine-grain concurrency. We describe the implementation of the service and show how to compose and map it onto a cluster. We experiment with the service to show how its behaviour can be adjusted to match the application and the underlying characteristics of the machine.

The advantage of a service-oriented approach is that it enforces low coupling between components with higher cohesion inside components. As a service, the ordered linked-list structure can be easily composed with application code and more generally it is an illustration of how complex data structures can be added to message-passing libraries and languages.

Keywords. service-oriented programming, function-level parallelism, distributed data structures, linked list, MPI-aware scheduler, MPI runtime, FG-MPI, MPICH2

Introduction

Process-oriented languages in general, and communication libraries like MPI in particular, do not have the rich collections of data structures present in today's object-oriented programming languages. The lack of data structure libraries for MPI makes it necessary for programmers to create their own and adds to the overall programming complexity of using message-passing languages. The difficulty in having to explicitly manage the distribution of the data is often pointed to as a major impediment to the use of message passing [1]. One approach that is at the heart of process-oriented systems [2,3] and many distributed systems [4,5] is to implement the data structure using its own collection of processes and provide the distributed data structure as a service. We call this service-oriented programming. A service-oriented approach enforces low coupling, because components interact only through message-passing and cannot directly access one another's data, and leads to higher cohesion inside components, because application and data structure management code no longer needs to be combined together. We investigate the use of a process-oriented design methodology together with a service-oriented programming approach (design pattern) to provide highly scalable, easy to use libraries for complex data structures.

Our interest in a service-oriented approach to programming arose in the context of our work on Fine-Grain MPI (FG-MPI) [6,7]. FG-MPI extends MPICH2 by adding an extra level of concurrency (interleaved concurrency) whereby every OS process comprises of several MPI processes, rather than just one. FG-MPI uses a coroutine-based non-preemptive threading model together with a user-level scheduler integrated into the middleware to make it possible to support thousands of processes inside an OS process to expose finer-grain concurrency with no change to the programming model. The support for having many light-weight

[1]Corresponding Author: *Sarwar Alam, Department of Computer Science, University of British Columbia, Vancouver, B.C. V6T 1Z4, Canada*; E-mail: `sarwar@cs.ubc.ca`.

MPI processes and the flexibility to map processes to functions[1] to execute concurrently inside an OS process, as well as in parallel across the cores and machines in a cluster makes it possible to consider a service-oriented approach to providing a distributed data structure. In this paper we use FG-MPI and our service-oriented approach to implement an ordered linked-list service.

We chose an ordered linked list because it is a common data structure, its linear structure makes it challenging to parallelise in an efficient way, and it highlights how FG-MPI makes it possible to design new types of MPI programs. Since most MPI implementations bind an MPI process to an OS process, one could implement a linked list as a collection of separate OS processes. Previous work by Iancu *et al.* [8] has shown that over-subscription of MPI processes to cores often leads to poor performance and because of this the usual practice is to execute one MPI process per core. This directly binds the number of MPI processes to the number of cores which results in the design of processes to match the machine rather than the available concurrency present in the algorithms. This leads to more coupled, less cohesive programs.

A second drawback to the previous approach is that when executing with one MPI process per core it is difficult to keep the application and service processes constantly busy. One common approach to avoid this problem is to use MPI in combination with Pthreads, or some other runtime, to combine the MPI application process with a thread implementing the concurrent data structure. This provides a fine-grain implementation, but at the expense of portability and predictability because the multiple models and runtimes interact in unpredictable ways. Other than MPI, there are concurrent data structures and APIs designed for multicore, like OpenMP, that can be used, but they do not scale to execute on clusters or efficiently scale to the hundreds and thousands of processor cores which is the intent of our design. Our approach to the design of an ordered linked-list service relies on FG-MPI and avoids the previously mentioned problems to provide a scalable data structure service to applications.

In this paper we give the design, implementation and evaluation of an ordered linked list data structure in FG-MPI. We make the following contributions:

- A demonstration of how the FG-MPI extensions to MPI makes it possible to follow, within MPI, a process-oriented design methodology where the resulting process structure reflects the problem rather than the machine. This makes it possible to add complex data structures to MPI by using a service-oriented approach.
- We provide a novel design of a pure message-passing implementation of an ordered linked list targeted to the type of multicore clusters that are common to high-performance and cloud computing environments.
- We introduce a variety of performance-tuning parameters and investigate the effect of these parameters with regards to tuning the service to the underlying characteristics of the machine and network to achieve good performance across varied machine sizes and communication fabrics.
- Unlike many MPI programs, our mapping of processes to OS processes on a multicore machine, makes it possible to expose a large amount of concurrency that can be predictably scheduled. We also introduce a free process sub-service to allocate and deallocate processes to dynamically adjust to varying loads and demands.
- We introduce shortcuts, a novel technique related to service discovery and relate this to trade-offs between data consistency and performance.

In Section 1 we briefly describe related work and how our approach differs from existing work in the area. A discussion of service-oriented programming and the support for it within the FG-MPI runtime is given in Section 2. In Section 3 we describe the design of the ordered

[1] C functions, named procedures

linked-list service, describing the implementation of the operations, trade-offs with respect to different consistency properties, and the introduction of parameters related to bounded asynchrony, trade-offs between parallelism and interleaved concurrency, adjustments to the granularity, and the mapping of the service onto the system. Shortcuts are discussed in Section 3.6. In Section 4 we describe how the service is mapped and composed with the application. In Section 5 we experimentally show how these parameters affect the behaviour of the service. Conclusions are given in Section 6.

1. Related Work

The focus of MPI is on performance and as a result it supports a wide variety of communication routines to match modern network protocols, interfaces and fabrics. In contrast to synchronous message-passing languages like occam-π [9], MPI programs make frequent use of non-blocking asynchronous communication to overlap computation and communication. The message latency between machines in a cluster is relatively high in comparison to on-chip memory-based messaging and this "send and forget" model of messaging makes it possible to off-load message delivery to the network interfaces and devices. The intent is to increase the network load to better utilise the network hardware and have more messages in flight to provide more work for each machine, thereby improving performance by mitigating the overhead of communication. As shown by Valiant [10], the amount of work needed per machine to overlap with communication depends on the latency and bandwidth characteristics of the network. This suggests that, for scalability, the messaging needs to be able to adjust to the characteristics of the network in much the same way that programs need to take into account the memory hierarchy. The tuning parameters introduced in our design come from a two-level view of the network and the need to adjust the messaging to the fast local and slower non-local communication mechanisms in a cluster of multicore machines. We view this as essential to the design of programs to scale to hundreds and thousands of processors and differentiates our work from languages and systems that execute on a single multicore machine.

A linked list is a simple well-known data structure that has been frequently implemented as a concurrent data structure [11,12,13]. We make use of some of the ideas in [13] (Chapter 9) for the design and correctness of our linked list. In particular the notion of atomic actions, hand-over-hand[2] and forward-only traversal of the list have their counterparts in our design, however, the design is significantly different in that the list elements are active processes where the data structure has control over the operations, not the application processes as in the case of concurrent data structures.

In cloud computing environments hash-based storage structures like Cassandra [14] (and key-values stores) are commonly used to allow for distributed access to the data. Although these systems are based on message passing they are optimised for coarse-grain access to large data. Our distributed ordered linked list provides similar types of access to data as Cassandra but with more of a focus on scalability and low-latency where we are considering finer-grain in-memory access to data rather than large blocks of data residing on disk. As well, as expected, distributed storage systems are designed for availability and reliability, which we do not consider. A notable advantage of our ordered linked-list structure is that because keys remain sorted it would be relatively easy to extend our operations to support range queries, which is inefficient on simple hash-based structures. More generally in Peer-to-Peer computing Distributed Hash Tables (DHT) provide a pure message-based implementation of a look-up service designed for Internet applications [15]. The primary focus of these

[2]The lock for the next item in the list is acquired before releasing the lock for the current list item.

systems is on scalability and availability, particularly in the case where processes in the DHT frequently leave and join (i.e., churn), which is substantially different from our use case.

2. Service-Oriented Programming

Our objective is to take advantage of FG-MPI's support for fine-grain processes to implement an ordered linked-list service. The service consists of a collection of processes with, in its finest granularity, one list node per process along with a message interface for applications to communicate with the service. In the following we give a brief description of FG-MPI and discuss the advantages of providing distributed data structures as a service.

An FG-MPI execution, $[P, O, M]$, can be described in terms of P the number of MPI processes per OS process[3], O, the number of OS processes per machine and M, the number of machines. A typical MPI execution is of the form $[1, O, M]$ where N, the total number of MPI processes as given by the "-n" flag of MPI's `mpiexec` command, equals $O \times M$. In FG-MPI we added an "-nfg" flag to `mpiexec` enabling one to specify $P > 1$, where $N = P \times O \times M$.[4] Figure 1 shows a $[4, 2, 2]$ mapping started with the command:

```
mpiexec -nfg 4 -n 4.
```

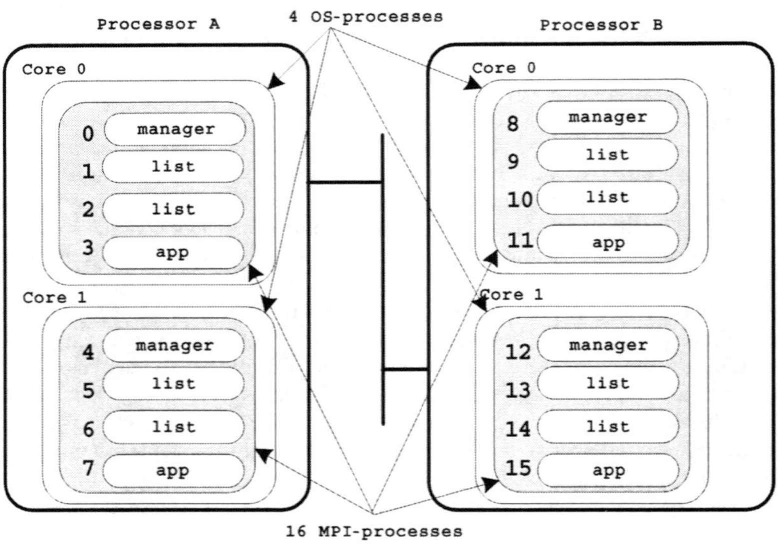

Figure 1. A cluster consisting of 2 machines each with 2 cores and a $[4, 2, 2]$ mapping with one manager, two list nodes, and an application process per OS process and 16 MPI processes in total. For the linked-list service, more list node processes will be added per core by increasing the `nfg` count (currently 4) in the `mpiexec` command.

The P co-located processes are full-fledged MPI processes that share the middleware inside the single address space of an OS process. FG-MPI uses coroutines to implement MPI

[3]We refer to these P MPI processes as *co-located* processes sharing a single address space. The terms *process*, *fine-grain process* and *MPI process* are used interchangeably. The term "OS process" is used to refer to an operating-system process.

[4]Both "`-n`" and "`-nfg`" can be used with `mpiexec` colon notation to create mappings with differing P's, O's and M's for each executable.

processes as non-preemptive threads with an MPI-aware user-level scheduler. The scheduler works in conjunction with the MPI progress engine to interleave the execution of the MPI processes [6]. Each MPI call provides a potential descheduling point for one co-located process to cooperatively yield control to another process. Each process, upon making an MPI call, will try and progress its own request[5], as far as possible, as well as any outstanding requests posted by other co-located processes. If it is unable to complete its call then it yields control to the scheduler, which selects another process to run. FG-MPI enables expression of function-level parallelism due to its fast context-switching time, low communication and synchronisation overhead and the ability to support thousands of MPI processes inside a single OS process [7]. This allows us to map MPI processes to functions instead of `main` programs (see Section 4). In summary, FG-MPI is integrated into the MPICH2 middleware and extends its execution model to support interleaved concurrency of finer-grain MPI processes inside an OS process.

In MPI programmers can, and frequently do, use non-blocking communication to switch between code segments to introduce "slackness" to overlap computation with communication or ensure a better computation to communication ratio. In essence, programmers are doing their own scheduling inside their program which adds to the complexity of the program and all but destroys program cohesion [6]. In FG-MPI programmers can take advantage of the integrated runtime scheduler to decompose their program into smaller processes that better reflect the program rather than code for optimising the messaging. In particular, for distributed data structures FG-MPI makes it possible to separate the data structure code from that of the application using that data structure.

As previously stated, we implement our distributed ordered-list structure as a distributed system service. We distinguish this type of library from the usual type of MPI library where the library computation is done by the same set of processes calling the library. For example, an MPI library for matrix operations first distributes the matrix and then all of the processes call the library routines to collectively perform the computation on their part of the matrix. Implementing more complex data structures on the same set of processes is difficult because of the need to intertwine the application code with the code for managing the data structure. As a result, implementations either use threads, as in implementations of tuple spaces [16] in MPI [17], or simply map the data structure onto its own set of resources as in C-MPI's implementation of a DHT for MPI [18]. FG-MPI provides an alternative way to map the application and data structure processes inside the same OS process.

This approach enhances portability, flexibility, predictability and controllability of the system. With regards to portability it relies on pure message passing in MPI and avoids the resource contention issues that arise in mixing MPI with threads [19]. FG-MPI enables one to expose more fine-grain concurrency and hence more parallelism to be potentially exploited. Each MPI process represents a "unit of execution" that can independently be replicated to scale up the program. As we describe in Section 4, we can flexibly map processes taking in account the size of the machine and to decide, relative to each other, which processes execute in parallel and which interleave their execution within a single core. The added predictability comes from the ability to use one core per OS process mapping to reduce the effect of OS noise [20] and the fact that an MPI-aware scheduler can provide more predictable interleavings. In addition, as described in Section 3, we can introduce a variety of parameters affecting the behaviour to give control over the granularity, number of outstanding messages and load-balancing.

These advantages could also be achieved by careful programming or re-factoring, but

---

[5]MPI defines `MPI_Request` objects that are used for a number of purposes. Requests here refer to send or receive request handles that are created internally as a result of MPI calls and are used to manage completion of MPI operations.

FG-MPI makes it easier to encapsulate program behaviour and externalise the parameters most affecting performance to match the application and machine. Although in FG-MPI there are some additional programming restrictions with regards to global and static variables[6] and overheads from the extra messaging costs, for many applications this is a reasonable trade-off for the increased programmability, portability and flexibility in exposing more fine-grain concurrency.

3. Ordered Linked-List Service

As a first step in the design of our ordered linked-list data structure service, we take a fine-grain approach where we represent each list node as an MPI process and store a single data item in that list node. Later in Section 3.5 we discuss a performance optimisation of the design to allow for multiple data items per list node, which lets the user relax granularity. By using FG-MPI we can develop and test this initial implementation with thousands of list items on one machine and even with millions of processes on a small cluster.

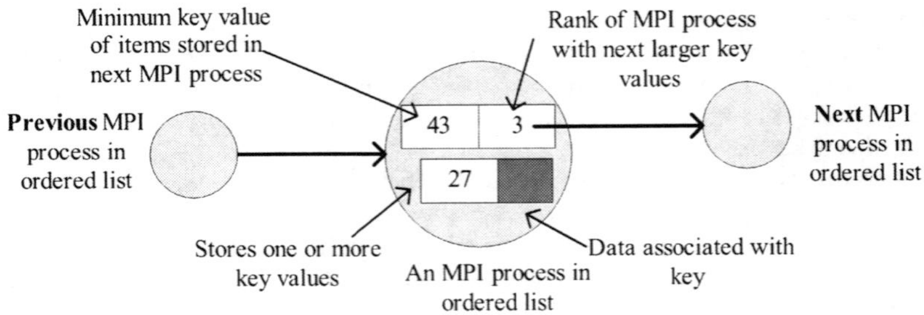

Figure 2. A node process (part of an ordered list)

As shown in Figure 2 each process (i.e., list node) in the list stores the following values:

a) the MPI process rank (i.e. identifier) of the next process in the list,
b) the key (or ordered list of keys in the multi-key case),
c) the data associated with the key, and
d) the minimum key (`min-key`) associated with the next process in the list.

The data stored inside each process is almost identical to that of the standard implementation of a linked list in a procedural language. The MPI process rank plays the role of a pointer to memory where a process can only communicate with the next process in the list. The only difference is item (d), the minimum key of the next process. This value is kept for correctness and performance of our list operations, which will be discussed in Section 3.1.

Our intent was to design a list potentially supporting millions of nodes with thousands of operations simultaneously accessing, adding and deleting from the list. In order to reason about the implementation and in particular to ensure freedom from deadlock we imposed the following two conditions:

a) processes initiate communication only in the forward direction, and
b) operations on the list process are atomic.

[6]The privatisation of global variables is commonly used in systems that allows multiple user-level threads to share the same address space and a number of tools are available to facilitate their privatisation [21,22].

Given a list of N processes, one can view the list as being comprised of a collection of nested servers where the i-th process is the client of the service provided by processes $i+1$ to N. According to our two conditions client i can only make requests to the service provided by process $i+1$. As well, the operation is atomic, implying that once a request has been made it completes and neither client i or the server at $i+1$ will accept new requests. In order to perform the request the client and server may engage in a series of communications as part of the request but this is assumed to terminate and all communications are local between a small fixed set of processes.

We claim that any request made to the head of the list eventually completes and thus the service is deadlock-free. The base case for server N is true because it depends only a series of local communications which are assumed to complete.[7] More generally, since list operations are atomic, we have that all requests by client $N-i$ to server $N-i+1$ complete and by induction it follows that requests by client $i=N-1$ (the first list process) complete. It also follows from the two conditions that requests to the list cannot jump over another request.

We take advantage of MPI message-passing semantics to enforce these two conditions. Every list process blocks on an `MPI_Recv()` with `MPI_ANY_SOURCE` (i.e., ready to receive a message from any source). Since a process's predecessor is the only process that knows its MPI rank, the predecessor is the only process that can initiate communication. Once this initial communication is made, the process learns the identity of its predecessor and the two can communicate back and forth to perform all local changes needed to complete the operation. Once both processes have completed the local operation they both block again on their respective `MPI_Recv()` routine. In addition MPI guarantees FIFO delivery of messages between two processes, which ensures that one message (operation) cannot overtake another inside the message queues; however special handling is necessary for operations where the process's predecessor changes (Sections 3.1.2, 3.1.3). In this case we need to ensure that all outstanding messages to a list process are received before switching that list node's predecessor.

3.1. List Operations

There are three types of requests that can be sent to our ordered linked-list service: FIND, INSERT and DELETE. Other operations such as UPDATE can be added but these three operations are sufficient to illustrate the basic properties of the service. Requests are of the form: operation type, source process rank, key-value, sequence number, and data (in the case of an insert operation). Application processes initiate requests and the process in the list performing the operation is the process that replies to the application process making the request.

In the following discussion we assume process A is one of the processes in the linked-list service and it is initiating contact with process B, the next process in the list. A third process, named C, with different roles for different operations is also present. We describe the implementation of the list operations in Sections 3.1.1, 3.1.2, and 3.1.3 together with Figure 3.

3.1.1. FIND Operation

As shown in step 1 of Figure 3(a), suppose that B has just received a FIND operation message from A. The message from A contains the operation, the key, the rank of the application node which made the request along with the application's sequence number. After receiving the message, B searches locally for the key in the message. Success is relayed back to the

[7] Server N acts as a sentinel and is configured with the NULL service, and any request generated by N signals failure which is relayed back to the application process.

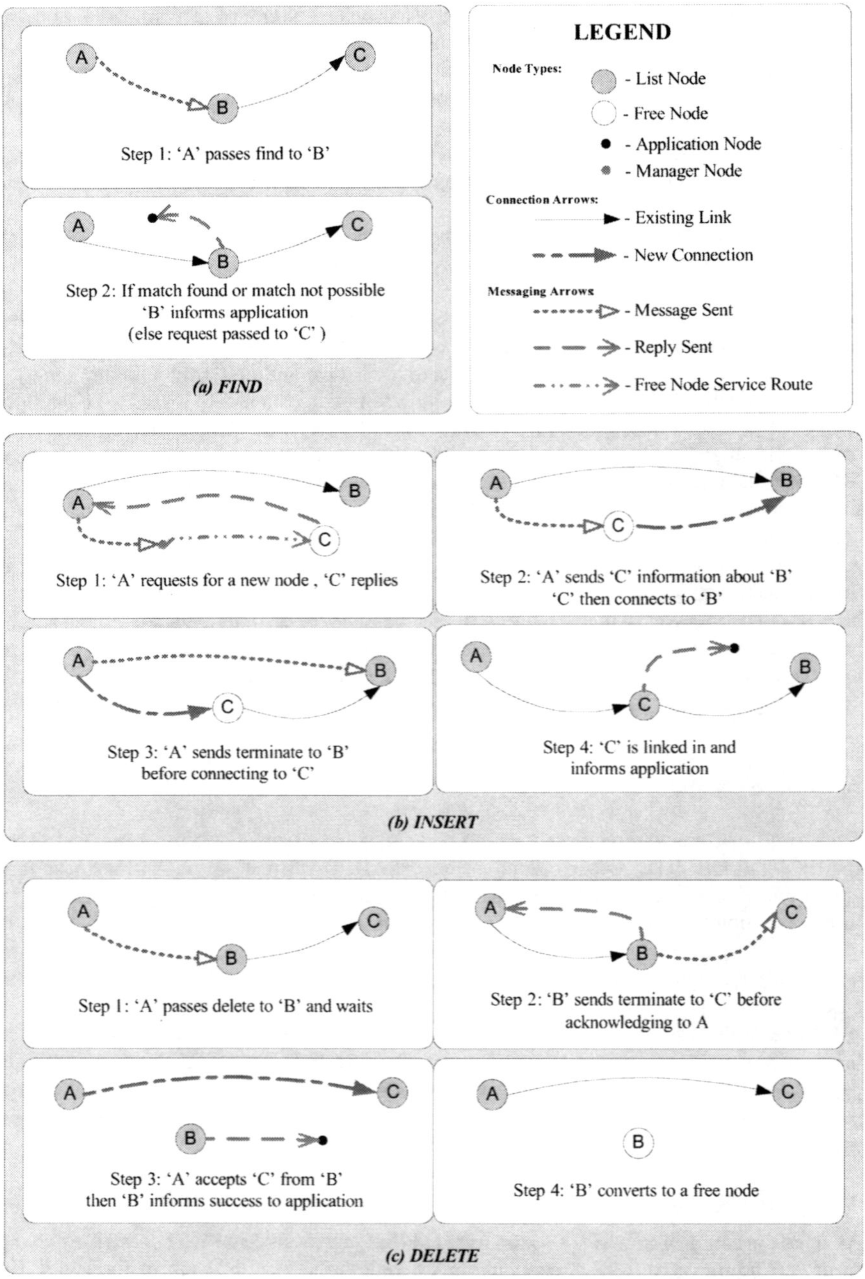

Figure 3. Implementation of the ordered linked-list operations FIND, INSERT, and DELETE. The arrows denote the different communications among processes A, B and C; three consecutive processes that participate in the operations.

application process, while failure makes B compare the key in the message with the `min-key` value. Recall that the `min-key` is the minimum key (or just the key, in the single key case we are describing) stored at B's successor C. If the key is less than `min-key` and the key is not the key stored at B, then the key does not exist in the list and B sends a message to signal failure back to the application process. If the key is the key stored at B, then B replies to the application process with a message containing the data associated with the key. Finally, if the key is greater than `min-key`, then B simply forwards the original message onto C.

3.1.2. INSERT Operation

As in the case of FIND, the INSERT message passes from list process to process until it reaches the location to be inserted. Assume we are currently at A where the key to be inserted is greater than the key inside A and is less than `min-key` of successor B. We need to insert a new process for this key between A and its successor B. Notice that `min-key` allows A to detect that a process has to be inserted after A, whereas without `min-key` it is process B that would need to insert a new process before itself, and need to communicate with A. For this to occur, while at same time satisfying the forward-only and atomicity conditions, every forwarded request would need a reply. Storing `min-key` makes it possible to asynchronously forward requests without having to wait for a reply. This ensures that the common case, forwarding requests, is done as quickly as possible.

As shown in step 1 of Figure 3(b), A sends a free-node request to a manager process (see Figure 1) for the free-node service (see Section 3.3). The free-node service, if successful, replies back to A with the MPI rank of an available free list process (process C). If there are no free processes, then the free-node service reports failure to process A, and in turn process A reports failure back to the application process making the INSERT request.

In step 2, after A receives the rank of C, A sends the INSERT operation, the `min-key` and rank information it has about B to C. In step 3 A synchronously sends[8] a terminate connection message to B which signals that there are no outstanding messages from A, and B can now accept a message from some process other than A. A updates its own information with C as its successor. Finally, in step 4, once C has received the message from A it can store the data and signal success back to the application process. A no longer has information about B and can only forward requests to C. As well, C is the only process that knows about B and thus the only one that can forward requests to B.

It is possible to overlap steps 1 with steps 2 and 3 if the free node service is able to receive the data for C from A and have node C instantiated with all of the needed data. In this case A only needs to communicate with the free-node service and with B, after which it can continue to process requests.

3.1.3. DELETE Operation

In the case of DELETE, as long as the key is less than `min-key` the request is forwarded. When the key equals `min-key` then A knows that the next process, B, is to be deleted. Like INSERT, `min-key` helps to maintain our two conditions since it makes it possible to detect when the next process needs to be deleted.

As shown in step 1 of Figure 3(c), A sends a delete message to B, the node to be deleted, and waits for a reply from B. In step 2, using synchronous send, B sends a disconnect message to its successor C, which signals to C that there are no outstanding messages from B. B notifies A that it is finished by replying to A with the information about C. In step 3, A updates itself so it now points to C and continues. During steps 3 and 4, B replies back to the application and notifies the free node service that it is now free, and B now behaves as a free

[8] `MPI_Ssend` is a synchronous send, and completes only after a matching receive has been posted.

process. In step 2, having A wait for the reply from B ensures that no other request can jump ahead of A (equivalent to hand-over-hand).

3.2. Application Programming Interface

An important part of the design is the implementation of the service interface. Applications first need to know how to discover the service. We configure the application processes with a pointer (i.e., MPI rank) to the head of the list, which does not store any list items. Later in Section 3.6, to improve response time, we augment this with local service discovery, but for now we only consider the case when application processes send all requests to the head process.

Although the list communication is deadlock-free, the composition of the list with the application processes introduces the following deadlock scenario. It is possible for there to be a communication cycle starting from (a) the application process making the request, to (b) the list nodes traversed, to (c) the final list process that sends the reply message back to (a). The message interface to the service must enforce the condition that the application be ready to receive all replies whenever they have outstanding requests to the service. This ensures that for this deadlock scenario, while all the list processes may be trying to send, the application process will first receive before sending; thus avoiding deadlock. There are also important performance consequences to receiving replies quickly, since at steady-state the in-flow of requests equals the out-flow of replies; each new request depends on the receipt of a reply.

As discussed further in Section 3.4, each application process can have a fixed number of outstanding requests. The replies to these requests can return out of order because the order depends on an item's location in the list; replies by processes towards the end of the list are likely to return after a later request for a list item near the head of the list. As mentioned in Section 3.1 request messages contain sequence numbers. The application service interface adds the sequence number. Sequence numbers are carried along as part of the request and returned to the process inside the reply message. We use MPI's `MPI_Irecv` to pre-post receive buffers for the replies for each outstanding request. The pre-posted receive buffers can be used, if necessary, as a hold-back queue [23] which can be managed by the application to re-order and return replies in the order of the requests. When all initial requests are sent to the head of the list, the list structure has the property that, since all requests are received in-order from the application process and no request can jump ahead of another one as requests traverse the list, we can maintain sequential consistency.

Moreover, although not done, if the head process replaced the sequence number of the request with the message-order timestamp, then all requests are totally ordered. The timestamp action can be viewed as a linearization point which implies the linearizability of executions, again assuming all requests are sent to the head process. Later in Section 3.6 we relax this condition.

3.3. Free Process Sub-service

MPI creates a static namespace at the initialisation of the MPI execution environment (`MPI_Init`), and assigns consecutive integer ranks to MPI processes in the range 0 to $N-1$. FG-MPI decouples the namespace of the MPI processes from that of the OS processes to allow for multiple MPI processes per OS process. We take advantage of FG-MPI's ability to expose large-scale concurrency to create a pool of free processes at the beginning of the program. This pool of free processes can be used to dynamically allocate and deallocate processes in response to the application load. As we discuss in this section, a free process is reincarnated as a list node process in case of an INSERT operation and released back to the pool on DELETE. This can be viewed as spawning of processes on a fixed namespace. The free processes are all blocked waiting for an allocate request and do not impose an overhead on the scheduler (see Section 4). The manager processes are able to dynamically load balance the allocation of free processes by communicating with each other. We discuss the details of the free process sub-service next.

At the start of a $[P, O, M]$ execution for our list service, three different types of MPI processes are launched per OS process, see Figure 1 and see Section 4 for the details on mapping. One of them is an application process, the second is a manager process and the remaining $P-2$ processes are ordered-list processes. Under this mapping the maximum size of the list is $(P-2) \times O \times M$ with $(P-2)$ nodes (i.e. ordered-list processes) per OS process. Initially the ordered list processes are enqueued in a local pool of $(P-2)$ free processes. Other mappings are possible and the only constraint is that whenever an OS process has an ordered-list process there must also be a manager process. This allows us to define the size of the list as well as control the distribution of list nodes to OS processes to balance the communication and work between co-located processes (interleaved) and non co-located processes (parallel).

On start-up each of the manager processes has a list of the list processes local to its OS process. It also has information about the ranks of manager processes in other OS processes. Managers use a pre-defined load-balancing strategy whose goal is to evenly distribute list processes to OS processes while trying to keep communication local between co-located processes. The trade-off between these two conflicting goals depends on the distribution of keys and overall workload and thus should be configurable.

In our experiments the keys are uniformly random from a 32-bit key space. We tested with varying workloads but all cases used the following simple load-balancing strategy. In the case of a free node request because of an INSERT operation the manager first attempts to satisfy the request locally. If there are no local free nodes, then the manager passes the allocation request onto the next manager where all managers have been configured into a ring and each knows the next manager in the ring. The manager who succeeds in allocating a free node replies directly to the list node making the request, allowing the INSERT to complete. If no free node is found by any of the managers after one round robin cycle, then the local manager signals failure back to the list node who made the request. In turn, the list node notifies the application process that the INSERT operation failed because the list is full. When the list is full or almost full, guaranteeing liveness is up to the application and the list service does not guarantee fairness. In the ring of manager processes deadlock is avoided by ensuring that each manager posts a receive buffer for receiving a message from the previous process before attempting to send a message to the next manager in the ring. Because the receive buffer is always pre-posted, MPI middleware can always deliver a message sent by one manager process to the next manager in the ring.

In the case of a DELETE, the free node registers itself back on the free node list. As described further in Section 3.6, locally we maintain a look-up table that processes can post to and consult. DELETE operations are fast, since they only require local operations, whereas INSERT can incur much larger delays when non-local communication is needed to satisfy the request.

Load-balancing is a good example of a strategy that impacts performance and needs to be configurable. The cost can depend on the workload, the relative cost of communication between local and non-local processes, the distribution of keys, and also the size of the list and P (the number of local processes per OS process). An advantage of this service-oriented approach is that load-balancing code is not scattered through the program but in one place and adding a new strategy only requires extending the manager processes inside the free node service.

3.4. Flow Control (W and K)

We define a configuration parameter W in the application API to control the flow of requests to the list. W is the maximum number of outstanding requests to the list service that an application process can have at any time. Let cw_i denote the current number of outstanding requests (i.e. number of requests minus number of replies) from process i. As previously mentioned, to avoid deadlock and improve performance, we require that whenever $cw_i > 0$, the application process must be able to receive a reply. The sum of all cw_i's at time T is the total load on the service. In the experiments in Section 5 we set W to be a constant for all application processes and increase the rate of requests together with W to reach steady-state where globally the flow of requests to the service matches the flow of replies from the service. This is used to measure the peak performance of the service. W cannot be set arbitrarily high, particularly when all requests go to the head of the list, since the saturation point tends to occur towards the head of the list and there are overheads in the middleware that increase, which result in decreased throughput. Thus peak throughput is achieved by carefully setting W. This parameter can be extended so as it can be changed at run time per application to maintain peak throughput.

In MPICH2, as in most MPI implementations, short messages are sent eagerly and buffered at the destination until matched by a receive. MPICH2, and as a result FG-MPI, has a global buffer pool that allocates space for eager messages in addition to other dynamic MPI objects. There is no flow control on eager sends and MPI simply aborts the execution when the buffer pool in the middleware of an OS process is exhausted. There is no simple technique for detecting these types of buffer problems in MPI, but it is possible to avoid them without resulting to completely synchronous communication [24]. To avoid these errors we implemented a *bounded asynchronous* scheme where for every K standard MPI sends (potentially eager sends) we send a synchronous `MPI_Ssend()` to ensure that there are no outstanding matching messages between the two processes. In the case of $K = 0$, every send becomes synchronous. The synchronous case is interesting because it is one technique to test for potential deadlock situations that are masked when buffering is available. Because MPI cannot guarantee the amount of buffering available for asynchronous communication the MPI standard defines a *safe* program as one that can execute synchronously. However, as was evident in our experiments, it is important to increase K as large as possible to overlap the computation and communication, while also ensuring there is always sufficient space in the buffer pool. This is all the more important in FG-MPI because there are more MPI processes sharing the middleware's buffer pool and thus a greater chance of exhausting it.

3.5. Granularity (G)

An obvious extension to our ordered list is to allow each list process to store more than one item. As previously described each process stores its local set of keys and the minimum key of its successor. It is simple to modify the list process to perform a FIND, INSERT or DELETE to search its local set of items to perform the operation. The only added complication occurs for INSERT since we can either add the item to the current node's set of items or insert the item by adding a new process. The INSERT operation was changed so that now the process at the insertion point (process A in Section 3.1.2) can decide to add the item to its own set of items or split the items between itself and a new process. This introduces another opportunity to add a load balancing strategy to attempt to optimise for the number of items to store locally. We used a simple strategy which split the items whenever the number of items exceeds a fixed threshold. For DELETE operations, if the delete key matches the `min-key` stored in a process, this implies either the next process is to be deleted, or when there are multiple items in the process, `min-key` needs to be updated. Both these cases can be handled by the reply from B to A in step 2 of the DELETE operation shown in Figure 3(c).

We call the maximum number of list items that a list node process can hold the *granularity*, G, of the list service. This can be adjusted dynamically. It is possible to set G initially to be small when the list is small and then increase it as the list size increases. Both G and P are helpful in reducing the idle time. Increasing G reduces the amount of messaging and therefore the amount of work done in performing requests. However, to maximise the opportunity for parallelism we need to distribute the list to the $O \times M$ processing cores.

3.6. Shortcuts

As mentioned earlier, the application processes send requests to the head of the linked list. This has an obvious disadvantage of creating congestion at the head of the list. When P is greater than one there are potentially active list processes co-located in the same address space. Rather than always send a request to the head, one optimisation is to have the application process first check locally. If the application process can look-up the key-values of the local list processes along with the associated rank, then it can send the request to the process whose key-value is closest to, but not greater than, the requested key-value. When none exist, it can always by default send the result to the head. One can view the local list processes as providing a set of possible shortcuts that the application process can use to reduce response time as well as reduce congestion for the head and processes towards the front of the list.

The disadvantage to the indiscriminate use of shortcuts is that there is no longer any guarantee on the order of operations since a later operation, on even the same item, could find a new shortcut allowing it to get ahead of an earlier request. However, for every outstanding request the application process can store the key-value of any shortcut that was used. This allows it to always choose the shortcut less than or equal to the earliest shortcut used by an outstanding request, thereby guaranteeing requests are performed in order. Likely the application process only cares about the order of requests to a given data item and gives more opportunity to use shortcuts since now earliest can be defined with respect to each individual key.

Given the existence of these shortcuts one can go further and make them available to the list processes themselves. There is no longer any guarantee on the order in which concurrent requests are done, even requests to the same key-value. Of course, the application process always has the option to wait for the reply before submitting the next request. This use of shortcuts by the list introduces an interesting type of probabilistic data structure where the shortcuts help balance requests and significantly reduce response time. The number of shortcuts, and hence the likelihood of finding good ones, depends on the size of the list and P. On average, assuming a uniform distribution, one expects a cost of N/P rather than $N/2$ communications to access N list processes distributed across all of the OS processes. As well, shortcuts help distribute the requests across the entire list of processes rather than at the front of the list. The managers also influence the quality of shortcuts. If managers only allocate free processes locally, then the shortcuts will tend to be local as well, whereas allocating non-locally results in better shortcuts but higher communication cost. The use of shortcuts, even without ordering, is good for the important use case where inserts and deletes are rare and the common operation is find.

Shortcuts are implemented by the manager process and service discovery is done by having each application process contact the local manager. We take advantage of the single address space of the OS process and the non-preemptive scheduling (i.e. atomicity) of processes to implement a shared look-up table. The list processes update the table when they are inserted or deleted and the application can query the table to find a suitable shortcut. Although this violates pure message-passing semantics, it can be viewed as part of the local execution environment of the processes similar to the `mpiexec` environment variables passed to the program at launch time or OS environment variables that can be read during execution.

Note that using shortcuts this way can lead to race conditions, where a message may be sent to a shortcut but the shortcut may be deleted or changed by the time the message reaches it. We are able to detect when a shortcut no longer exists and we modified the list and free processes to report shortcut failure back to the application when this occurs. We are also able to turn off the use of shortcuts and the application can re-submit the request with shortcuts off for that particular request.

We use the terms *total-ordering* for the implementation that does not use shortcuts, *sequentially-consistent* for the one where only application processes use shortcuts, and *no-consistency*[9] for the version that allows use of shortcuts by both the application and the list processes. We extended our service discovery to include both these shortcut semantics.

4. Mapping Issues

In this section we describe how the application, manager and list processes are mapped to OS processes (as an example see Figure 1). The `mpiexec` command performs a 1-dimensional mapping of MPI processes to different cores and machines. FG-MPI provides a second dimension to the mapping of multiple MPI processes within the same OS process. Just as `mpiexec` binds OS processes to cores, we define a function `FGmpiexec`, called from `main`, that binds co-located MPI processes to different functions. The binding of processes to functions can be defined in a general way through a user-defined function that takes a process's `MPI_COMM_WORLD` rank as its input parameter and returns a pointer to the function that MPI process will be executing. Note that each of these functions, which the processes are bound to, are written as regular MPI programs, beginning with `MPI_Init` and ending with `MPI_Finalize` calls. The number of co-located processes is specified by the `nfg` parameter to `mpiexec` and as shown in Figure 1, the MPI process ranks are assigned in consecutive block ranges of size `nfg`.

Listing 1 shows an example of binding MPI processes to three different types of functions per OS process: `LocalManagerNode` function for the manager process, `AppNode` function for the application process and `ListNode` function for the list processes. Listing 1 contains two user-defined mapping functions; `binding` and `map_lookup`. The `binding` function, as previously mentioned, takes the process rank as input and returns the corresponding function pointer. The `map_lookup` mapper function takes a string as its third parameter and maps that to a binding function. In this example the `str` parameter is not being used, however, its purpose is to allow selection of different binding functions at runtime through the `mpiexec` command line. The FG-MPI runtime system inside each OS process is initialised through a call to the `FGmpiexec` function, which takes `map_lookup` as a parameter and spawns the MPI processes. For the linked-list example this allows us to interleave the application and data structure node processes within the same OS process.

Depending on the operations on the list service, the list processes may change state from "free" to "active" and vice versa. As mentioned in Section 3.3, initially all the $P-2$ list processes add themselves to the local pool of free list processes. Adding to the free list is done by each of the list processes executing a `MPI_Recv` call. Since this is a receive call for which no matching messages have arrived, the receive call cannot complete and FG-MPI's scheduler places all these list processes in its blocked queue. In the free state, these list processes do not take up any additional resources other than the small amount of memory used to store state. These list processes move to "active" state and become part of the list service when they are allocated by the free process service, which unblocks them by sending a message to them.

[9] No-consistency means there is no guarantee on the order in which requests are performed. Note, the list itself always properly orders the items.

```
/******* FG-MPI Boilerplate begin *********/
#include "fgmpi.h"
/* Forward declarations of the three
 * functions that the MPI processes are bound to. */
int ListNode(int argc, char** argv);
int LocalManagerNode(int argc, char** argv);
int AppNode(int argc, char** argv);
FG_ProcessPtr_t binding(int argc,char** argv,int rank)
{
    if (rank == starting_rank_of_co-located_processes)
      return (&LocalManagerNode);
    else if (rank == last_rank_of_co-located_processes)
      return (&AppNode);
    else
      return (&ListNode);
}
FG_MapPtr_t map_lookup(int argc,char** argv,char* str)
{
    return (&binding);
}

int main( int argc, char *argv[] )
{
    FGmpiexec(&argc, &argv, &map_lookup);
    return (0);
}
/******* FG-MPI Boilerplate end *********/
```

Listing 1. An example of MPMD mapping for the linked list.

FG-MPI makes it possible to combine those processes implementing the service with the application processes using the service. As a service this makes it possible to separate the data structure code from the application code as well as having the ability to flexibility map application and service processes in a variety of ways.

5. Experiments

In this section we experiment with different parameters of the ordered linked-list service and show how varying them affects the performance. Recall that operations on a linked list are inherently non-scalable and our objective here is to demonstrate the working of a service-based approach and the flexibility in tuning of its parameters. In all the experiments reported below we report the steady-state throughput (i.e., number of operations per second) of the ordered linked list. The INSERT operation was initially used to build the list to capacity and then two different workloads were used to measure the throughput. One workload consisted of an equal number of INSERT, DELETE and FIND operations and the second consisted of FIND operations only. The two workloads produced similar results with a slight difference in performance, therefore, we report the results for the second workload only.

For the experiments, the test setup consisted of a cluster with 24 machines connected by a 10GigE Ethernet interconnection network. Each of the machines in the cluster is a quad-core, dual socket (8 cores per machine) Intel Xeon® X5550, 64-bit machine, running at 2.67 GHz. All machines have 12 GB of memory and run Linux kernel 2.6.18-308.16.1.el5.

5.1. Consistency Semantics

Figure 4 shows the throughput achieved with the three consistency semantics described in Section 3.6, i.e., total-ordering, sequentially-consistent and no-consistency. The size of the

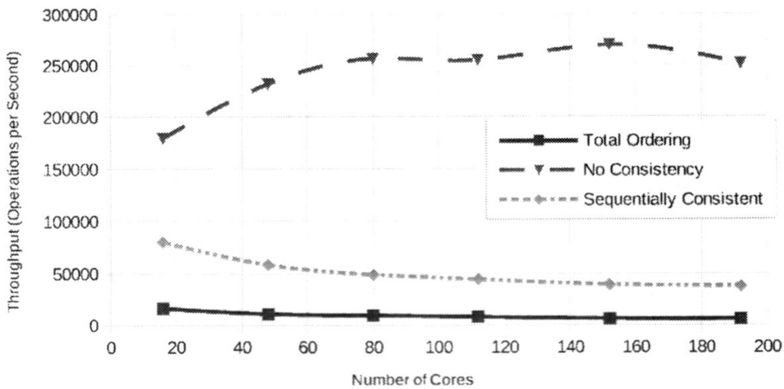

Figure 4. Throughput achieved with different consistency semantics for a fixed list size, evenly distributed over the number of cores=$(O \times M)$.

linked list was set to 2^{20} in this experiment and the number of list nodes were evenly distributed over the number of cores $(O \times M)$, by setting $ListSize \div (O \times M) = P \times G$, with P set to 10. As expected, the throughput with no-consistency is the highest followed by that for sequential-consistency and then total-ordering. In Figure 4, as we move from left to right along the x-axis, the list is becoming more distributed and the number of items being stored per list node is decreasing, while keeping the list size constant. The no-consistency semantics benefits from more distributed nature of the list because the requests are spread over more cores. One could argue that, in the probabilistic sense, selecting the optimal shortcut for the no-consistency semantics takes a maximum of $\mathcal{O}(O \times M)$ steps. Operations with the other two types of semantics i.e., total ordering and sequential consistency have to traverse the list to reach the correct list node and on the average take $\mathcal{O}(O \times M \times P)$ steps assuming G is small enough to neglect its cost. Total ordering also suffers from congestion in the head of the list, which is why its curve is, as expected, the worse of the three.

5.2. *G versus P*

Figure 5 shows how performance changes when interleaved concurrency (P) is increased while reducing granularity (G), keeping the list size and the number of cores fixed. Since the number of cores is fixed, the overall amount of parallelism is fixed and the figure shows the trade-off between interleaved concurrency versus coarser-grain processes with more list items inside a process. Figures 5(a) and (b) show the trade-off between G and P with sequentially-consistent and no-consistency shortcut semantics. In Figure 5(a), as expected, it is better to reduce P and increase the granularity, G.

Parameters G and P define the first two levels in the communication hierarchy of the machine. In terms of traversing the list, from fastest to slowest, there is traversal (a) inside a process, (b) between processes inside the same OS process, (c) between processes on the same machine, (d) between processes on separate machines. There is a significant difference in communication costs between these levels (a) tens of nanoseconds, (b) less than a microsecond, (c) a few microseconds, and (d) on the order of 50 microseconds. Making G large and P small is analogous to keeping all data in the primary cache rather than the secondary cache. Figure 5 is a simple illustration of the trade-offs. Since the focus of the paper is on the design of the library we did not explore the use of different applications or additional load-sharing strategies for the manager processes to better exploit the hierarchy.

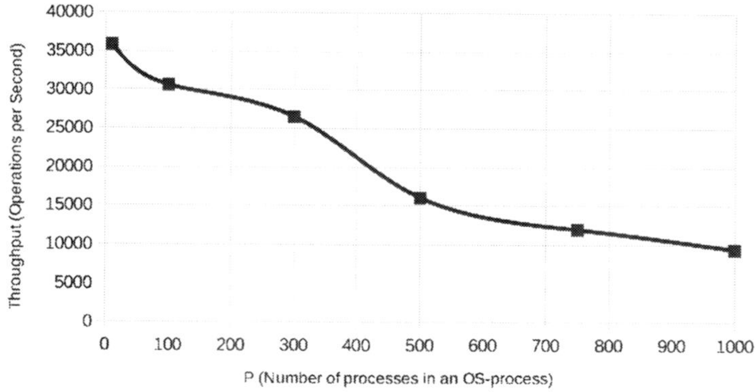

(a) Sequentially-consistent.

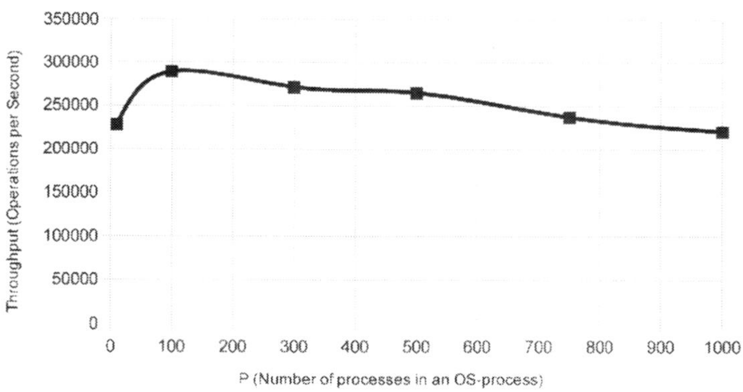

(b) No-consistency.

Figure 5. Effect of changing P, while keeping $P \times G$ constant. The list size and the number of cores are fixed. Semantics used are (a) sequentially-consistent and (b) no-consistency. (ListSize = $G \times P \times O \times M$=$2^{20}$ and cores=$O \times M$=176).

There are interesting questions that arise with regards to efficiently managing communication costs across the communication hierarchy. On a list of size N, assuming uniformly distributed keys, we have that $N = G \times (P \times O \times M)$ where the expected number of hops is $\frac{1}{2}(G + P \times O \times M - 1)$. With regards to latency (response time) the optimal distribution is to have $P = 1$, $G = N/(O \times M)$ and map OS processes consecutively from one machine to the next to keep communication as local as possible. But for throughput, increasing G limits the rate at which requests enter the system, since when viewed as a pipeline the search inside the list of size G at the FIND location delays all requests behind the current one. Although this depends on the location of the FIND, in steady-state it is likely to occur often and have more of an effect closer to the head of the list where the load is higher. Intuitively, larger G reduces the capacity of pipeline and speeds up requests while increasing P increases the capacity of the pipeline but slows down requests. We believe the value of G in Figure 5(a) was never large enough to see this effect and provide a rationale for increasing P in the sequentially-consistent case.

The throughput shown in Figure 5(b), with the use of shortcuts inside the list of processes, is significantly larger than that of Figure 5(a). P determines the number of possible shortcuts inside each OS process. Once there are more than a few shortcuts there is a high probability of the existence of a shortcut to a closer OS process. As a result, the expected number of hops begins to depend on the number of OS processes rather than the size of the list. There is a diminishing return for adding more shortcuts since there is less chance of it being of value. As Figure 5(b) shows the first 100 shortcuts increase performance after which more shortcuts is adding more overhead than benefit as P increases.

5.3. W and K

Parameter W specifies the number of outstanding requests that each application can submit to the list service. Increasing the value of W increases the load on the system. K is a throttling parameter that each list node uses to specify how many requests can be forwarded down the list, before having to wait for a previously forwarded message to complete. In the perfectly balanced case, when W equals P, each list node receives exactly one application request. When W exceeds P, each list node progresses multiple application requests up to a maximum of K. The smaller the value of K, the higher the throttling effect on the flow of requests through the list service. Figure 6 shows the effect of different values of K on the throughput.

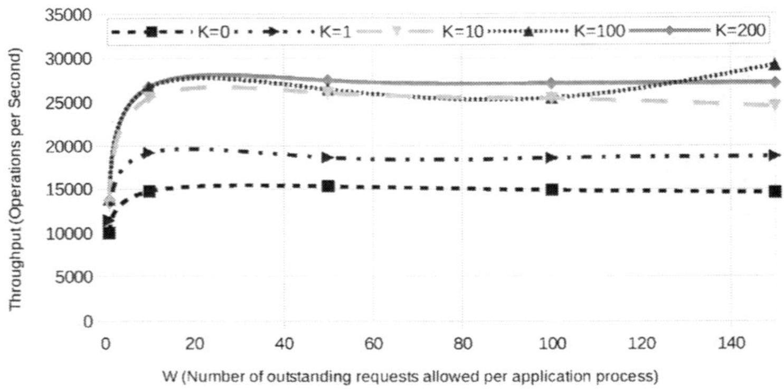

Figure 6. Throttling effect of K on the throughput while increasing the outstanding requests (W) by the application processes. List size used is 2^{20}. $P = 100$. $O \times M = 192$ cores. Sequentially-consistent semantics are used in this experiment.

The effect of small K clearly limits the throughput. Increasing K increases fluidity in the network for request propagation. There is a limit to the rate at which a process can forward request, and once K exceeds this limit it no longer has any effect (it's at full throttle). This limit depends on computation and communication characteristics of the machine. K gives us a control over the flow inside the service and it should be tuned to the communication characteristics of the underlying network.

6. Conclusions

In this paper we discussed a service-oriented approach to the design of a distributed data structures and presented an ordered linked-list service as an example. The implementation of a distributed data structure as a service uses its own collection of processes which makes it possible to separate the data structure code from the application code resulting in reduced coupling and increased cohesion in the components of the system.

In parallel computing, more so than in other areas, there is the additional challenge that the performance of the library must be portable across a wide range of machines sizes and communication fabrics. Scalability across machines and communication fabrics requires the addition of tunable parameters to control resources based on the characteristics of the machine and networks. In our design we introduced a variety of these parameters to control computation/communication overlap (K), interleaving versus parallel (P), load on the data structure (W) and process granularity (G).

The service-oriented approach to the design of an ordered linked list is substantially different from existing work. It borrows ideas from work on concurrent linked lists but is based on pure message-passing and does not require support for shared memory. There is some commonality with distributed systems in cloud environments but with a focus on scalability and latency, and not availability and reliability. Our ordered linked list implementation can execute on a single multicore but more importantly it can also potentially scale to machines with thousands of cores to support sorting and querying for large datasets that can take advantage of the large amounts of available memory.

The experiments demonstrated the ability to scale to an ordered linked list with over a million data items with up to 192,000 MPI processes executing on 192 cores. As expected, for a list of size N, the $O(N)$ communications per request makes the one item per process impractical for large lists. However, we showed that better performance is possible and depends on the level of consistency needed and trade-offs between G, the number of list items stored inside a process, and P, the number of processes inside an OS process. The values of P, O, G and N are all runtime or dynamic parameters that do not require re-compilation of the program making it possible to flexibly map an application onto varying types of cluster architectures without a need to re-design or recompile the library.

In conclusion, by using a service-oriented approach we are able to create an ordered linked-list library that can be used within MPI programs. The linked list is a relatively complex data structure and our design decouples the code needed to implement the list from the application and encapsulated the code related to performance inside the library while exposing a simple set of parameters for the application programmer or user to flexibly tune the performance of the system. This supports our view that a service-oriented approach to distributed data structures provides an interesting design space for large scale, high performance, distributed systems. It lies at the intersection of parallel computation and distributed system design, and well-suited for the types of cluster of multicore machines used in high performance and cloud computing environments.

Acknowledgements

The authors wish to thank the anonymous reviewers for their valuable comments. We gratefully acknowledge NSERC Canada for their support of this research.

References

[1] B.L. Chamberlain, D. Callahan, and H.P. Zima. Parallel programmability and the Chapel language. *Int. J. High Perform. Comput. Appl.*, 21:291–312, August 2007.

[2] Carl G. Ritson and Peter H. Welch. A process-oriented architecture for complex system modelling. *Concurrency and Computation: Practice and Experience*, 22:965–980, March 2010.

[3] P.H. Welch. Process Oriented Design for Java: Concurrency for All. In H.R.Arabnia, editor, *Proceedings of the International Conference on Parallel and Distributed Processing Techniques and Applications (PDPTA'2000)*, volume 1, pages 51–57. CSREA, CSREA Press, June 2000.

[4] Mike Burrows. The chubby lock service for loosely-coupled distributed systems. In *Proceedings of the 7th symposium on Operating systems design and implementation*, OSDI '06, pages 335–350, Berkeley, CA, USA, 2006. USENIX Association.

[5] Daisuke Tsukamoto and Takuo Nakashima. Implementation and evaluation of distributed hash table using MPI. *Broadband, Wireless Computing, Communication and Applications, International Conference on*, 0:684–688, 2010.

[6] Humaira Kamal and Alan Wagner. An integrated fine-grain runtime system for MPI. *Computing*, pages 1–17, 2013.

[7] Humaira Kamal and Alan Wagner. Added concurrency to improve MPI performance on multicore. In *Parallel Processing (ICPP), 2012 41st International Conference on*, pages 229 –238, Sept. 2012.

[8] C. Iancu, S. Hofmeyr, F. Blagojevic, and Yili Zheng. Oversubscription on multicore processors. In *Parallel Distributed Processing (IPDPS), 2010 IEEE International Symposium on*, pages 1 –11, april 2010.

[9] Peter H. Welch and Frederick R. M. Barnes. Communicating mobile processes: Introducing occam-pi. In Ali Abdallah, Cliff Jones, and Jeff Sanders, editors, *Communicating Sequential Processes. The First 25 Years*, volume 3525 of *Lecture Notes in Computer Science*, pages 712–713. Springer Berlin / Heidelberg, April 2005.

[10] Leslie G. Valiant. A bridging model for parallel computation. *Commun. ACM*, 33(8):103–111, August 1990.

[11] Mikhail Fomitchev and Eric Ruppert. Lock-free linked lists and skip lists. In *Proceedings of the twenty-third annual ACM symposium on Principles of distributed computing*, PODC '04, pages 50–59, New York, NY, USA, 2004. ACM.

[12] Timothy L. Harris. A pragmatic implementation of non-blocking linked-lists. In *Proceedings of the 15th International Conference on Distributed Computing*, DISC '01, pages 300–314, London, UK, UK, 2001. Springer-Verlag.

[13] Maurice Herlihy and Nir Shavit. *The Art of Multiprocessor Programming*. Morgan Kaufmann Publishers Inc., San Francisco, CA, USA, 2008.

[14] Avinash Lakshman and Prashant Malik. Cassandra: a decentralized structured storage system. *SIGOPS Oper. Syst. Rev.*, 44(2):35–40, April 2010.

[15] R Morris IonStoica, David Karger, M Frans Kaashoe, and Hari Balakrishnan. Chord: A scalable peer-topeer lookup service for internet applications. In *SIGCOMM01*, pages 27–31.

[16] David Gelernter. Generative communication in Linda. *ACM Trans. Program. Lang. Syst.*, 7(1):80–112, January 1985.

[17] L. M. Silva, J. G. Silva, and S. Chapple. Implementing distributed shared memory on top of MPI: The DSMPI library. In *Proceedings of the 4th Euromicro Workshop on Parallel and Distributed Processing (PDP '96)*, PDP '96, pages 50–, Washington, DC, USA, 1996. IEEE Computer Society.

[18] J. M. Wozniak, B. Jacobs, R. Latham, S. Lang, S. W. Son, and R. Ross. C-MPI: A DHT Implementation for Grid and HPC Environments. (Preprint ANL/MCS-P1746-0410), April 2010.

[19] William Gropp. MPI 3 and beyond: Why MPI is successful and what challenges it faces. In Jesper Larsson Träff, Siegfried Benkner, and Jack J. Dongarra, editors, *EuroMPI*, volume 7490 of *Lecture Notes in Computer Science*, pages 1–9. Springer, 2012.

[20] Kurt B. Ferreira, Patrick Bridges, and Ron Brightwell. Characterizing application sensitivity to OS interference using kernel-level noise injection. In *Proceedings of the 2008 ACM/IEEE conference on Supercomputing*, SC '08, pages 19:1–19:12, Piscataway, NJ, USA, 2008. IEEE Press.

[21] Photran. An Integrated Development Environment and Refactoring Tool for Fortran. Available from `http://www.eclipse.org/photran`.

[22] Elsa: An elkbound based C++ parser. Available from `http://www.cs.berkeley.edu/~smcpeak/elkhound`.

[23] G.F. Coulouris, J. Dollimore, T. Kindberg, and G. Blair. *Distributed Systems: Concepts and Design*. International computer science series. Addison-Wesley, 2011.

[24] Alex Brodsky, Jan Bækgaard Pedersen, and Alan Wagner. On the complexity of buffer allocation in message passing systems. *Journal of Parallel and Distributed Computing*, 65(6):692 – 713, 2005.

Communicating Process Architectures 2013
P.H. Welch et al. (Eds.)
Open Channel Publishing Ltd., 2013

Scalable Performance for Scala Message-Passing Concurrency

Andrew BATE [1]

Department of Computer Science, University of Oxford, UK

Abstract. This paper presents an embedded domain-specific language for building massively concurrent systems. In particular, we demonstrate how ultra-lightweight cooperatively-scheduled processes and message-passing concurrency can be provided for the Scala programming language on the Java Virtual Machine (JVM). We make use of a well-known continuation-passing style bytecode transformation in order to achieve performance that is several orders of magnitude higher than native JVM threads. Our library is capable of scaling to millions of processes and messages on a single virtual machine instance, and our runtime system will detect deadlock should it occur. Message-passing is over 100 times faster than Erlang, and context switching is 1000 times faster than native Java threads. In benchmarks, the performance of our library is close to compiled code.

Keywords. concurrency, Scala, message passing, lightweight threads, JVM

Introduction

Concurrent programming is often considered error-prone and difficult to scale. However, it is our belief that these problems are due to the prevalence of *shared-memory concurrency* and not due to the multiple threads of control.

Hardware is becoming increasingly parallel, with the number of cores taking the place of clock speed as the specification of first interest when describing a processor [1,2]. The massively parallel architectures of the future are unlikely to have uniform memory access [3], and thus shared memory will not be a good abstraction. Message-passing concurrency, on the other hand, models the world as a collection of independent parallel processes that share nothing and interact only by passing messages [4].

Message-passing programs can be built from millions of extremely lightweight processes that can run on a single processor, a multi-core processor, or on a network of processors [5]. Thus message-passing is appropriate for both local and distributed algorithms [6,7]. In contrast, the idioms of shared-memory concurrency are substantially different from those of distributed programming [6,8].

By exposing the potential parallelism as small well-defined and encapsulated logical tasks, existing programs built using message passing are able to scale with the increasing parallelism of the hardware on which they are run [9,10,11]. Experience has shown that this is not equally true of programs built with shared-memory concurrency.

It is also extremely difficult to prove correctness or to reason about fairness and efficiency in the presence of fine-grained locks and access to shared data structures [12]. In particular, reasoning about memory corruption and deadlock in shared memory systems is diffi-

[1] Corresponding Author: *Andrew Bate, Department of Computer Science, University of Oxford, Oxford, OX1 3QD, UK.*; E-mail: `andrew.bate@cs.ox.ac.uk`.

cult. For message passing, there exist well-developed mathematical models and robust tools for formal verification of design [13].

In this paper, we present an embedded domain-specific language (DSL) for message-passing concurrency, called CSO_2, which provides a judicious combination of the ideas from Hoare's CSP [4,13] and Milner's π-calculus [14] for the Scala programming language. In a similar way to the original CSO [15,16] and JCSP [17,18], it follows many of the principles of the occam-π [19,20] programming language, exchanging compiler-enforced safety for programmer checked rules, but allowing the model to be used in a mainstream programming language. In contrast to these previous approaches, however, our library offers low process management and communication overheads, all without any modification to the Java Virtual Machine (JVM).

In order to make message-passing concurrency and process-oriented programming a compelling alternative to shared-memory concurrency, we need fast inter-process communication and lightweight threads. The JVM, arguably the most widely deployed and targeted virtual machine, supports many languages, including Java [21], Scala [22], and Clojure [23]. However, unlike in Erlang [5] or occam-π [19], the JVM does not natively support lightweight threads or message-passing: a JVM thread corresponds to a thread provided by the operating system [24]. Thus, Java threads are too heavyweight to assign per HTTP connection in a web server or per transaction in a database.

Our library provides lightweight and fast cooperatively scheduled user-level processes. The speed of context-switching of CSO_2 processes is far superior to a mainstream concurrent programming language, and close to compiled native code. We achieve this performance by transforming the bytecode produced by the Scala compiler using a well-known continuation-passing style (CPS) transformation [25,26,27,28]. This transformation is necessary in order to achieve good performance when many more JVM threads are used than there are available processors. We also provide a lightweight runtime system, which includes support for deadlock detection, similar to that available for the Go programming language [29].

The syntax of our embedded DSL is inspired by that of CSO [15]. In our experience, its syntax is sufficiently concise and constructs are in a close correspondence to their CSP counterparts. Feedback from taught courses at the University of Oxford has reinforced to us its suitability.

Outline of the Paper

It is our intention that this paper be self-contained. In Section 1, we introduce the syntax of CSO_2 and, in Section 2, provide our rationale for implementing it as an embedded DSL. In Section 3 we describe the post-compilation CPS bytecode transformation. We describe our user-level scheduler in Section 4, report on our implementation of message-passing channels in Section 5, and provide details of our implementation of runtime deadlock detection in Section 6. In Section 7 we benchmark the performance of our concurrency library in three key aspects: (i) the ability to create many processes, (ii) speed of process creation, and (iii) speed of message passing. A survey of related work, considerations for possible future work, and concluding remarks are provided in Sections 8 through 10.

1. Introducing the Syntax of CSO_2

Owing to space constraints, we introduce only the principal constructs of CSO_2. However, we will demonstrate their use on a parallelised implementation of Quicksort. Since our syntax is similar to CSO [15], this section may be skipped should the reader already possess familiarity with this syntax.

1.1. Simple Processes

A process is constructed using the notation:

```
def ExampleProcess(x1: T1, ..., xn: Tn) = proc (name) { body }
```

where name is an optional[2] String used to identify the process, body is a sequence of statements (called the *process body*) to be executed when the process is run, and x1, ..., xn are the arguments to body with the respective types T1, ..., Tn. Any of the Ti may be ports (discussed below in Section 1.3).

If P is a process, then **run** P executes P in the current thread, whereas **fork** P executes P in parallel and allows the caller to continue.[3]

Our library is capable of multiplexing millions of these processes onto as many JVM threads as there are available hardware threads, in the manner described in Section 3. Each JVM thread is in turn mapped onto a kernel thread [24].

1.2. Parallel Composition

The parallel composition of processes is achieved using the || operator. For example, the composition of the four processes P1, P2, P3 and P4 is given as follows:

```
val comp = P1 || P2 || P3 || P4
```

A composition of parameterised processes P(0), P(1), ..., P(n) is denoted by:

```
|| (i <- 0 to n) (P(i))
```

where n may be determined at runtime.

If Q = P1 || P2 || ... || Pn, then **run** Q causes each of the processes P1, P2, ..., Pn to be run in parallel. The process Q terminates when all of the components Pi of its composition have terminated.

It is important to reaffirm the distinction between a thread and a process: a *process* is a lightweight object of type Process that is a specification of runtime behaviour, whereas a *thread* is a JVM thread running the body of some process. For example, the statement **run** (P || P) results in the execution of a parallel composition of two separate instances of the body of process P.

1.3. Ports and Channels

Processes communicate via channels. A channel on which values of type T may be read or written is an instance of a subclass of Chan[T] and consists of one output port of type ![T] and one input port of type ?[T]. Processes are parametrised with the ports on which they may read or write data, as described in the previous section.

Input ports and output ports are named from the perspective of the process. Therefore, an output port provides a method !(T): Unit to write values to the channel. Similarly, an input port provides a method ?(): T to read values from the channel.

Channels are either *synchronous* or *buffered*. For a synchronous channel, termination of the execution of a ! at the output port is synchronised with the termination of the execution of

---

[2] If name is not supplied, then the name of the process defaults to that of the method where it is defined. For example, **def** P = **proc** { ... } is equivalent to **def** P = **proc** ("P") { ... }.

[3] In the original CSO syntax, one would write P() to run P. However, forgetting to write the double parentheses was a common source of bugs, especially since the Scala type system distinguishes between a method definition with no formal parameter list and one with an empty list.

a corresponding ? at the input port. A buffered channel, however, does not synchronise any execution of ! with ?.

A OneOne[T] channel is a synchronous channel where no more than one process at a time may access its output port or its input port. Similarly defined are the channels ManyOne[T], OneMany[T], and ManyMany[T], where the naming convention is the restriction on sharing of the output port followed by that of the input port. Each of these channels have the following buffered counterparts: OneOneBuf[T](n), ManyOneBuf[T](n), OneManyBuf[T](n) and ManyManyBuf[T](n), respectively, where n is the buffer capacity.

Our library supports changing process topology by the communication of ports, as is available in occam-π [19], JCSP [30], and CSO [15]. This is equivalent to the mobility of channel names in the π-calculus [14].

Finally, a channel is closed by invoking its close method. Closing a channel enforces the contract that the channel will never be read from or written to again.

1.4. Alternation

A *guarded output event* is a construct of the form (guard &&& outport) –!–> { cmd }, where guard is a Boolean expression, outport is the output port of some channel, and cmd is a statement. We say that the event is *ready* when outport is ready to perform a ! at that instant and guard evaluates to true.[4] A *guarded input event* is an expression of the form (guard &&& inport) –?–> { cmd } whose semantics is defined analogously.

An **alt** consists of a collection of guarded events:

```
alt ( (guard1 &&& port1) --> { cmd1 }
    | ...
    | (guardn &&& portn) --> { cmdn }
    )
```

where each (guard &&& port) ––> { cmd } can be either a guarded input or output event.

A process executing an **alt** waits until it finds an event that is ready and then executes the corresponding cmd, after which the **alt** terminates. Note that it is the cmds that are responsible for performing the reading from or writing to a channel.

1.5. Example: Quicksort using Recursive Parallelism

We illustrate the use of the CSO_2 library with the application of the recursive parallelism pattern to Quicksort, whereby recursive method calls in the sequential code are replaced by a composition of parallel processes with the same effect [31]. In the base cases, the result is computed directly.

The base case of the Quicksort process is when the input channel contains no data, in which case we just close the output channel. Otherwise, if the input channel is non-empty, we choose the first element as the pivot, and recursively use two copies of QSort to sort the data that is less than the pivot, and greater than or equal to the pivot, respectively.

```
def QSort(in: ?[ Int ], out: ![ Int ]): Process = proc {
  attempt {
    val pivot = in?
    val toHigher, toLower, fromHigher, fromLower = OneOne[Int]

    def Partition = ... // As defined below
```

[4] We have chosen for our terminology to deviate slightly from that of Sufrin [15] for sake of clarity.

```
    // Compose the system and run it
    run ( Partition || QSort(toHigher, fromHigher) || QSort(toLower, fromLower))
  }{
    out.close // Close once all data has been received
  }
}
```

The construct **attempt** P Q executes P, and if P throws a Stop exception (for example, if a channel is closed), then it executes Q.

To perform the partitioning of elements, we define a nested process called Partition that is responsible for passing data to the correct recursive copy of the QSort process, and then reading back the sorted data from the recursive processes.

```
def Partition = proc {
  // Partition the data received around the pivot
  repeat {
    val n = in?
    if (n < pivot) toLower ! n else toHigher ! n
  }
  // Close channels to recursive processes once the final input is received
  toHigher.close; toLower.close

  // Now output the results
  Copy(fromLower, out); out ! pivot; Copy(fromHigher, out)
  out.close
}
```

where the construct **repeat** { body } will repeatedly execute the statements of body until it invokes an operation on a closed channel, and Copy is an auxiliary process that repeatedly copies values from in to out:

```
def Copy(in: ?[Int], out: ![Int]) = repeat { val n = in?; out ! n }
```

Of course, in practice, one would use limited recursive parallelism, whereby recursive processes are only spawned to a maximum depth; however, we have elided the logic necessary to achieve that for brevity.

2. Why an Embedded Domain-Specific Language?

In this section, we motivate our decision to design an embedded DSL [32] for Scala, as opposed to implementing (yet another) standalone programming language.

The historically prevalent approach has been to build standalone languages, with their own custom syntax. This has the advantage that the syntax and semantics of the language can be designed specifically for message passing. Go and occam-π are notable examples of this approach [19,29].

Conversely, implementing a new standalone language is a significant undertaking, involving a separate parser and compiler, and perhaps an interactive editor too. Considerable effort is spent implementing the common features of a general purpose programming languages, including variables, definitions, conditionals, and so on. The custom syntax for these common features also increases the learning curve of the new language.

The alternative approach to is implement message-passing concurrency as an embedded DSL within an existing host language, and thus retaining as much as possible of the existing syntax and language features, without having to go to the trouble of defining a separate language. The DSL is then implemented as a library of definitions written in the host language, and familiarity with its syntactic conventions can be carried over to the DSL. Furthermore, this approach overcomes need to implement compiler optimisations for the sequential code in addition to the message passing code. In some benchmarks, the lack of compiler optimisations for sequential code has led to reduced performance for occam-π [6].

However, there are some downsides. It requires great care to preserve the abstract boundary between the DSL its host, and useful error messages may be more difficult to implement. In spite of this, it turns out that the Scala programming language is particularly well suited for hosting embedded DSLs, in particular because of four language features: higher-order functions, a rich type system supporting type inference, macros supporting compile-time metaprogramming, and a lightweight syntax supporting customisation.

Taken together, these features have allowed us to design an embedded DSL with syntax close to what one might provide for a corresponding standalone language. For the majority of the message-passing constructs in our DSL, we hope that the syntax bears close resemblance to its occam-π counterpart.

3. Lightweight Processes

In this section we describe our implementation of lightweight processes. We begin by giving an overview of the runtime system before presenting the bytecode transformation, with additional details of the scheduler and message-passing channels provided in later sections.

Our user-level scheduler is responsible for multiplexing possibly millions of Process objects onto as many JVM threads as there are hardware threads. In this way, we view each java.lang.Thread as a virtual processor. Each process is cooperatively scheduled.

The idea is as follows: Each process has an internal execute() method which is invoked by the user-level scheduler whenever the process is to begin running from a waiting state. Further, each process has an internal pause() method that moves the process from a running state to a paused state, and an internal resume() method that moves the process from a paused state to a waiting state; see Figure 1.

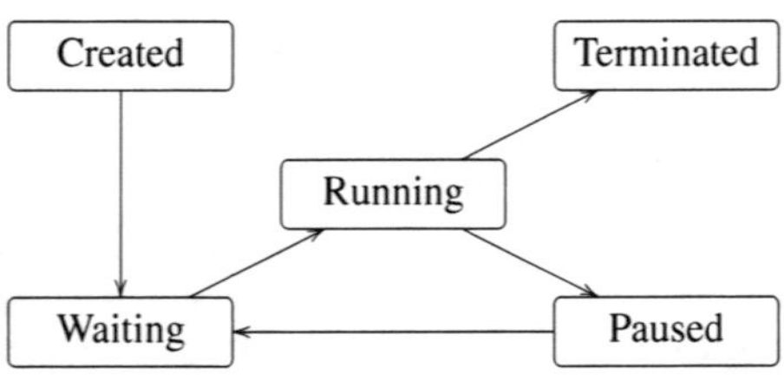

Figure 1. The various process states defined by the user-level scheduler with arrows indicating the possible transitions between states.

For example, when a process attempts to write to a buffered channel that is full, the pause() method of that process will be called. Later, when space is available, another process will invoke the corresponding resume(). Intuitively, pause() and resume() are similar to wait() and notify(), respectively, except that pausing does not block the underlying thread.

It is our user-level scheduler that is responsible for automatically managing the stacks of our processes. This cooperative scheduling is achieved through a *continuation-passing style*

(CPS) transformation and so works by rewriting the bytecode that is produced by the Scala compiler,[5] as is described in Section 3.1.

Optimisations employed in our implementation are discussed in Section 3.2. Difficulties arising in the implementation owing to restrictions placed upon us by the JVM are reported in Section 3.3, and difficulties caused by Scala's functional style are reported in Section 3.4. Performance considerations are discussed in Section 3.5, and issues relating to shared memory are detailed in Section 3.6.

3.1. Bytecode Transformation

The transformation described in this section is applied statically to the bytecode generated by the Scala compiler for the source program.

In accordance with previous work [25], we say that a method is *pausable* if it either directly or transitively (with respect to the call graph) invokes the pause() method.

The first step in the post compile-time transformation is to annotate each pausable method with the Pausable annotation. Furthermore, if a method overrides another method annotated as Pausable, then that method must also be annotated as Pausable. This is necessary because methods annotated as such will be subsequently transformed as described below, and this transformation involves the modification of method signatures. Clearly, the application of these annotations requires whole-program analysis.

Next, the resulting bytecode is further transformed using a well-known CPS transformation [25,26,27,28]. The technique we employ is a simple variant of single-shot delimited continuations [33] and largely identical to that of Srinivasan [25], and we give a brief account of that technique here.

A continuation represents the remainder of a computation from a given point in a program; hence, by using CPS, we can effectively suspend a process at the user level in order to resume it later. We use a restricted form of continuations that always transfer control back to their caller but maintain an independent stack.

The signature of every method annotated as Pausable is transformed to include an extra parameter: the continuation object, known as a *Fiber* [34,35]. Each instance of a process has a Fiber associated with it. Whenever pause() is called from within such a method, the isPausing flag of the Fiber is set. This flag is later cleared by the scheduler when another process calls the corresponding resume().

In addition to the extra argument, each method annotated as Pausable has a so-called *prelude* block injected at the start, and each call site of a method annotated as Pausable is surrounded by both *pre-* and *post-call* sections. This is illustrated in Figure 2.

The purpose of these blocks can be understood as follows: Each pre-call block captures the current state of the activation frame, including the values of local variables and the program counter, and stores them in the Fiber passed as argument. In the corresponding post-call block, the isPausing flag of the Fiber is inspected: If unset, execution can continue unaffected. Otherwise, the process is pausing and hence the current method must return immediately (possibly with a dummy return value; see Section 3.3). This continues recursively all the way up until the internal execute() method of the process has been popped off the stack. Afterwards, the process is no longer running and is now paused.

The process will move from the paused state to the waiting state when another process calls the corresponding resume(). It then waits to be scheduled, at which point the user-level scheduler invokes the execute() method once again.

[5] We have developed a plugin for the Simple Build Tool (SBT), the de facto build tool for Scala projects, that launches the bytecode rewriter after compilation. Both the library and this plugin are available from the project website: `http://www.cso.io`.

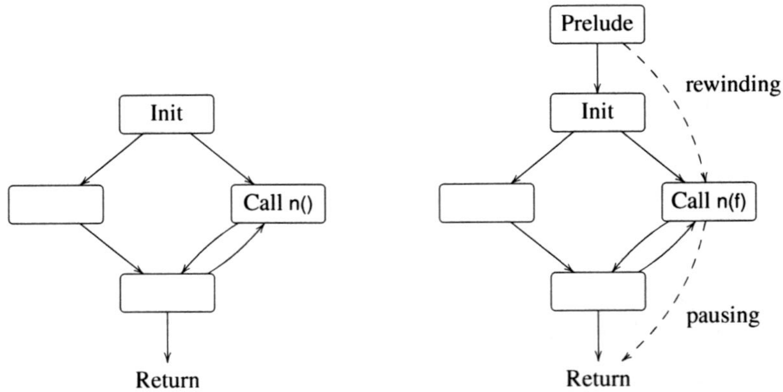

(a) Before transformation. (b) After transformation.

Figure 2. A sample control flow graph of a pausable method that calls another pausable method, before (a) and after (b) the transformation. The transformation adds the prelude basic block (as described in the text) and two possible jumps in control flow (shown by dashed edges) for each pausable method n: one from the prelude to the call site of n that is to be followed when n is being restored onto the stack, and another from the call site of n to the end of the method that is to be followed when n is pausing.

This causes the process to be wound back onto the thread, one activation frame at a time, as follows: Each prelude section inspects the Fiber, detects that the stack is being rewound, restores the local state of the method, and jumps to the call site of the method that most recently triggered the pause. Thus the prelude section contains a sequence of conditional jumps to each call site of a method annotated as Pausable. After each frame has been restored, the process is in the running state once again and can continue execution where it left off.

Since all necessary logic is injected at the bytecode level, our library manages the stack with no changes to the source code. Thus our library is entirely portable, and is agnostic of any particular JVM implementation. For example, our approach is equally targetable towards any of the Oracle Hotspot, Oracle JRockit, IBM J9, or Google Dalvik virtual machines.

Previous work has shown that this technique, known as *stack-ripping*, allows for faster stack switching than any implementation based on the throwing and catching of exceptions [34,25].

3.2. Implementation Details and Optimisations

The bytecode rewriter is written in Scala and makes use of the ASM 4 library for bytecode transformation [36] which is the library used by the Scala compiler for bytecode generation.

In our implementation, we follow the ideas of Srinivasan [35], and optimise the transformation described in the previous section as follows: On the return of a method marked as Pausable, the caller examines the Fiber to determine whether the isPausing flag is set. If so, then the caller will store the local state needed upon resumption in the Fiber only if this has not already been done on an earlier pause/resume cycle, and then itself return control to its caller. This is more efficient then storing the state eagerly in the pre-call section.

Hence, if a method pauses for a second time, the states of intermediate frames are already stored in the Fiber and are not stored again. This explains the need for the four cases in the post-call section of Figure 3: the caller's behaviour depends on both whether the callee is pausing or returning normally, and whether the local state of the caller has or has not already been captured on an earlier pause/resume cycle.

```
void a() throws Pausable {




   x = ...

   b(); // b is pausable














   c(x); // c is not pausable
}
```

(a) Before transformation.

```
         void a(Fiber f) {
           switch (f.pc) { // prelude
             case 0: goto START;
             case 1: goto CALL_B;
           }
  START:   x = ...
 CALL_B:   f.down(); // pre-call
           b(f);
           f.up();   // post-call
           switch (f.status) {
             case NOT_PAUSING_NO_STATE:
               goto RESUME;
             case NOT_PAUSING_HAS_STATE:
               restore state;
               goto RESUME;
             case PAUSING_NO_STATE:
               capture state;
               return;
             case PAUSING_HAS_STATE:
               return;
           }
 RESUME:    c(x);
         }
```

(b) After transformation.

Figure 3. Example method before (a) and after (b) the insertion of logic for suspending and resuming the stack, using the optimisations described in the text. The transformation is on bytecode, hence the presence of `goto` statements. Example adapted from [35].

Another optimisation in this technique over traditional CPS transformations is that, because only those method signatures annotated as Pausable are rewritten, only those call sites that correspond to pausable method calls (i.e. those that may involve stack switching) need be transformed and added to the prelude. This allows external libraries that do not use CSO_2 constructs (including the Scala standard library) to be used without modification (but certain restrictions apply, as detailed in Section 3.4).

We also use two important optimisations employed by Srinivasan [35]: firstly, we use live variable analysis to store in the Fiber only the values of variables which are to be used again after the restoration point, and secondly we do not store the values of constants or variable aliases in the Fiber.

Evaluation has shown that the runtime overhead of the additional code injected to implement the prelude and all pre- and post-call blocks is low, and is certainly outweighed by the increased performance in context switching (see Section 7).

3.3. *Living with the JVM Verifier*

The JVM does not permit a method to inspect the stack frame of its caller, thus necessitating the design described in this paper. Nonetheless, as reported in Section 7, the performance of this technique is far in excess of Java primitives.

Furthermore, the JVM verifier requires that the operand stack is always the same size and contains the same types of values, irrespective of the code path that is taken to reach any particular point [37]. Therefore it is necessary to push onto the stack as many (dummy) constants of the expected types in the correct order as is required before issuing the jump in the prelude. For example, we push `iconst_0` whenever an `int` is expected.

After transformation, it is necessary to recompute the stack map table of each modified class file. These tables were introduced with Java 6 in order to speed up the runtime verification of classes by caching type information known to the compiler. The table records the type of values of local variables and the types of each operand stack slot immediately before specific bytecode instructions [37].

The `jsr` bytecode instruction, used to jump to a local subroutine defined within the body of a method, has caused implementation difficulties for previous bytecode rewriting frameworks [35,38]. However, the Scala compiler and all modern Java compilers no longer emit the `jsr` instruction (i.e. those that target bytecode version 51.0 and higher [37]), and thus our bytecode rewriter does not include logic for handling `jsr`.

3.4. Interaction with Functional Expressions

The Scala programming language offers both imperative and functional language features [22]. As of the current version of Scala (2.10.x), functional expressions, including for comprehensions, are implemented as standard library method calls. For example,

```
for (i <- 0 until n; j <- i until n) println(i)
```

is compiled to:

```
intWrapper(0).until(n).foreach(i: Int => intWrapper(i).until(n).foreach(j: Int => println(i)))
```

A difficulty arises when pausable methods are called from within anonymous functions passed to such standard library methods or external libraries. This would require those external methods to already have the Pausable annotation and the CPS transformation applied.

Our (partial) solution is to rewrite commonly used functional expressions, such as for comprehensions, as the equivalent imperative code at compile time using Scala macros. Since macros define abstract syntax tree transformations, they operate at the abstract source level rather than at the object code level, and so should be preferred wherever possible over bytecode transformation. For example, if we change the above expression to:

```
for (i <- 0 until n optimized; j <- i until n optimized) println(i)
```

our code is now compiled to:

```
var i: Int = 0
while (i < n) {
  var j: Int = i
  while (j < n) { println (i); j += 1 }
  i += 1
}
```

where i and j are fresh variable names. As a side benefit, the performance of the source program is optimised.

Nevertheless, there will always be pathological cases where such ad-hoc program transformations are unable to avoid delegation to library routines. In these cases, the transformation attempt will fail. Accounts of real user experience are needed to determine whether such rewritings alone are sufficient in practice. An alternative design, which would lift all aforementioned restrictions, would employ bytecode rewriting at runtime using a custom class loader. However, this could impact on performance.

3.5. Performance Considerations

Support for tail-call optimisations in the Scala compiler, which are unavailable in all Java compilers of which we are aware, are of great benefit in this setting: the storing and restoring of call stacks will obviously degrade performance when recursion is deep or call chains are large; tail-call optimisations can transform some recursive functions into equivalent iterative code, thus eliminating repeated passing of continuations.

Moreover, in process-oriented programming, systems are often built by the composition of processes, each of which have a specific well-defined task [39]. Thus, the lengths of call chains, and hence the depths of activation frames that are to be unwound and rewound, are expected to be shorter if process-oriented patterns are employed in users' programs. Therefore we expect the techniques outlined above to perform well in practice. Of course, however, benchmarks and case studies of real systems are needed.

3.6. Support for Shared Memory

Although we are primarily concerned with message-passing concurrency, for the reasons outlined in the introduction, our library also allows for shared-memory concurrency. This requires some additional changes to the bytecode in the post-compilation stage: If a pausable method calls another pausable method from within a synchronized block, then all usages of that object's monitor must be replaced with a newly introduced explicit lock which sets the thread affinity of the process to the current thread, and any call to wait(), notify() or notifyAll() must be replaced with the equivalent to calls to pause() and resume(). These changes are necessary to ensure that, after a call to resume(), a process is rewound back onto the same thread when it holds a lock.

4. Scheduling

The user-level scheduler is provided by our small runtime system. The scheduler manages the states of processes as described in the introduction of Section 3 and illustrated in Figure 1. We have experimented with two different scheduler algorithms, as reported below. The relative performance of these two schedulers is considered in Section 7.

Empirical evaluation has shown us that the best scheduling performance is achieved when one scheduler is responsible for each JVM thread (as opposed to one scheduler multiplexing processes across many threads). Thus, each thread has its own scheduler in an effort to improve cache affinity. Furthermore, previous work shows us that throughput is increased when the number of JVM threads matches the number of hardware threads on the system: more threads will likely result in a high context switching overhead, and fewer threads could result in higher cache misses due to the increased possibility of a thread being rescheduled on a different core [40].

4.1. Naïve First In, First Out Scheduler

The implementation of the FIFO scheduler is quite simple: the running process occupies the JVM thread; the processes that are paused have called pause() but no corresponding resume() has yet been invoked; and the processes that are waiting are placed in a queue. Processes enter the queue when a resume() is called, and leave once the JVM thread is unoccupied. In this scheme, processes do not migrate between threads when a resume() is invoked.

The scheduler is fair in that processes are scheduled in first in, first out order. One could experiment with different scheduling disciplines by specifying a process priority function and using a priority queue.

```
function WriteToBufferedChannel(msg)
    // The Enqueue operation atomically stores a message in the ring buffer,
    // returning FALSE iff the buffer was full.
    while not Enqueue(msg) do
        // The Pause operation causes the current stack to unwind.
        Pause
    r := GetPausedReader
    Resume(r)
```

Algorithm 1. Writing to a buffered channel.

4.2. Multi-Core Batch Scheduler

This scheduler is a simplified implementation of that provided by the occam-π runtime. It is intended that in a future release, we will provide a scheduler that implements both the work-stealing algorithm and all of the runtime heuristics detailed in [6].

The scheduler instance maintains a *run queue*, which is a linked list of *batches*. Each batch is a linked list of processes. The scheduler executes each batch by moving the processes it contains to the *active queue*. The *dispatch count* is calculated as a constant multiple of the batch size, bounded by a fixed constant. The dispatch count is decremented each time a process is taken from the active queue and executed. When the dispatch count reaches zero (or the active queue is empty), the processes in the active queue form a new batch which is appended to the end of the run queue (if non-empty), and the head of the run queue is moved to the active queue for execution. Since these data structures are manipulated only by the current thread, they can be maintained without locking.

When a process is resumed, and the dispatch count is non-zero, it is added to the end of active queue of the rescheduling thread, otherwise it is added to a new batch which is appended to the end of the run queue of the rescheduling thread. In this way, a process can migrate between threads during execution (unless prohibited by an explicit affinity setting, as detailed in Section 3.6).

The use of batches addresses the issue of cache thrashing that can occur with process-oriented designs [6]. By partitioning the run queue, we reduce the size of the working set for each thread, which should fit within the processor cache. Performance is improved by repeated execution of batches within the cache.

5. Channel Communication

Our library, CSO_2 provides all of the message-passing capabilities of the original CSO library [15], including synchronous and buffered channels, extended rendezvous, barriers, and port type variance. In particular, it provides a CSP-style generalized **alt**, borrowing the provably-correct implementation of Lowe [16]. Future work will consider how the performance of this primitive can be optimised in our setting of cooperative yielding.

Intuitively, the implementations of channels are as usual, except with calls to wait() and notify() replaced with calls to pause() and resume() on the appropriate processes. Owing to space constraints, we have illustrated this for the case of buffered channel communication only, in Algorithms 1 and 2.

For a single-core machine, since there will be only one JVM thread onto which all processes are cooperatively scheduled, it is safe to eliminate the locking used in the ENQUEUE and DEQUEUE operations because there will be no preemption.[6]

[6]We did experiment with several lock-free algorithms for buffered channel communication, but were unable

```
function READFROMBUFFEREDCHANNEL
    // The DEQUEUE operation atomically fetches a message from the ring buffer,
    // returning NIL if the buffer was empty.
    msg := DEQUEUE
    while msg = NIL do
        // The PAUSE operation causes the current stack to unwind.
        PAUSE
        msg := DEQUEUE
    r := GETPAUSEDWRITER
    RESUME(r)
    return msg
```

Algorithm 2. Reading from a buffered channel.

Our library makes use of the specialization of primitives on the parametric polymorphic types of channels. In Scala, type specialization allows the compiler to generate primitive-specific versions of classes that have type parameters [41]. This feature is similar in spirit, although not identical, to that of C++ templates. Therefore the performance of communicating primitives is optimised in CSO_2 compared to CSO. For example, this feature is of particular use when building a concurrent solution to numeral integration.

One last feature of interest is the possibility of optimising *extended rendezvous* in our framework. Given a channel c that can communicate values of type T and a function f : T ⇒ U, an extended rendezvous is of the form c?f, and is evaluated by pausing the caller until a value v can be read from c and then computing f(v) before allowing the writer process to continue. However, this is existentially equivalent to f(c?) provided f does not perform any channel operations nor accesses shared memory. Since channel operations are tagged in the bytecode, we are able to optimise the former into the latter. The user needs to be aware however that with these optimisations switched on, should f perform any I/O, then this may occur in a different order.

6. Runtime Deadlock Detection

Our runtime system also includes deadlock detection for global deadlocks, similar to that available for the Go programming language [29]. Informal benchmarking showed that this feature adds little to no overhead to the runtime performance since our scheduler records those lightweight processes currently paused.

Once all processes are paused, we run the deadlock detection algorithm shown in Algorithm 3, which is due to [42]. In the algorithm, we say that process u is waiting for process v when u is attempting to communicate with v across some channel. The runtime reports the cycle of ungranted requests that caused the deadlock. For example, the output provided for a deadlocking run of the dining philosophers problem is as follows:

```
Deadlock detected! The cycle of ungranted requests is:
Philosopher0 -!-> Fork4          Fork0 -?-> Philosopher0
Philosopher1 -!-> Fork0          Fork1 -?-> Philosopher1
Philosopher2 -!-> Fork1          Fork2 -?-> Philosopher2
Philosopher3 -!-> Fork2          Fork3 -?-> Philosopher3
Philosopher4 -!-> Fork3          Fork4 -?-> Philosopher4
```

to find an algorithm that outperformed the use of locks. We conjecture that this is a result of instantiating only as many JVM threads as there are hardware threads, and thus the contention for locking is low. This is a topic for future work, especially as the number cores on commodity hardware increases.

```
// Input: contention network G = (V, E) where
//   V is the set of nodes representing processes, and
//   E contains all edges (u, v) such that u is waiting for v.
// Returns: TRUE if some sub-network is deadlocked.
function CHECKDEADLOCK(nodes,edges)
    interiorNodes := {u | (u, v) ∈ edges}
    // The leaves are the processes not waiting for any other process
    leaves := nodes \ interiorNodes
    // If a non-empty graph contains no leaf nodes then all nodes are waiting
    if (nodes ≠ ∅ ∧ leaves = ∅) then return true
    else
        // Construct the adjacency list
        Adj := {u ↦ {v | (u, v) ∈ edges} | u ∈ nodes}
        // Maintain set of visited nodes (processes known not to be deadlocked)
        visited := leaves
        // Starting from the leaves, perform a backwards mark and sweep
        q := MAKEQUEUE(leaves)
        while q ≠ ∅ do
            next := DEQUEUE(q)
            for all n ∈ Adj [next] do
                if n ∉ visited then
                    visited := visited ∪ {n}  // Mark n as deadlock free
                    ENQUEUE(q,n)
        // Graph is acyclic iff we have visited all nodes
        return nodes = visited
```

Algorithm 3. Deadlock Detection [42].

where the left-hand column shows the ungranted write requests and the right-hand column shows the ungranted read requests.

This is a level of diagnostics not available without great effort when using native Java threads and shared memory, and is of particular importance given that such software errors are inherently difficult to reproduce.

7. Performance Evaluation

In this section we evaluate the performance of our library against competing solutions. The source codes for the two benchmarks reported here and for each language/library tested are available from `http://www.cso.io`.

Erlang is a concurrent message-passing functional language [5], and Go is a compiled language that natively supports message-passing concurrency [29]. JCSP is a library which provides CSP-style concurrency primitives for Java [17], and occam-π is a concurrent programming language [19]. All are therefore suitable candidates for benchmarking our library against. The Scala Actors library is the language designers' recommendation for concurrency in Scala [22], and is thus also a suitable candidate.

The first performance benchmark measures the total time for n processes to each communicate a single message to every other process, for several values of n. Thus each test involved $O(n^2)$ messages in total.

The benchmarks were performed on a server with two 3.06 GHz Intel Xeon X5667 processors (each with four cores) and 96GB RAM running Windows Server 2008 R2 Enterprise x64, with JVM 1.7 update 21, Scala 2.10.1, JCSP 1.1-rc4, Erlang R16B and Go 1.0.3. For the benchmarking of occam-π we installed CentOS 6.4 and KRoC 1.6.0 on the same hardware. Fifty times were recorded for each experiment, after allowing virtual machines to warm up. Some Erlang experiments reported very high times, which we believe, having inspected the memory graph in the task manager, were due to the garbage collector. These times were excluded; the variances of the remaining results in all experiments were small enough to be ignored. The times reported in Figure 4 are the mean wall times in milliseconds and include the time for context switching and process start-up.

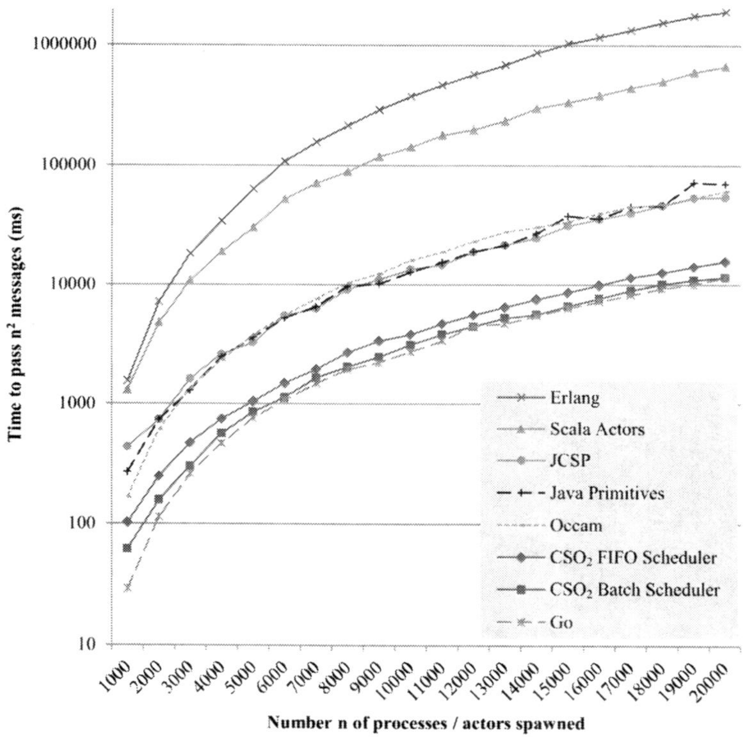

Figure 4. Benchmarking the performance of our CSO_2 library, using both the FIFO and batch scheduler, against several alternatives for speed of message passing. Time is the mean wall time in milliseconds for fifty runs on a logarithmic scale (lower is better).

The results show that our library and Go scales linearly in n^2 when communicating n^2 messages amongst n processes, and that our library achieves the best performance when using the batch scheduler.

From the graph, we can read off that the performance of the internal mechanism of the library, whereby processes are unwound from the stack whenever they need to pause and rewound back onto the stack when they are later able to continue, is almost as good as compiled Go code and well in excess of 100 times faster than Erlang.

When we inspected the raw data we found that our library, and also Go, provides exceptionally consistent timings even under heavy loads. In the case of our library, we attribute this both to the uniform way in which methods that perform concurrency operations are transformed in our framework and to the reliability of the JVM, particularly the engineering effort to reduce pauses during garbage collection [43].

We also timed the performance of both synchronous and buffered channel communications between exactly two processes, both with and without the use of an **alt** at either end of the channel. We found that the number of communications per second achieved by our library was equal to that of the original CSO library. Therefore we conclude that the chief utility of CSO_2 is when the programmer requires many more processes then there are available processors.

In the second benchmark we measured the mean wall time in milliseconds for n processes, whose communication graph formed a ring, to communicate a single message around the ring 300 times. As before, fifty times were recorded for each experiment, after allowing the JVM to warn up. The results are shown in Figure 5. It can be seen that in this benchmark the performance of our library is far in excess of Java primitives. However of greatest interest is the influence of the scheduler on the runtime performance: It demonstrates that grouping processes into cache affine batches and allowing processes to migrate between worker threads can result in greater throughput.

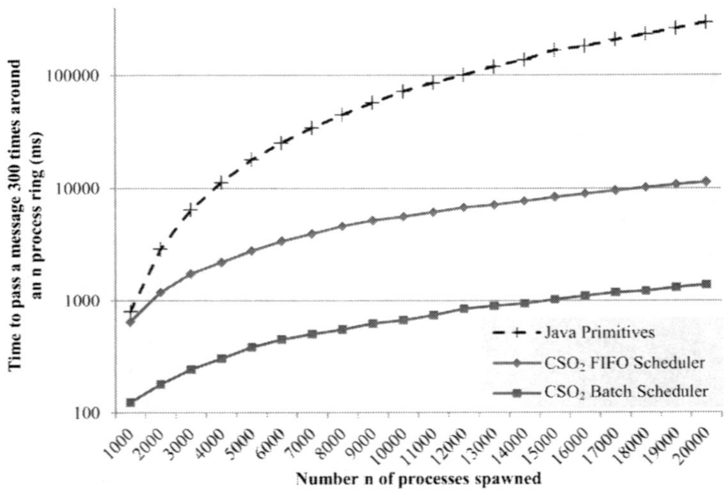

Figure 5. Benchmarking the performance of the batch scheduler against both the FIFO scheduler and Java primitives. Time is the mean wall time in milliseconds for fifty runs on a logarithmic scale (lower is better).

One should not place too much reliance on micro-benchmarks; however, this evaluation demonstrates that our library is capable of achieving excellent performance. Whilst we seek real experience with CPU-intensive applications, we feel that it is a safe conjecture that our library provides better performance than any other competing message-passing solution on the JVM.

8. Related Work

The work presented here builds on the ideas of [25,44,35,26,27,28]. A good survey of related techniques appears in [35].

Srinivasan implemented a similar continuation-passing style transformation for the Java Kilim actors library [35]. One significant difference is that Srinivasan requires that pausable methods be explicitly annotated in the user's source code. Our view is that these additional annotations clutter the source code of methods that need to perform concurrency operations regardless, and in Scala where lambda functions are supported as first-class citizens, this is not always possible. We do acknowledge, however, that the use of explicit annotations does alert the programmer to the potential cost of message passing and stack switching [25].

Thus our approach is similar to Apache Javaflow [27], which transforms a method if it can reach an operation that causes the owner thread to suspend. However, in contrast to our approach, Javaflow unnecessarily transforms all methods reachable from that method, irrespective of whether or not those methods perform thread suspending operations. This can lead to a substantial increase in the size of the bytecode.

There have been a number of projects to develop lightweight threads at a lower level than we have investigated here. One example is the Capriccio project [45], a modified version of POSIX threads that is capable of scaling to hundreds of thousands of preemptively-scheduled user-level threads. The Native POSIX Thread Library also offers lightweight threads with greatly improved performance over the original LinuxThreads implementation [46].

Pettyjohn et al. [47] demonstrate how continuations can be supported in environments that allow neither stack inspection nor manipulation, such as the JVM or the CLR. However, their approach relies on tail-call optimisations, which are unavailable on the JVM (as of version 1.7, but are available in the Scala compiler). They also use exceptions to capture state and their transformations result in loops being split into several method invocations, which is likely to be expensive on the JVM [34]. Furthermore, since their transformation changes access modifiers to increase visibility of private methods, the transformed code is unsuitable for linking against by other developers.

Of particular importance is occam-π and the Kent Retargetable occam Compiler [19]. Occam natively supports lightweight threads and combines many of the features of both CSP and the π-calculus. The runtime overheads are low enough to support millions of concurrent processes. The occam-π scheduler groups processes into cache affine batches of communicating processes using runtime heuristics and process migration. It has been demonstrated that these techniques can lead to superior communication times [6].

9. Future Work

One interesting future direction for our library would be to allow the developer to define a user-level scheduler. These have proven to be of particular use in the performance tuning of applications that deal with database transactions, through the use of an earliest-deadline-first scheduler [25].

A formal verification of design for our library, as has been done for core components of CSO [16] and JCSP [48], is needed, and this is a consideration of future work.

In the long term, support for lightweight threads on the JVM at the execution engine layer is needed. The JVM does not permit any form of stack manipulation, and thus we have been forced to follow the design described in this paper. Whilst our approach is close in performance to a compiled concurrent programming language, it is clear that direct access to memory would allow for more efficient support of continuations. A lower-level implementation of these techniques would also allow for more efficient jumps to call sites within methods, as arbitrary jumps are disallowed by the bytecode verifier (as explained in Section 3.3). Support for lightweight threads in the JVM could improve context switching by a further two orders of magnitude [25].

An important area of work is to support distributed programming. Such an extended library would provide a uniform interface for both local concurrent and distributed programming, allowing systems to scale from one machine to many, with no architectural changes to the user's code.

Another area of future work will be in an improved runtime, supported with the use of a debugger. This will largely be a port of our existing work described in [42]. We believe that the performance and usefulness of this debugging tool will be even greater in this setting, owing to the cooperative yielding of user processes. In particular, we would expect a reduction in the reordering of atomic actions caused by the overhead of the instrumentation of the debugger.

Finally, it should be relatively straightforward, if not development intensive, to bring many of the features of Erlang [5], such as process-linking and supervisor processes to our library, with good runtime performance.

10. Conclusions

We have presented an embedded domain-specific language for building massively concurrent systems using Scala on the JVM. The technique we have described here offers performance that is over 100 times faster than Erlang, and context switching that is 1000 times faster than competing libraries and native JVM threads. The performance of our library is also close to that of compiled Go code.

The traditional threading facilities available on the JVM are tied to kernel resources, which limits their scalability and the efficiency of context switching. Using the techniques described in this paper, we are able to map millions of processes onto only as many threads as there are hardware threads. We achieve this by a CPS transformation of bytecode produced by the Scala compiler and through the use of a user-level scheduler that supports process migration. The approach described in this paper is entirely portable and can target any JVM. Our runtime system also supports deadlock detection, reporting it to the user should it occur.

We have demonstrated that message-passing concurrency can be realised in both a portable and a pragmatic way. We have done so by leveraging an existing body of work, and extending it to full message-passing concurrency and functional programming while maintaining excellent performance.

Acknowledgements

I would like to thank Gavin Lowe for his valuable feedback on an early draft of this paper. I would also like to acknowledge Sriram Srinivasan for his work on high performance Java Actors whose ideas we have followed, and Bernard Sufrin whose work on CSO lead us to choose Scala as the host language.

References

[1] Matt Gillespie. Preparing for the Second Stage of Multi-Core Hardware: Asymmetric (Heterogeneous) Cores. White paper, Intel Corporation, 2008.

[2] R. M. Ramanathan. Intel® Multi-Core Processors: Making the Move to Quad-Core and Beyond. White paper, Intel Corporation, 2006.

[3] Intel® QuickPath Architecture. White paper, Intel Corporation, 2008.

[4] C.A.R. Hoare. *Communicating Sequential Processes*. Prentice-Hall, London, 1985.

[5] Joe Armstrong. *Making Reliable Distributed Systems in the Presence of Software Errors*. PhD thesis, The Royal Institute of Technology, Stockholm, Sweden, December 2003.

[6] Carl G. Ritson, Adam T. Sampson, and Frederick R. M. Barnes. Multicore Scheduling for Lightweight Communicating Processes. In John Field and Vasco Thudichum Vasconcelos, editors, *Coordination Models and Languages, 11th International Conference, COORDINATION 2009*, volume 5521 of *Lecture Notes in Computer Science*, pages 163–183. Springer, June 2009.

[7] A. W. Roscoe, C. A. R. Hoare, and Richard Bird. *The Theory and Practice of Concurrency*. Prentice Hall PTR, Upper Saddle River, NJ, USA, 1997.

[8] Joe Armstrong. *Programming Erlang: Software for a Concurrent World*. Pragmatic Bookshelf, 2007.

[9] M. McCool, J. Reinders, and A. Robison. *Structured Parallel Programming: Patterns for Efficient Computation*. Elsevier Science, 2012.

[10] András Vajda. *Programming Many-Core Chips*. Springer, 1st edition, 2011.

[11] James Reinders. *Intel Threading Building Blocks: Outfitting C++ for Multi-core Processor Parallelism*. O'Reilly Media, 1st edition, July 2007.

[12] Edward A. Lee. The Problem with Threads. *Computer*, 39(5):33–42, May 2006.

[13] A.W. Roscoe. *Understanding Concurrent Systems*. Springer-Verlag, New York, NY, USA, 1st edition, 2010.

[14] Robin Milner. *Communicating and Mobile Systems: the π-Calculus*. Cambridge University Press, New York, NY, USA, 1999.

[15] Bernard Sufrin. Communicating Scala Objects. In Peter H. Welch, S. Stepney, F.A.C Polack, Frederick R. M. Barnes, Alistair A. McEwan, G. S. Stiles, Jan F. Broenink, and Adam T. Sampson, editors, *Communicating Process Architectures 2008*, pages 35–54, September 2008.

[16] Gavin Lowe. Implementing Generalised Alt. In Peter H. Welch, Adam T. Sampson, Jan Baekgaard Pedersen, Jon Kerridge, Jan F. Broenink, and Frederick R. M. Barnes, editors, *Communicating Process Architectures 2011*, pages 1–34, June 2011.

[17] Peter H. Welch, Neil C.C. Brown, James Moores, Kevin Chalmers, and Bernhard H.C. Sputh. Integrating and Extending JCSP. In Alistair A. McEwan, Steve Schneider, Wilson Ifill, and Peter H. Welch, editors, *Communicating Process Architectures 2007*, pages 349–369, July 2007.

[18] Peter Welch, Neil Brown, James Moores, Kevin Chalmers, and Bernhard Sputh. Alting Barriers: Synchronisation with Choice in Java using JCSP. *Concurrency and Computation: Practice and Experience*, 22(8):1049–1062, June 2010.

[19] Peter H. Welch and Frederick R. M. Barnes. Communicating Mobile Processes: Introducing occam-pi. In Ali Abdallah, Cliff Jones, and Jeff Sanders, editors, *Communicating Sequential Processes. The First 25 Years*, volume 3525 of *Lecture Notes in Computer Science*, pages 712–713. Springer Berlin, April 2005.

[20] Frederick R.M. Barnes, Peter H. Welch, and Adam T. Sampson. Barrier Synchronisation for occam-pi. In Hamid R. Arabnia, editor, *Proceedings of the 2005 International Conference on Parallel and Distributed Processing Techniques and Applications (PDPTA'05)*, pages 173–179, Las Vegas, Nevada, USA, June 2005. CSREA Press.

[21] J. Gosling, B. Joy, G.L. Steele, G. Bracha, and A. Buckley. *The Java Language Specification, Java SE 7 Edition*. Addison-Wesley, 1st edition, 2013.

[22] Martin Odersky, Lex Spoon, and Bill Venners. *Programming in Scala: A Comprehensive Step-by-Step Guide*. Artima Incorporation, USA, 2nd edition, 2011.

[23] C. Emerick, B. Carper, and C. Grand. *Clojure Programming*. O'Reilly Media, 1st edition, April 2012.

[24] JDK 1.1 for Solaris Developer's Guide. Technical report, Sun Microsystems, Inc., 901 San Antonio Road, Palo Alto, CA 94303-4900, U.S.A., February 2000.

[25] Sriram Srinivasan. Kilim: A server framework with lightweight actors, isolation types and zero-copy messaging. Technical report, University of Cambridge, Computer Laboratory, February 2010.

[26] Stefan Fünfrocken. Transparent Migration of Java-based Mobile Agents. In Kurt Rothermel and Fritz Hohl, editors, *Mobile Agents*, volume 1477 of *Lecture Notes in Computer Science*, pages 26–37. Springer Berlin Heidelberg, 1998.

[27] JavaFlow: Apache Commons project for Java comunications.
`http://commons.apache.org/sandbox/commons-javaflow`.

[28] Tatsurou Sekiguchi, Hidehiko Masuhara, and Akinori Yonezawa. A Simple Extension of Java Language for Controllable Transparent Migration and its Portable Implementation. In *Proceedings of the Third International Conference on Coordination Languages and Models*, COORDINATION '99, pages 211–226, London, UK, 1999. Springer-Verlag.

[29] The Go Programming Language Specification.
`http://golang.org/ref/spec`.

[30] Kevin Chalmers, Jon Kerridge, and Imed Romdhani. Mobility in JCSP: New Mobile Channel and Mobile Process Models. In Alistair A. McEwan, Steve Schneider, Wilson Ifill, and Peter H. Welch, editors, *Communicating Process Architectures 2007*, pages 163–182, July 2007.

[31] G.R. Andrews. *Multithreaded, Parallel, and Distributed Programming*. Addison-Wesley, 1st edition, 2000.

[32] Marjan Mernik, Jan Heering, and Anthony M. Sloane. When and How to Develop Domain-Specific Languages. *ACM Computing Surveys*, 37(4):316–344, December 2005.

[33] Carl Bruggeman, Oscar Waddell, and R. Kent Dybvig. Representing Control in the Presence of One-Shot Continuations. In *Proceedings of the ACM SIGPLAN 1996 conference on Programming Language Design and Implementation*, volume 31 of *PLDI '96*, New York, NY, USA, May 1996. ACM.

[34] Atul Adya, Jon Howell, Marvin Theimer, William J. Bolosky, and John R. Douceur. Cooperative Task Management Without Manual Stack Management. In *Proceedings of the General Track of the annual conference on USENIX Annual Technical Conference*, ATEC '02, pages 289–302, Berkeley, CA, USA, 2002. USENIX Association.

[35] Sriram Srinivasan. A Thread of One's Own. In *Workshop on New Horizons in Compilers*, 2006.

[36] Eric Bruneton. *ASM 4.0: A Java bytecode engineering library*. OW2 Consortium, September 2011.

[37] T. Lindholm, F. Yellin, G. Bracha, and A. Buckley. *The Java Virtual Machine Specification, Java SE 7 Edition*. Addison-Wesley, 1st edition, 2013.

[38] Xavier Leroy. Java Bytecode Verification: Algorithms and Formalizations. *Journal of Automated Reasoning*, 30(3-4):235–269, August 2003.

[39] Adam T. Sampson. *Process-Oriented Patterns for Concurrent Software Engineering*. PhD thesis, University of Kent, October 2010.

[40] Kuo-Yi Chen, J.M. Chang, and Ting-Wei Hou. Multithreading in Java: Performance and Scalability on Multicore Systems. *IEEE Transactions on Computers*, 60(11):1521–1534, 2011.

[41] Iulian Dragos and Martin Odersky. Compiling Generics Through User-Directed Type Specialization. In *Proceedings of the Fourth Workshop on the Implementation, Compilation, Optimization of Object-Oriented Languages and Programming Systems*, ICOOOLPS '09, pages 42–47, New York, NY, USA, 2009. ACM.

[42] Andrew Bate and Gavin Lowe. A Debugger for Communicating Scala Objects. In Peter H. Welch, Frederick R. M. Barnes, Kevin Chalmers, Jan Baekgaard Pedersen, and Adam T. Sampson, editors, *Communicating Process Architectures 2012*, pages 135–154, August 2012.

[43] David Detlefs, Christine Flood, Steve Heller, and Tony Printezis. Garbage-First Garbage Collection. In *Proceedings of the 4th international symposium on Memory management*, ISMM '04, pages 37–48, New York, NY, USA, 2004. ACM.

[44] Sriram Srinivasan and Alan Mycroft. Kilim: Isolation-Typed Actors for Java. In Jan Vitek, editor, *Proceedings of the 22nd European conference on Object-Oriented Programming*, ECOOP '08, pages 104–128, Berlin, Heidelberg, July 2008. Springer-Verlag.

[45] Rob von Behren, Jeremy Condit, Feng Zhou, George C. Necula, and Eric Brewer. Capriccio: Scalable Threads for Internet Services. In *Proceedings of the Nineteenth ACM Symposium on Operating Systems Principles*, SOSP '03, pages 268–281, New York, NY, USA, October 2003. ACM.

[46] Ulrich Drepper and Ingo Molnar. The Native POSIX Thread Library for Linux. Technical report, RedHat, Inc, 2003.

[47] Greg Pettyjohn, John Clements, Joe Marshall, Shriram Krishnamurthi, and Matthias Felleisen. Continuations from Generalized Stack Inspection. In *Proceedings of the Tenth ACM SIGPLAN International Conference on Functional Programming*, ICFP '05, pages 216–227, New York, NY, USA, 2005. ACM.

[48] Peter H. Welch and Jeremy M. R. Martin. Formal Analysis of Concurrent Java Systems. In Peter H. Welch and Andrè W. P. Bakkers, editors, *Communicating Process Architectures 2000*, pages 275–301, September 2000.

Communicating Process Architectures 2013
P.H. Welch et al. (Eds.)
Open Channel Publishing Ltd., 2013

ProcessJ: A Possible Future of Process-Oriented Design

Jan Bækgaard PEDERSEN [a,1] and Marc L. SMITH [b]

[a] *Department of Computer Science, University of Nevada Las Vegas, NV, USA*
[b] *Computer Science Department, Vassar College, Poughkeepsie, NY, USA*

Abstract. We propose ProcessJ as a new, more contemporary programming language that supports process-oriented design, which raises the level of abstraction and lowers the barrier of entry for parallel and concurrent programming. ProcessJ promises verifiability (e.g., deadlock detection), based on Hoare's CSP model of concurrency, and existing model checkers like FDR. Process-oriented means processes compose, unlike thread-based or asynchronous message-passing models of concurrency; this means that programmers can incrementally define larger and larger concurrent processes without concern for undesirable nondeterminism or unexpected side effects. Processes at their lowest, most granular level are sequential programs; there are no global variables, so no race conditions, and the rules of parallel composition are functional in nature, not imperative, and based on the mathematically sound CSP process algebra. Collectively, these ideas raise the level of abstraction for concurrency; they were successful once before with the occam language and the Transputer. We believe their time has come again, and will not go away, in this new age of multi-core processors. Computers have finally caught up with CSP and process-oriented design. We believe that ProcessJ can be the programming language that provides a bridge from today's languages to tomorrow's concurrent programs. Learning or teaching the programming model and language will be greatly supported through the educational part of the proposed project, which includes course templates and an online teaching tool that integrates in-browser programming with teaching material. Our efforts are encouraged by the forthcoming 2013 IEEE and ACM curricula guidelines, which for the first time include concurrent programming as a core knowledge area at the undergraduate level.

Keywords. ProcessJ, process-oriented programming

Introduction

Over the last decade, a new architecture model has been introduced to the world of computing: multi-core processors. As speeds of conventional single-core processors plateaued, a new approach had to be taken to continue the performance increase so desperately sought by CPU manufacturers and users.

Today, it is not uncommon to find 32-core or even 64-core processors, and there is no reason to believe that this trend will not continue. Although the idea of using multiple CPUs to solve large problems is not new — message passing and shared memory architectures have been around for a long time — the reality of multiple cores on a single socket (chip) sharing resources is. Meanwhile, programming such architectures to make good use of the multiple

[1]Corresponding Author: *Jan Bækgaard Pedersen, University of Nevada Las Vegas, 4505 Maryland Parkway, Las Vegas, NV, 89154, United States of America.* Tel.: +1 702 895 2557; Fax: +1 702 895 2639; E-mail: `matt.pedersen@unlv.edu`.

cores available is difficult, time consuming, and error prone. On a quad-core processor, sequential code cannot use more than 25% of the processing power. If more is needed, there has to be concurrently executable code, and lots of it. For there to be a chance of keeping the cores busy with useful work, there needs to be an order of magnitude more concurrency in the software than is available in the hardware: the principle of parallel slackness, first discussed by Valiant [1].

Asynchronous message-passing programming (e.g., MPI_Send/MPI_Recv), which is typically fairly coarse-grained parallelism, will not make optimal use of the computing resources; multi-threaded programming, which requires locks on shared resources, is a complicated and often dangerous game. Debugging such programs is, at best, hard; even worse, because of possibly rare race conditions, programs that during testing seem fine, might deadlock or livelock if scheduled differently. In addition, trying formally to argue about or prove properties of lock-based programs is rarely achievable.

Despite all these difficulties, concurrency is needed to make full use of computing resources. We believe we have a better model that is more appropriate and much safer. This programming model for multi-core processors does not require locks and semaphores, etc.; it does not require the programmer to code explicitly for N cores; it reduces the number of potential errors associated with the concurrent part of the program by means of a solid foundation in mathematics as well as automatic verification and formal model checking; and it is highly efficient and easily teachable. In this paper we propose a complete suite, from programming language through online tools to IDEs that we believe could serve as a possible way to reintroduce process-oriented programming to the general computing community in the twenty-first century.

1. Existing Programming Models and Languages

Let us briefly point out some of the problems and issues we perceive to exist in current programming models that could be used to program multi-core CPUs. We focus on the threads and locks model, asynchronous message passing, and languages (besides occam [2,3] and occam-π [4,5,6,7,8,9,10]) that are inspired by CSP [11,12,13,14], but have not broken into the mainstream.

1.1. Thread and Lock Model

Shared memory initially seemed like a good idea: treat data as a shared resource between concurrent threads of execution. This model provides a way for threads to communicate with each other. However, this solution creates even greater problems: nondeterminism, race conditions, and the need for mutual exclusion. Without shared memory, these problems don't exist.

Access to shared memory requires the use of locks, semaphores, and mutexes, etc. Such synchronisation primitives are hard to master and get right; reasoning about a program with just a few locks quickly becomes a complicated task, and even very capable programmers get it wrong. For example, a solution to the Santa Claus problem, written in Java with locks [15], was shown to be incorrect [16].

Programs written using the thread-and-lock model are error prone; even if the error does not show up in tests, such programs are hard to reason about or even prove correct. Testing is challenging enough for sequential (deterministic) programs, and its importance is well established. There are even techniques for testing nondeterministic, randomised sequential programs, which include repetition and testing for a range of expected results. However, concurrency presents another layer of complexity due to runtime scheduling-based nondeterminism: race conditions, or even deadlocks, can remain hidden throughout long, extensive periods of system testing, and not reveal themselves until unfavorable scheduling occurs after a system is in production.

Although all cores on a multi-core CPU share the main memory, it is not necessary to use shared memory primitives (explicitly) to program it. We argue that a model with synchronous channel (point-to-point) communication with no shared data is a much safer alternative; actually, it is so safe that we can formally prove that our programs never deadlock or livelock, and race conditions can never happen, ever!

1.2. Asynchronous Message Passing

Coarse-grained parallelism, as found in asynchronous message-passing programming, is not entirely suitable for a multi-core architecture either. Often, such systems are developed with a certain architecture size in question, frequently using a master/worker model as well as the embarrassingly parallel programming paradigm. There is no inter-process communication, and changing the size of the number of processors causes load imbalance. Reasoning about such things as buffer allocation is difficult or impossible [17,18]. In addition, the asynchronous nature of a program execution makes reasoning about synchronisation — and ultimately the concurrent behavior of the whole system — extremely complicated, if at all possible. What we need is fine-grained parallelism without the explicit use of locks, so that the model is safer to program in, and without asynchronicity, so that programs are easier to reason about. This, we believe, can be achieved through process-oriented design [19], built on the solid foundation of the process algebra CSP. Process Oriented Design uses synchronous channel communication, no locks, and no shared data. Each process is in full control of its own data; thus, no race conditions can ever occur. The following three points are important for optimal utilisation of multi-core processors; we believe they are provided by process-oriented design and programming with the system we describe in this paper.

- Architecture size independence: To fully utilise all the computing resources of a multi-core system, a concurrent program should consist of more processes than cores, at least if the code being implemented is not embarrassingly parallel.
- Verification: process-oriented design is based on the mathematical formalism of CSP and the π-calculus [20], which means that we can formally reason about the programs (the communication parts, at least) through the rigorous rules of the underlying mathematical model. We also can use tools, such as the model checker, FDR 2 [21], to prove properties about the code, including lack of deadlock or livelock.
- Efficiency: Extremely efficient runtime systems for multi-core architectures exist that support process-oriented programming [22,23], and we plan to make use of them.

These design principles and programming models were utilised on the Transputer, a CPU developed by INMOS, a British semiconductor firm, in the 1980s [24], these models were not mature for general computing at the time. The Transputer architectures were not like the multi-core architectures of today. As hardware matured, it became obvious that it is time to revive, re-invent and further develop the process-oriented programming/design model. In this paper, we describe the ProcessJ suite, including the design of a new language, tools for development (IDEs and graphical 'modular programming' interfaces), and built-in verification. In addition, we will develop teaching materials to make the model available to students, including a process component repository, online cloud storage, and a programming-in-a-browser tool. This combines online access to the compiler, and allows code to be run in the browser, with the possibility of incorporating it in a Web page with video clips as well as web-page text and pictures, something which is gaining momentum in the online teaching world.

1.3. Other CSP-Inspired Languages

occam is not the only language with channels and CSP-inspired semantics. One such language is XC [25], developed by Douglas Watt and David May (who, of course, was the designer of occam) at XMOS. As its name implies, XC has C-like syntax, with CSP-style extensions designed to take advantage of the XMOS processor architecture. XC programs are written for the XMOS architecture, and require XMOS emulators to run on other hardware platforms. Another CSP-inspired language is Go [26], developed by Robert Griesemer, Rob Pike, and Ken Thompson at Google. Go borrows the idea of channels from CSP, and has *goroutines* (coroutines) that communicate over these channels. Go channels are synchronous by default, but *are* buffered and can be specified as asynchronous, at which point the program becomes more Actor-inspired [27] than CSP-inspired.

2. Process-Oriented Design, CSP and Formal Verification

Building a programming model and language on top of a mathematical formalism like CSP provides an extremely robust foundation. We can actually prove things about the code we write, even without running it. Static proofs can be constructed for the absence of deadlocks or livelocks. A concurrent system can be verified before it is implemented, and different scenarios of behavior can be explored; best of all, no testing for concurrency errors is ever necessary. This alone should reduce the amount of time and money spent on testing; furthermore, we actually know more about the behavior of the code, after using the model checker, than we would by only testing it. Therefore, more confidence can be placed on the products developed.

The process-oriented programming model is a perfect fit for much of the technology being developed today, including robotics, UAVs, and other component-based systems. Many systems that would benefit from a multi-core CPU consist of semi-autonomous sub-systems that, if programmed using the standard thread-and-lock model, easily become an entangled mess that cannot be reasoned about. Unless Object Oriented programming models are significantly changed to behave exactly like the Actor model [27,28], with each object running in its own thread, there is not much to be gained [29,30,31].

In a process-oriented system such as the one we propose, sub-systems can be programmed independently, verified and tested; the composition of such sub-systems can be verified as well. Since CSP has composable semantics, so will the processes in the language. For example, there is no need for explicitly testing if an aircraft's control system fails to respond, and deadlocks, if the auto pilot is switched off at the same time as the flaps are extended. Such problems can be formally proven to be absent or present before the model is implemented.

To better appreciate the proposed project, it is worth focusing on what process-oriented programming and design is. The basic tenet is that a program consists of a number of processes, possibly executing concurrently; however, each executes in their own thread of control and each is 100% in control of its own state (data). Processes communicate through directed channels that are connected to a sender and a receiver – or multiple senders and receivers, in the case of shared channel ends. All communication is synchronous, something that enables us to reason and prove properties about our programs. Processes can be composed in sequence, e.g., statements in Java or C using the ";" operator; or in parallel, e.g., using the PAR construct in occam-π. We can model such a language — or at least the concurrency and the communication — by using the mathematical formalism CSP developed by Hoare [11,12], and extended by Roscoe [13]. FDR 2 [21], a formal model checker, can be used to statically verify properties about our programs.

Process composition facilitates *local reasoning*, that is, reasoning about a process at any level of composition. Local reasoning permits reasoning about a process, in isolation, without having to reason about the other processes it contains, or interacts with. Contrast this with Java's thread-based method invocation semantics, as described by Welch [29]. Threads of execution weave in and out of many objects as methods are invoked from inside the bodies of other methods. Objects are at the mercy of any other object that has a reference to it, and there is nothing to stop passing as a parameter the reference of one object to another (aliasing). Objects are passive; they are not in control of their own destiny. Methods may be invoked at any time, regardless of whether the object is ready. To reason about object behavior is to follow the spaghetti trails of threads of execution in and out of multiple objects; that is, one must reason about many objects to reason about the behavior of one object. This is not so with process-oriented design.

In addition to what we already described, CSP supports the idea of external choice. When executing a receive call on a channel, since the communication semantics is synchronous, a process is blocked until the sender is available to transfer data; however, this might not always be what a process wants. An external choice operator, often referred to as an alternation, is the ability to choose between several events — typically input, but also could be timeouts or barriers — and to nondeterministically choose from one of the events that are ready.

CSP reifies nondeterminism in a concurrent program by making external choice available explicitly through alternation. By avoiding alternation in a concurrent program, the program remains deterministic! In other words, process-oriented concurrent programs are deterministic by default! By giving the programmer the choice of when she needs to explicitly introduce alternation into a program, the programmer controls when a program may be nondeterministic!

With these seemingly simple primitives — parallel composition, synchronous communication, and alternation — we can write quite powerful concurrent code; even better, we can reason about it and prove properties, freedom from deadlock and livelock.

3. New Languages, Education and Teaching

One obstacle faced by new languages is that the general public does not use them. This poses a serious risk. However, we believe that since the idea of teaching concurrent computing is becoming more and more common, and since both the IEEE and the ACM are including parallel programming in their core curriculum [32], we believe that if a language is packaged with the right tools — integrated development environments, graphical programming tools, and online teaching tools — as well as a comprehensive library of support material (lesson plans, notes, slides, etc.), there is a much greater chance for a language to be used in a teaching setting. The more people who learn it, the greater the chance it will make its way into industry.

For this reason, to support teaching process-oriented design and programming in the classroom or to anyone who has an interest, we propose a project that integrates both a technical part — writing the software to support the new language and its tools — and an educational part that leverages an online approach. The project we propose includes developing a new process-oriented language, IDE/GUI tools, online programming tools, lecture plans, course descriptions, notes, and other support materials.

4. Proposed Work

The project we propose consists of two different yet intertwined parts: a technical and an educational part. In the technical part, an appropriate language and the aspects associated with implementing it will be developed. These aspects are not just about the compiler, but also include such topics as graphical programming environments, runtime systems, and automatic generation of scripts for code verification as well as an online process repository and cloud storage, enabling easy process reuse and sharing. A JavaScript runtime and a compiler back-end is proposed to interface with in-browser programming, an emerging trend in online education. A tool that could be helpful in implementing this is Emscripten [33], a source-to-source (LLVM to JavaScript) compiler, which translates LLVM bytecode, typically created from compiling C or C++, to JavaScript. This would be an obvious solution (at least as a proof of concept) as we plan to produce C++ code from the ProcessJ Compiler.

The educational part primarily will be concerned with developing materials supporting the teaching and learning of techniques related to process-oriented design and programming through the use of the ProcessJ language and the associated tools as well as the integrated Web Online Teaching Tool. Such materials can include lectures notes, slides, handouts, exercises, and supplementary materials. A book, we believe, will be a necessity, and videos would incorporate nicely into the online education experience.

Figure 1 illustrates the proposed ProcessJ suite. Boxes in light blue correspond to projects that will be undertaken as part of the proposed work, and boxes in gray are existing tools/resources that will be utilised in the project. Figure 1 contains three different types of arrows signifying what a sub-project produces, what it is based on, and which other sub-projects it uses; these three different types represent output, implemented interfaces, and inheritance.

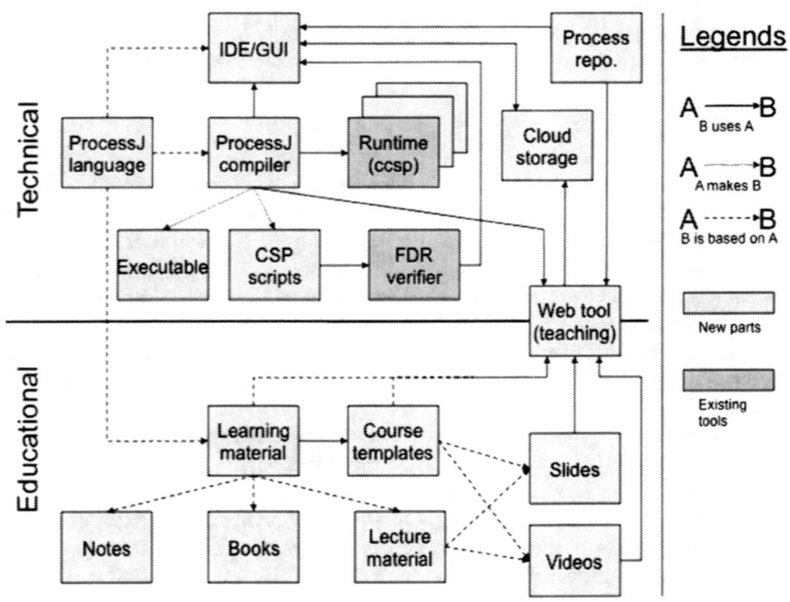

Figure 1. The overall proposed work of the project.

5. Technical Aspect

The technical part of this project is concerned with the actual software development, specification, and implementation. The following steps are to be undertaken:

- Language specification: Study existing popular languages (like C/C++ and Java) and develop a language specification that takes the familiar syntax and semantics of these languages into account for all similar syntactic constructs; that is, the syntax and semantics of a "while-statement" in ProcessJ is the same as in C/C++ or Java. New primitives required for concurrency — such as process, channels, barriers, and alternations — will have the same syntactic feel; however, the semantics will be that of CSP.
- Compiler construction: Implement the compiler for ProcessJ using modern compiler writing techniques and analysis. At first, we will utilise the CCSP/C++CSP [34,35,36] existing runtime systems to provide an efficient and reliable execution environment, later adding new runtimes or target architectures, such as clusters. One separate runtime in JavaScript is already included in the project.
- Graphical Programming Environment / Integrated Development Environment: Develop a graphical programming environment that supports the compositional and naturally layered approach that suits process-oriented design, including "programming by pictures" [37,38] and code reuse through a "LEGO catalog" method [39]. This IDE should interface with the compiler and the formal model checker, FDR 2, as well as an online repository for sharing ProcessJ code.
- Auto-generated Program Specification: The compiler should automatically generate CSP specifications of a program in machine readable form (CSP_M), enabling automatic verification of the absence or existence of deadlocks, livelocks, etc. [40]; furthermore, the support of specifying assertions about the program in order for the model checker to check should be supported [41].
- New back-end/runtime systems: Additional back-ends of the compiler targeting different execution architectures, such as clusters, using MPI, and the Java Virtual Machine, possibly using JCSP [19,42,43]. As mentioned we have already planned a runtime in JavaScript and a back-end code generator that produces JavaScript. Other interesting runtime systems include the LUNA CSP-Capable Execution Framework [44] and the light weight process MPI system [45] from the University of British Columbia.
- Online Repository: To encourage code reuse and code sharing, an online process source repository is planned, similar to the well-known app stores for mobile phones.
- Cloud Storage: To fully integrate the development and sharing of code in both the IDE and the online teaching tool (See next bullet point), a cloud storage can serve as a central hub for (private) code storage, making access available from any place with a network connection.
- ProcessJ Online Teaching Tool: A tool to tie the educational and the technical aspects of this project together is a Web based programming interface that can be placed on Web pages along with text, video, pictures, etc. This uses the JavaScript runtime, mentioned earlier, to facilitate execution of the compiled ProcessJ code directly in the user's browser. Such a tool will allow anyone to program in ProcessJ, and to execute code on any platform that contains a browser that supports JavaScript, albeit not as fast as on a multi-core architecture.

In the following subsections, we investigate each aspect in greater detail. In addition, we specify what experiences we have with each aspect topic, any work that has been done in the area, and what remains to be done.

5.1. Language Specification

There are many reasons why a language is not successful. Some languages are hard to learn and master, some have hard-to-understand semantic models, and others have poor syntax. Naturally, this assessment could be subjective, and subject to arguments. Others believe that languages who "made it" did so because they filled a void at just the right time in history.

We believe that now is the time for process-oriented programming to become popular; however, rather than make the mistakes of other languages developed throughout history, we are not going to make up a whole new set of syntactic rules, but rather take advantage of the knowledge that programmers already possess. Our aim is to produce a language specification that is syntactically similar to languages like C/C++ and Java, and which has the same semantics for all the well-known sequential language constructs, such as loops and variables. Then, we plan to introduce new process-oriented concurrency primitives — channels, communication, barriers, and mobile processes / channel-ends with their respective CSP/π-calculus semantics — while retaining a syntax that would be familiar, or at least reasonably obvious, to C/C++ and Java programmers.

CSP semantics apply to concurrency primitives like channel, channel-communication, barrier, and alternation (i.e., choice of which channel, out of many that may be possible, to engage). The semantics of the π-calculus apply to the mobility aspects of the language (mobile processes and mobile channel ends). Mobility is an area in which we are very interested and on which we have worked in the past. Our approach differs from the one (currently) taken by occam-π [4,5], for example, in the way we implement interfaces to the mobile processes. The occam-π approach uses a single interface for each mobile process, possibly requiring dummy actual-arguments for some parameters that might not be referenced; we propose a polymorphic approach, where each resumption of the mobile process uses a different interface, something we have (theoretically) investigated in [46,47] and intend to include in ProcessJ. At this point, we have a draft version of the ProcessJ language, which was used in [48]. Figure 2 shows a simple *producer-consumer* system written in ProcessJ, whose behaviour we shall return to in Section 5.4.

What remains to be done to complete this task includes developing a solid model for a package structure similar to Java's package model, finalising a reasonable syntax for structured literals, and determining how heap allocated values should be handled, for example whether to have explicit "free" or not, and if not, what sort of garbage collection the language should support. Although not part of the language directly, time must be spent on integrating into the ProcessJ language support for automatic verification; that is, an accompanying specification language must be developed. Some work already has been done on this incorporation with our research colleagues at the University of Kent at Canterbury [40,41].

5.1.1. Aliasing and Usage Checking

One of the holy grails (assuming we can have more than one) of occam is the non-aliasing semantics. This is something we strongly believe needs to be discussed, and perhaps rethought. One the one hand, no aliasing is an extremely nice feature: it prevents us from making an awful lot of mistakes, some potentially real bad. On the other hand, non-aliasing also prevents us from implementing a lot of very useful data structures (for example, circular lists).

We firmly believe that for a language to have a real impact, it is important that it is general enough to allow the programmer to actually write the data structures that are needed for the algorithms implemented. If non-aliasing within data structures prevents the language from being fully utilised, then perhaps aliasing should be allowed in certain circumstances; but the programmer should be aware that, that in those circumstances, some things cannot be checked by the compiler. Not giving the programmer the freedom to design the data structures the application needs hampers the usability of the language, which is not an attraction.

```
proc void Producer (chan<int>.write out) {
    int x = 42;
    while (true) {
        while (x < 1000) {
            out.write (x);              // write to channel
            x++;
        }
        while (x > 0) {
            out.write (x);              // write to channel
            x--;
        }
    }
}

proc void Monitor (chan<int>.read in) {
    int last = in.read ();              // read channel
    while (true) {
        int x;
        x = in.read ();                 // read channel
        if (x == last) {
            ... system failure detected
        }
        last = x;
    }
}

proc void main() {
    chan<int> c;
    par {                               // in parallel do ...
        Producer (c.write);
        Monitor (c.read);
    }
}
```

Figure 2. A two process *producer-consumer* system in ProcessJ. The Monitor process checks that the `Producer` never sends the same number twice in succession (i.e., that `Producer` does not violate such a condition in its specification). A CSP model, verifying that this system failure will not happen, is shown in Figure 4.

We hope this topic will form the basis of a heated debate in the process oriented community — it is something that needs to be considered. It touches upon the age old issue with how strict compilers should be (how restrictive are the semantics of the language?) and how much responsibility can be safely given to the programmer. We strongly believe that a middle ground, with the possibility of support from the compiler when needed, is the best approach. We note that Barnes' design of Guppy [49] (a *spin-off* language from occam-π) and the discussion proposals from Welch [50] (in these proceedings) incline to this view.

As an important side note to aliasing is the issue of parallel usage checking; having the compiler guarantee no memory location is accessed concurrently in a way that might create a race hazard is a strong semantic check. It is an extremely useful feature to have, but unfortunately, it becomes even more difficult if aliasing is allowed. We have not yet decided on where to draw the line with respect to aliasing, but parallel usage checking is something that will impact this decision. However, part of the future research of this project is to establish a balance between semantic strictness of the language and the usability; both are extremely important subjects, and several papers on this subject alone could be written. Enough to say for now that we are aware of the issue and it will play an important role in defining the semantics of the language and the implementation of the compiler.

5.1.2. APIs

Arguably, one of the reasons occam has not enjoyed more widespread adoption (apart from its Transputer roots and indentation syntax, which was not fashionable until Python came along) is its lack of APIs for certain common utilities. For example, utilities that provide the ability to interact directly with the underlying OS and network, as well as access to common collections classes (arrays and records are not the only data structure programmers desire).

The CSP model, at its core, is composed of *sequential* processes. The ProcessJ API should permit access to processes written in native code (e.g., C for systems calls), as well as provide collections classes. We've previously discussed (in Section 5.1.1) the challenges and necessity of supporting aliasing. Perhaps wrapping such recursive data structures in a collections API is one way forward. Regardless, APIs must exist for a language to be useful, and they must be extensible and allowed to evolve with the feedback of the language's user community.

5.2. Compiler

A language without a compiler is, of course, a dead language; therefore, we propose the development of a new compiler written in C/C++ for speed, which, besides the regular parsing, scope, and type checker, etc., will produce code that can be linked with the CCSP/C++CSP runtime [35,36,22]; this is a highly efficient multi-core runtime system for process-oriented languages, developed and maintained at the University of Kent at Canterbury. We have authored papers [51,52,53,40,41,54,9] and done research with this group in the past, and we are actively working with them now. We anticipate a close collaboration between our two institutions and the Kent group. It should be noted that many of the people who have worked on several occam-π compilers — Fred Barnes and Peter Welch (KRoC [55]) as well as Neil Brown (CHP [56]) — are still associated with this group, something we view as an asset for support in the development process.

As a proof of concept, a Java version of the ProcessJ compiler was implemented by a UNLV M.Sc. student [48]; this implementation in Java was slow and extremely incomplete; it contained virtually no semantic analysis. As stated, a new compiler written in C++ is needed, and two research undergraduate students at UNLV currently are working on this project. We have a working scanner and parser as well as parse trees; however, no semantic analyses or code generation have been implemented. Part of the implementation of the compiler will include automatically producing CSP scripts; this will be described in Section 5.4.

5.3. Graphical Programming Interface / Integrated Development Environment

Today, many popular programming languages, such as Java, C#, and Objective C, have very sophisticated programming environments: NetBeans for Java, Eclipse for Java and many other languages; Visual Studio for C#; and Xcode for Objective C. We believe that providing good programming support in terms of an integrated development environment can help boost a language's availability and reduce its learning curve. We have worked on a prototype for such an IDE for occam-π, called Visual occam [37] (see Figure 3). This prototype was by no means a finished product, but it taught us important lessons about what such an IDE must contain.

Since ProcessJ is a process-oriented language, the underlying model consists of processes that are composed in parallel, and which communicate through the use of channels. This model is a perfect fit for a graphical programming tool, in which processes and channels as well as barriers are visualised. Figure 3 shows a number of processes and channels. A graphical tool like this can support a programmer in developing and maintaining a library of processes that can be reused; we often refer to this as the programmer's "LEGO catalog" because processes, like LEGO blocks, serve as building blocks for bigger structures: LEGO blocks click together, and processes compose by the means of channels and barriers. Once a programmer has a process, which itself can contain several levels of other concurrently composed processes, he or she can add such a process to the catalog of existing, reusable processes and, at a later date, simply drag it back into the IDE and reuse it.

In keeping with this idea, the online process source repository (Section 5.5) goes hand in hand with an IDE; it should integrate flawlessly with the "LEGO catalog" functionality of the IDE, allowing the programming to move already written processes from the repository to the local "LEGO catalog". Since verification is one of the pillars of process-oriented programming and development, the IDE should be able to interface with the compiler and also with the verification module (FDR 2). It should interpret the output of the verifier and relate it back to the source code, if possible. In other words, it should be possible to perform all the actions necessary from within the IDE; this ranges from process design at the source level, process layout (the boxes on the middle canvas in Figure 3), through to verification in the model checker (FDR 2). No work has been done yet specifically towards this sub-project. However, many important lessons were learned in the development of Visual occam, and we believe that we have a good idea of the pitfalls and what to do better in order to produce a highly usable IDE for ProcessJ. For a proof of concept that this kind of feeding back from the output of FDR to source level is possible, see the ASD:Suite by Verum [57]. FDR was integrated into their tool set and if a check for deadlock or livelock fails, the trace leading to it is presented in terms of a message sequence on the input model and its states and events.

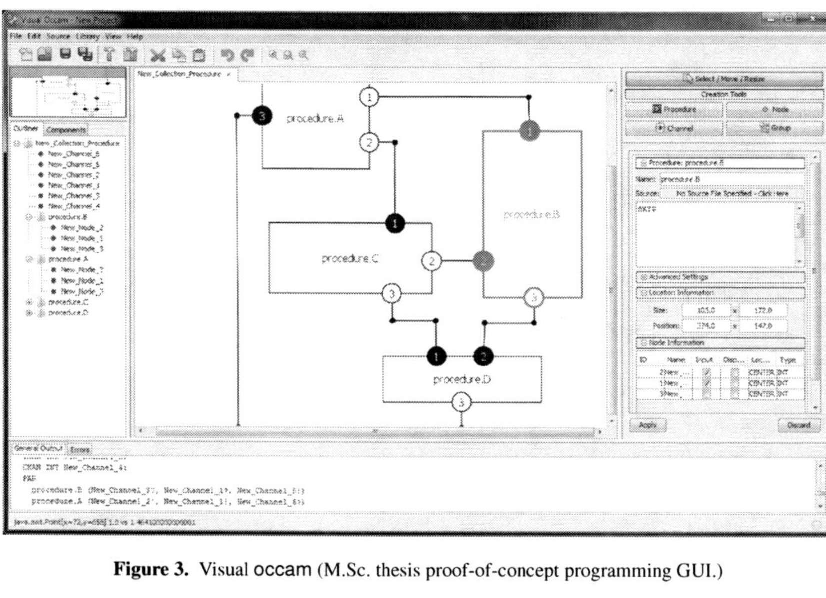

Figure 3. Visual occam (M.Sc. thesis proof-of-concept programming GUI.)

5.3.1. Documentation

The graphical process diagrams can also be used as part of the documentation, perhaps along with a JavaDoc style of comments, to generate comprehensive (online) documentation; this is an area we have not considered extensible, but which will become fairly important for the usability and learning curve of ProcessJ and its tools.

5.4. CSP Specifications

As argued earlier, one of the strengths of a language based on a mathematical formalism like CSP is the ability to prove properties of the programs. ProcessJ will be based on CSP process algebra — and also the π-calculus for mobility (although the FDR tool does not currently support verification of mobility). Thus, a one-to-one correspondence between runnable code and verifiable specification will exist. This is further discussed in [40,41].

We believe that such a mapping can be automated and added to the ProcessJ compiler, like a new target language. Rather than targeting the CCSP/C++CSP runtime, we will produce CSP_M, which is the machine-readable form of CSP accepted by the model checker FDR 2. For example, Figure 4 shows the CSP model of the producer-consumer system in Figure 2, with an assertion that the system is deadlock-free added at the end.

```
Producer (out) = ProducerUp (out, 42)

ProducerUp (out, n) =
  if n < 1000 then
    c!n -> ProducerUp (out, n + 1)
  else
    ProducerDown (out, n)

ProducerDown (out, n) =
  if n > 0 then
    c!n -> ProducerDown (out, n - 1)
  else
    ProducerUp (out, n)

Monitor (in) = in?x -> MonitorLast (in, x)

MonitorLast (in, last) = in?x ->
  if x == last then
    STOP                          -- system failure detected
  else
    MonitorLast (in, x)

channel c : {0..1000}

Main = (Producer (c) [| {|c|} |] Monitor (c)) \ {|c|}

assert Main :[ deadlock free [F] ]
```

Figure 4. CSP_M specification for the *producer-consumer* system in Figure 2.

FDR verifies this assertion. This means that the `STOP` in `Monitor` can never happen, since otherwise there would be deadlock. In turn, this means that consecutive numbers output by `Producer` are never the same, since otherwise the `STOP` would have been reached. Therefore, in systems incorporating this `Producer` process, there is no need to program the run-time checks implemented by `Monitor` in any processes taking its production — we have *verified* that `Producer` fulfills this specification. So, production code can safely dispense with such run-time monitoring. This is a powerful win from formal analysis and deadlock checking.

CSP is powerful enough to prove properties such as *trace refinement* (i.e., that the event sequences the *refined* process can perform are event sequences of the original, usually called the *specification*) and *failure refinement* (i.e., that the sets of events the *refined* process can refuse, in any of its states, are sets of events also refused by the *specification*). Trace refinement establishes *safety*: only specified event sequences can happen. Failure refinement establishes *liveness*: if the specification says that certain events will happen (i.e., that they will *not* be refused), then they *will* happen in the refined process!

These are strong properties of the process algebra, which we think should be accessible from the ProcessJ language, so that verifictaion becomes part of normal programming practice. Such an approach should greatly narrow the chasm that currently exists between implementation and verification language. If the two are not significantly alike, proving that the formal model is a correct abstraction of the implemented code becomes almost as daunting as the verification process itself. We believe that by narrowing this gap and automating the process, there is a much greater chance that verification will be used more frequently.

As a side note, we have taught verification using FDR 2 and $\mathrm{CSP_M}$ in courses. Students find these tools which we use to verify programs almost as easy to understand as the programs themselves. The ultimate goal is to reach a state where verification is just programming, and we think we can do that.

No work pertaining to the proposed project has been undertaken to date; however, early investigation has begun. Currently we are working on a research paper with the Kent group concerning this subject that should lay a solid foundation for completing this sub-project.

5.5. The ProcessJ Process Repository

As described in Section 5.3, in addition to a local LEGO catalog within the IDE, we propose an online repository—much like the Apple App-store and Google's Android Market, where developers can share source code for processes they have written. Such a repository should integrate into the IDE as well as the online tool set, and also have a searchable/browsable Web interface. Users can upload their ProcessJ code for other programmers to use. Processes are divided into categories so they can be browsed either through a Web interface (directly or through the ProcessJ Online Teaching Tool, allowing for transfer from the repository to the user's own private cloud storage—See Section 5.8) or through the IDE. Such a repository should of course be searchable. Furthermore, we propose the possibility of associating $\mathrm{CSP_M}$ verification certificates with uploaded code. That is, when a user uploads a piece of source, each process can have CSP code associated with it, and in addition typical things like deadlock, livelock etc. can be marked as 'checked'. This enables other users to know about certain properties of the code before they download and use it.

The process repository does not require any development of new tools, but simply an implementation using HTML and also a database. However, we believe it forms a key feature in making ProcessJ a common language: the more libraries and publicly available code, the easier the development gets of new code. No work has been done on this either; however, this can be undertaken as a completely independent project from much of the other development that is required for the project.

5.6. Multiple Back-Ends/Architectures

As proposed in Section 5.2, the first iteration of the ProcessJ compiler will target the CCSP/C++CSP runtime; that is, it will generate C or C++ code that gets compiled and linked with the runtime using the gcc or g++ compiler. The CCSP/C++CSP runtime is specifically developed and optimised for multi-core architectures; however, as already seen, we might need to generate code for different targets/back-ends. We have already mentioned two: $\mathrm{CSP_M}$ for use in the FDR 2 model checker and JavaScript for the ProcessJ Online Teaching Tool. To make process orientation more available to other platforms/targets/architectures, we have considered a number of possible other back-ends:

- Instead of generating C/C++ and linking with the CCSP/C++CSP runtime, we could generate Java and link with the JCSP [43] library. This would make process-oriented programming available on architectures that do not support the CCSP/C++CSP runtime; not particularly efficient, because it is built upon Java threads.
- Clusters that often use MPI [58] (and C) could also run process-oriented code. The compiler would then generate C with MPI calls rather than CCSP/C++CSP calls. A particularly interesting possibility is the MPI implementation from The University of British Columbia [45] which supports much finer grained levels of parallelism than the regular MPI implementations.
- Other virtual machines could be targeted, such as the LLVM [59]. In [22], a translator for ETC (Extended Transputer Code) to LLVM was developed.

Countless other targets exist, but we will focus on the MPI and the JCSP targets.

5.7. The ProcessJ Online Teaching Tool (Technical Part)

The recent success of massive open online courses (MOOCs) in attracting student interest from around the world is impossible to ignore. With the use of recent technologies incorporated into major Web browsers under the umbrella term of HTML5, it is now possible to offer highly interactive online computer science courses, which run entirely within a student's Web browser.

By producing a ProcessJ compiler back-end and runtime system that targets JavaScript, ProcessJ code can be executed directly within the Web browser environment. The core of the ProcessJ Online Teaching Tool will consist of four main components:

- The code editor/debugger interface for writing ProcessJ programs in the browser.
- A ProcessJ to JavaScript compiler back-end, which will run on a server and compile programs on demand.
- The ProcessJ JavaScript runtime system, which will provide concurrent execution of processes in the single-threaded JavaScript interpreter using co-operative multitasking.
- A ProcessJ API for the HTML5 canvas for interactive graphical programs.

In addition to the online programming environment, the ProcessJ Online Teaching Tool will provide access to learning materials and provide a framework for automated grading of students' programming exercises. With these features in combination, self-contained online classes can be produced that can be accessed by students in any place and at any time. Figure 5 shows a mock-up of the ProcessJ Online Teaching Tool. We discuss similar browser-based IDEs and online learning environments (for other languages) in Section 6.3.

5.8. Online Cloud Storage

Since an integral part of the ProcessJ suite is the online programming environment, which enables a programmer to develop ProcessJ code from any Web browser, the issue of code storage arises. To make available a user's code in any Web browser (on any computer with a Web browser that supports JavaScript and is connected to the Internet), code cannot be stored locally (at least not as the only location for the code).

We propose an online cloud storage (in the style of Dropbox [60], though not user managed to the same extent), that integrates the online programming environment, and the integrated development environment as well as the Online Repository.

A user logs into his or her cloud storage and the code is immediately available to the in-browser development environment; similarly, logging in within the IDE will make all the cloud storage code available.

Furthermore, the cloud storage makes getting code from and putting code into the Online Process Library an easy task. Cloud storage enables additional possibilities, and we envision

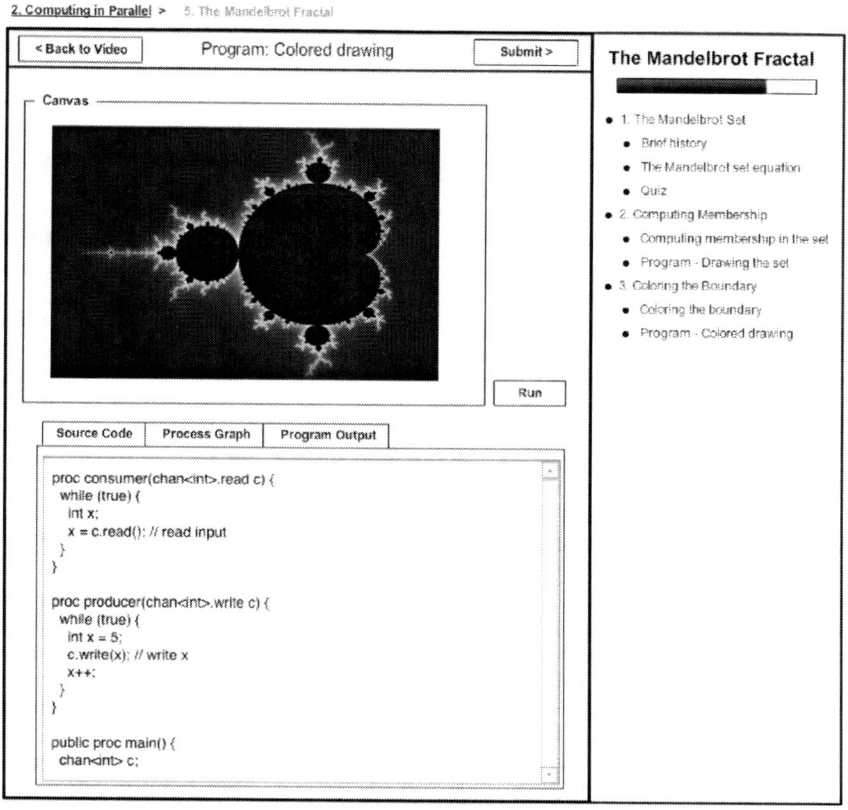

Figure 5. A Web mock-up of the ProcessJ Online Teaching Tool (PJOTT).

many possible extensions. For example versioning and sharing. As the project develops, more extensions will surely be included.

One possible implantation of this could be on top of Google Drive, using the Google Drive SDK [61]; alternatively, GitHub [62] could be utilised through the graphical user interface, or even the ProcessJ website. There are many ways to implement online cloud storage, and implantation wise, it might be a great idea to allow a plug and play situation, where the user can decide which system he wants to use (Google Drive, GitHub, Dropbox etc.)

6. Educational Aspect

Every ten years, an ACM-IEEE Computer Society joint task force puts out an updated document containing guidelines for Computer Science curricula; the latest version, Computer Science Curricula 2013 (CS2013) [32], is due later this year. The latest draft (Ironman Draft Version 0.8) includes information about important (and exciting!) changes for the curriculum, especially regarding what topics are considered “core”. The last time full guidelines were published in 2001 [63], and in interim updates [64,65], parallel computing wasn't identified as its own Knowledge Area within the Body of Knowledge. The story is different this time around: in 2012, the joint task force stated, when questioned about new topical areas that should be added to the Body of Knowledge, survey respondents indicated a strong need to add the topics of Security as well as Parallel and Distributed Computing . . . CS2013 includes these two new KAs (among others): Information Assurance and Security, and Parallel and Distributed Computing [32]. Concurrency should not be a last resort to gain performance, but should be taught as a fundamental model of interaction between processes. Supporting such a way of thinking both from an educational side as well as a programming-language side, for example, the use of a process-oriented language, makes concurrency as easy to approach for students just beginning their programming careers as any other sequential construct; one might argue that fully understanding state and mutation is even harder than synchronously communicating processes.

It is not a surprise that parallel computing has been reified in CS2013; Moore's Law has plateaued over the last decade, replaced by a focus on Amdahl's Law [66] to promote the development of ever-larger multi-core processors. As discussed in [67,68], the free lunch is over. In other words, in order to continue enjoying increased performance, we have to learn to harness the power of multi-core processors. We can no longer count on a faster processor coming out next year. Our challenge is to write better parallel programs and also to teach the next generation of programmers and computer scientists how this can be done.

Assuming we are completely successful in developing ProcessJ and its supporting tools, that wouldn't be enough to call it a day. We must provide a compelling set of materials for courses and course modules for faculty to use in the classroom. We must provide a path for teachers and professors to learn this material well enough that they can teach it to their students. To help get the word out, once the ProcessJ compiler is ready for broader use, we plan to offer workshops at parallel computing conferences, including the formerly named *Supercomputing (SC)* conference, now named the *International Conference for High Performance Computing, Networking, Storage and Analysis*; and the *International Parallel and Distributed Processing Symposium (IPDPS)*. Once the ProcessJ Online Teaching Tool is ready, we plan to offer workshops at such computer science education venues as SIGCSE and one or more of the regional CCSC conferences. We have some experience, through the CPA community, in presenting during teaching tracks and tutorials at conferences.

6.1. Learning Materials

Even as parallel processing enjoys a higher profile and recognition as a new Knowledge Area and core topic within the CS2013 guidelines, much of what people know about the topic is either limited — parallel and distributed computing is not everyone's area of expertise — or misunderstood — what is currently popular or being promoted regarding multi-core programming is not the only approach to solving these problems. Some people may think that threads and locks as well as shared memory is the only approach to multi-core programming. Others may know about asynchronous message passing, but not synchronous message passing.

The CSP model and process-oriented design, despite its long history, is still not in the mainstream, and what people think they know about it is wrong. It is our responsibility to provide learning material that both educates and re-educates our prospective audience regarding Process Oriented Design, while accommodating the different learning styles and environments of today's faculty and students. We will provide sample lecture notes, assignments, and video tutorials that present individual topics as well as demonstrate the use of the ProcessJ compiler and online teaching tool. These materials will be accessible online through a Web browser accessing the ProcessJ Web pages. We also will provide an online community space for teachers and practitioners to ask questions and exchange ideas, following the model of BlueJ's Blueroom [69].

The biggest obstacle we must overcome in our learning materials is the long-standing reputation that concurrent programming is hard, error prone, difficult to get right, and virtually impossible to reason about. For example, Muller and Walrath, in an article by Sun Microsystems on Threads and Swing [70] state:

> *"If you can get away with it, avoid using threads. Threads can be difficult to use, and they make programs harder to debug."*

This may, in fact, be a well-deserved reputation in the mainstream of concurrent processing, but it doesn't apply to process-oriented design. The challenges of unwanted nondeterminism, race hazards, and deadlock do not go away; they simply don't enter the picture in the first place. Our materials must illustrate that process orientation is the right model and abstraction for modeling our parallel world.

We have some very elegant, time-tested approaches to teaching process-oriented design, using Welch's 'LEGO' catalog method approach to processes [39]. We have used such libraries of processes in our courses for the last decade of teaching our own concurrent programming courses, and will implement this library for the new ProcessJ language. These programs and exercises build parallel programming skills on top of students' existing sequential programming skills so that parallel composition becomes just another programming abstraction, like sequential composition, conditional statements, recursion and iteration. Because processes compose, process-orientation is a natural way to deal with complexity, including underlying parallelism. This approach also illustrates how we write parallel programs at much finer-grained level of parallelism than prevailing conventional wisdom currently believes is feasible.

6.2. Course and Module Templates

The CS2013 guidelines are broken down into Knowledge Areas that do not necessarily correspond to full courses. This allows for maximum flexibility for departments to design their curricula according to their particular levels of staffing, expertise, and numbers of students. As mentioned earlier, Parallel and Distributed Computing is a new Knowledge Area, but not all programs will decide to add a dedicated course in parallel programming. One option is to distribute parallel programming topics among several existing courses, for example, Algorithms, Operating Systems, Computer Architecture, and (we hope) introductory courses. This means we need to provide building blocks that can be put together into a single course or spread out among courses in existing curricula; the more self-contained we can make these building blocks, the more flexibility faculty and departments have in using the materials we provide. We will offer modules that can be used in a variety of courses, including algorithms, architecture courses, operating systems courses, and software engineering. We will use the CS2013 guidelines as a source for categorising different modules; in this way, faculty and departments can most easily use our ProcessJ materials in conjunction with the CS2013 document to revise their curricula to accommodate increased coverage of parallel processing material. We also will recommend different possibilities for a dedicated course in parallel processing, stressing the value of process-oriented design.

6.3. The ProcessJ Online Teaching Tool (Educational Part)

With the ProcessJ Online Teaching Tool (PJOTT), the technical and educational parts of this project intersect. PJOTT will provide students with a browser-hosted interface to access ProcessJ and its planned online resources. It also will make using ProcessJ in the classroom more convenient for faculty with limited resources because it won't be necessary to arrange for the installation and configuration of another programming language or IDE in their labs. Browser accessibility means faculty and students can focus on process-oriented design from

Day One. This intersection of the technical and educational provides an important feedback loop for improving the tool and its interface for students, based on ongoing feedback. Another advantage of providing a browser-based IDE is that changes to PJOTT are centralised on the server, and don't require pushing out changes for the users. Each time the browser loads PJOTT, users access the latest version. This is good, because it takes time, effort, and user feedback to develop a well-engineered interface that follows pedagogically sound guidelines for online learning tools. What good is a lot of information if it is not organised and presented in an intuitive manner?

Other languages now have browser-based online learning playgrounds. Some examples include GoLang [71], Codecademy [72], and CodingBat [73]. While GoLang specialises in the Go language, Codecademy and CodingBat support learning multiple languages. These online learning environments have some similarities but different missions and goals. CodingBat provides a set of problems to be solved, and an editor with a run button for users to try to solve them. The teaching takes place elsewhere; CodingBat is a place to practice one's Java or Python coding skills. Codecademy provides courses for a larger set of languages using tutorial style lessons. These tutorials including an editor for users to type in their solutions each step along the way, and check for correctness. Codecademy attracts both students and teachers, providing a framework for both. Codecademy also provides a playground for several languages for users that just wish to have access to these languages from their browser. GoLang is a portal to all things Go-related, including providing an online editor and interpreter, and examples of programs that can be run from the browser. These existing online learning resources provide proofs of concept for some of what we plan to do for the PJOTT; however, we envision a deeper and more comprehensive integration with the teaching materials.

7. Conclusions and the Environment

We would like to conclude our paper, but we also feel it would be helpful to comment on the environment of this proposed work. Hoare introduced through CSP the notion of reasoning about the meaning of a computation through reasoning about its trace of observable events recorded by an Olympian observer. These observable events were offered by the *Environment*, and engaged in by the processes composed in a process network. Process-oriented design as a programming paradigm also exists within a larger environment, and if it is to succeed and become widely adopted, the process-oriented community must find a compelling way to communicate its ideas within this environment: the greater concurrency and high performance computing community.

7.1. Conclusions

We have presented what we believe to be a complete system for a new process-oriented programming system that is extremely well suited to take full advantage of today's multi-core architectures. This system includes a new process-oriented language, ProcessJ, rooted in the mathematical process algebra CSP. Programs written in ProcessJ are verifiable with respect to their concurrency. In addition, a programming environment and online code libraries and online storage are proposed as well as an online teaching tool, which allows for in-browser programming.

From an educational point of view, we propose creating course descriptions, support material, notes, and slides; in other words, anything needed to teach a course about process-oriented programming and design at the college level. The online teaching tool utilises these techniques to make such course material available to anyone with a Web browser. We believe this can be a very exciting project; the modularity of it means that a great many people and institutions can be involved with the development at the same time. We strongly believe that if the entire community pulls in the same direction, towards the completion of a suite like described in this paper, we stand a much better chance of 'reviving' process-oriented programming for the masses.

7.2. The Environment

This paper also formed the basis for an NSF proposal for around $750,000; however, the proposal was not funded, but three reviews were returned, and since this is a paper about a possible future of process-oriented programming, we would not do it justice without reporting the major issues found by the reviewers. We believe this can not only help shed some light on how to approach this project, but also give some ideas about what the scientific community (who are interested in concurrent programming) in general thinks about process-oriented design.

It seems that the scientific computing community thinks only in terms of HPC and shared memory, and if they ever deviate from that, MPI is as far as they are willing to venture. One of the comments was:

> *"At times, the proposal is talking about CSP as a panacea. Unfortunately, this (in my opinion) reveals ignorance of the richness of both the problem space (types of applications) and the solution space (known models for parallel or concurrent computation), and thus undermines the credibility of success. For example, try writing a well-performing parallel matrix multiplication in CSP."*

We believe that such statements actually shows the lack of understanding by the reviewer (and thus we might conclude by the broader community as a whole) both as to what CSP is (it's not a programming language!) and that *noone* is proposing a panacea. We are not suggesting that all problems be solved with process-oriented design, but that the tasks for which process oriented design is tailored may be solved by using it — and that the breadth of application be explored. Apart from the fact that a version of matrix multiplication could be implemented extremely efficiently using a systolic array (which could be programmed easily using process-oriented programming techniques), we are not proposing that process-oriented design and programming should be utilised for all scientific and HPC programming; but we are advocating that it be used for embedded system like *setups*, where either the hardware directly maps to processes or, on shared memory systems, where the problem naturally decomposes to processes.

We believe there is a perception that after deriding parallel computing as not necessary to perpetuate the prophesy of Moore's Law for 30+ years, now that we have hit the wall for clock speed and multicore has come along, they are going to deride concurrency and make it a scapegoat if it now cannot perform that service. But we know there are limitations to speed-up, which gives rise to the speed-up myth (solve a problem of size N on one processor in time N means we should be able to solve a problem of size N on P processors in time N/P). Some problems are more amenable to speed-up than others. They look at matrix multiplication as an embarrassingly parallel (or more politically correct these days, "pleasingly parallel") algorithm; but that is based on a shared memory model of concurrency. Process-oriented design and programming is not a shared memory model, although it works extremely well on shared-memory platforms [22,23]! If your problem is matrix multiplication, and you have the cores, you may use a shared memory model — but communicating processes may also work (and be simpler). However, most problems do not have embarrassingly parallel solutions and their needs are for interesting and rich patterns of interaction between the sub-components that will solve them. Process orentation is looking to address this more general class of problem, as well as those more regular systems whose needs for memory vastly exceeds that which can be supported by a single memory bus fought over by a host of processor cores.

Programming-with-pictures is an extremely important approach to process-oriented design, and its strengths must be emphasised in the future. A lack of understanding of the importance of composability is shown by comments such as:

> *"The inclusion of a graphical interface / IDE is a nice idea, but orthogonal as a research problem."*

and:

> *"Graphical IDEs have been around forever, but have never made it beyond the fringe. They tend to survive only if supported by a monopolistic proprietary owner."*

Finally, it seems that the community does understand the importance of verification, and they do seem familiar with the strengths of CSP, but they often fall short with comments like:

> *"Can't you just use Google's Go?"*

In conclusion to these reviews and comments, we believe that the following issues must be emphasised in the future to better promote process-oriented programming and design:

- Process-oriented programming and design is *not* trying to be a panacea; we do not propose implementing all algorithms in this fashion, but we do propose using the design and programming techniques to those problems that naturally fit the paradigm; these include, but are not limited to, embedded systems and safety critical concurrent systems, such as control systems for machinery, aircraft, etc.
- Reusability and composability are important; and communicating processes, in particular, can be separately reasoned about, verified and reused.
- The relationship between verifiability, composability and code reuse does not seem to be recognised and must be emphasised.

Acknowledgments

The authors wish to thank the editors and the many helpful comments and references provided by our reviewers. Their careful comments and valuable feedback helped improve the content and presentation of this paper.

References

[1] Leslie G. Valiant. A Bridging Model for Parallel Computation. *Communications of the ACM*, 33(8):103–111, August 1990.

[2] Dick Pountain and Michael D. May. *A Tutorial Introduction to occam Programming*. McGraw-Hill, Inc., New York, NY, USA, 1987.

[3] Michael D. May. CSP, occam and Transputers. In Ali E. Abdallah, Cliff B. Jones, and Jeff W. Sanders, editors, *25 Years of CSP*, volume 3525 of *Lecture Notes in Computer Science*, pages 75–84. Springer Verlag, April 2005.

[4] Peter H. Welch and Frederick R. M. Barnes. Communicating Mobile Processes: introducing occam-π. In Ali E. Abdallah, Cliff B. Jones, and Jeff W. Sanders, editors, *25 Years of CSP*, volume 3525 of *Lecture Notes in Computer Science*, pages 175–210. Springer Verlag, April 2005.

[5] Frederick R. M. Barnes and Peter H. Welch. Communicating Mobile Processes. In Ian East, Jeremy Martin, Peter H. Welch, David Duce, and Mark Green, editors, *Communicating Process Architectures 2004*, volume 62, WoTUG-27 of *Concurrent Systems Engineering Series, ISSN 1383-7575*, pages 201–218, Amsterdam, The Netherlands, September 2004. IOS Press. ISBN: 1-58603-458-8.

[6] Peter H. Welch and Frederick R. M. Barnes. Mobile Barriers for occam-π: Semantics, Implementation and Application. In Jan F. Broenink, Herman W. Roebbers, Johan P.E. Sunter, Peter H. Welch, and David C. Wood, editors, *Communicating Process Architectures 2005*, volume 63, WoTUG-28 of *Concurrent Systems Engineering Series*, pages 289–316, Amsterdam, The Netherlands, September 2005. IOS Press. ISBN: 1-58603-561-4.

[7] Peter H. Welch and Frederick R. M. Barnes. A CSP Model for Mobile Channels. In *Communicating Process Architectures 2008*, volume 66, WoTUG-31 of *Concurrent Systems Engineering Series*, pages 17–33, Amsterdam, The Netherlands, September 2008. IOS Press. ISBN: 978-1-58603-907-3.

[8] Peter H. Welch. *An occam-π Quick Reference Guide*. Programming Languages and Systems Research Group, University of Kent, `https://www.cs.kent.ac.uk/research/groups/plas/wiki/OccamPiReference/`, 2011.

[9] Peter H. Welch and Jan B. Pedersen. Santa Claus: Formal Analysis of a Process-oriented Solution. *ACM Trans. Program. Lang. Syst.*, 32(4):37, April 2010.

[10] Peter H. Welch, Kurt Wallnau, Adam T. Sampson, and Mark Klein. To Boldly Go: an occam-π Mission to Engineer Emergence. *Natural Computing*, 11(3):449–474, September 2012. `http://dx.doi.org/10.1007/s11047-012-9304-2`.

[11] Charles A. R. Hoare. Communicating Sequential Processes. *Communications of the ACM*, 21(8):666–677, August 1978.

[12] Charles A. R. Hoare. *Communicating Sequential Processes*. Prentice-Hall, 1985.

[13] Andrew W. Roscoe. *The Theory and Practice of Concurrency*. Prentice Hall, 1997.

[14] Andrew W. Roscoe. *Understanding Concurrent Systems*. Springer, 2010.

[15] John A. Trono. A New Exercise in Concurrency. *SIGCSE Bulletin*, 26(3):8–10, 1994.

[16] Mordechai Ben-Ari. How to solve the Santa Claus Problem. *Concurrency: Practice and Experience*, 10(6):485–496, 1998.

[17] Alex Brodsky, Jan B. Pedersen, and Alan Wagner. On the Complexity of Buffer Allocation in Message Passing Systems. *Journal of Parallel and Distributed Computing*, 65(5):692–713, June 2005.

[18] Jan B. Pedersen, Alex Brodsky, and Jeffrey Sampson. Approximating the Buffer Allocation Problem Using Epochs. *Journal of Parallel and Distributed Computing*, 68(9):1263–1282, September 2008.

[19] Peter H. Welch. Process Oriented Design for Java – Concurrency for All. In *PDPTA 2000*, volume 1, pages 51–57. CSREA Press, June 2000. ISBN: 1-892512-52-1.

[20] Robin Milner. *Communicating and Mobile Systems: the Pi-Calculus*. Cambridge University Press, 1999. ISBN-10: 0521658691, ISBN-13: 9780521658690.

[21] Formal Systems (Europe). *FDR2 (Failures, Divergence, Refinement) model checker*. `http://www.fsel.com/software.html`.

[22] Carl G. Ritson, Adam T. Sampson, and Frederick R. M. Barnes. Multicore Scheduling for Lightweight Communicating Processes. In John Field and Vasco Thudichum Vasconcelos, editors, *Coordination Models and Languages, COORDINATION 2009, Lisboa, Portugal, June 9-12, 2009. Proceedings*, volume 5521 of *Lecture Notes in Computer Science*, pages 163–183. Springer, June 2009.

[23] Carl G. Ritson, Adam T. Sampson, and Frederick R. M. Barnes. Multicore Scheduling for Lightweight Communicating Processes. *Science of Computer Programming*, 77(6):727–740, 2012. Revised from LNCS (5521) paper with same name.

[24] Hamid R. Arabnia. The Transputer Family of Products. In A. Kent and J.G. Williams, editors, *Encyclopedia of Computer Science and Technology*, 1998.

[25] XMOS. XC Programming Guide. `http://www.xmos.com/published/programming-xc-xmos-devices`. [Online; accessed 01-August-2013].

[26] Google. The Go Programming Language. `http://golang.org/doc/`. [Online; accessed 01-August-2013].

[27] Carl Hewitt, Peter Bishop, Irene Greif, Brian Smith, Todd Matson, and Richard Steiger. Actor induction and meta-evaluation. In *In ACM Symposium on Principles of Programming Languages*, pages 153–168, 1973.

[28] Carl Hewitt. Viewing control structures as patterns of passing messages. *Artificial Intelligence*, 8(3):323364, June 1977.

[29] Peter H. Welch. Java Threads in the Light of occam/CSP. In Peter H. Welch and André W.P. Bakkers, editors, *Architectures, Languages and Patterns for Parallel and Distributed Applications, Proceedings of WoTUG 21*, volume 52 of *Concurrent Systems Engineering*, pages 259–284, Amsterdam, The Netherlands, April 1998. WoTUG, IOS Press. ISBN: 90-5199-391-9.

[30] Peter H. Welch and Jeremy M. R. Martin. A CSP Model for Java Multithreading. In *Proceedings of the 5th International Symposium on Software Engineering for Parallel and Distributed Systems (PDSE 2000)*. IEEE Computer Society, jun 2000.

[31] Peter H. Welch and Jeremy M. R. Martin. Formal Analysis of Concurrent Java Systems. In Peter H. Welch and André W. P. Bakkers, editors, *Communicating Process Architectures 2000*, volume 66, WoTUG-31, pages 275–301, Amsterdam, The Netherlands, sep 2000. IOS Press. ISBN: 978-1-58603-907-3.

[32] ACM/IEEE-CS Joint Task Force for Computing Curricula. Computer Science Curricula 2013, Ironman Draft (Version 0.8). Technical report, November 2012. `http://ai.stanford.edu/users/sahami/`

CS2013/, [Online; accessed 01-August-2013].

[33] Alon Zakai. Emscripten. Technical report. http://emscripten.org, [Online; accessed 01-August-2013].

[34] Neil C.C. Brown. C++CSP2: a Many-to-Many Threading Model for Multicore Architectures. In Alistair A. McEwan, Steve Schneider, Wilson Ifill, and Peter H. Welch, editors, *Communicating Process Architectures 2007*, volume 65, WoTUG-30 of *Concurrent Systems Engineering Series*, pages 183–205, Amsterdam, The Netherlands, "July" 2007. IOS Press.

[35] Neil C. C. Brown and Peter H. Welch. An Introduction to the Kent C++CSP Library. In Jan F. Broenink and Gerald H. Hilderink, editors, *Communicating Process Architectures 2003*, WoTUG-26, Concurrent Systems Engineering, ISSN 1383-7575, pages 139–156, Amsterdam, The Netherlands, September 2003. IOS Press. ISBN: 1-58603-381-6.

[36] James Moores. CCSP – a Portable CSP-based Run-time System Supporting C and occam. In B.M. Cook, editor, *Architectures, Languages and Techniques for Concurrent Systems*, volume 57 of *Concurrent Systems Engineering series*, pages 147–168, Amsterdam, The Netherlands, April 1999. WoTUG, IOS Press. ISBN: 90-5199-480-X.

[37] Mikolaj Slomka. Visual Occam: High Level Visualization and Design of Process Networks. Master's thesis, Department of Computer Science, University of Nevada Las Vegas, 2010.

[38] Maarten M. Bezemer, Robert J. W. Wilterdink, and Jan F. Broenink. Design and Use of CSP Meta-Model for Embedded Control Software Development. In Peter H. Welch, Frederick R. M. Barnes, Kevin Chalmers, Jan B. Pedersen, and Adam T. Sampson, editors, *Communicating Process Architectures 2012*, pages 185–199. Open Channel Publishing, August 2012. ISBN 978-0-9565409-5-9.

[39] Peter H. Welch. Concurrency Design and Practice. http://www.cs.kent.ac.uk/projects/ofa/sei-cmu/, November 2007. A course on concurrency and process-oriented design, based on occam-π.

[40] Peter H. Welch, Jan B. Pedersen, Frederick R.M. Barnes, and Carl G. Ritson. Self-Verifying Concurrent Programming. Presentation to IFIP Working Group 2.4, September 2011. https://www.cs.kent.ac.uk/research/groups/plas/wiki/IFIP_WG24.

[41] Peter H. Welch, Jan B. Pedersen, Frederick R.M. Barnes, Carl G. Ritson, and Neil C.C. Brown. Adding Formal Verification to occam-π. In *Communicating Process Architectures 2011*, volume 68 of *Concurrent Systems Engineering Series*, page 379, Amsterdam, The Netherlands, June 2011. IOS Press. *Endnote address*. ISBN 978-1-60750-773-4.

[42] Peter H. Welch, Neil C.C. Brown, James Moores, Kevin Chalmers, and Bernard Sputh. Integrating and Extending JCSP. In Alistair A. McEwan, Steve Schneider, Wilson Ifill, and Peter Welch, editors, *Communicating Process Architectures 2007*, volume 65 of *Concurrent Systems Engineering Series*, pages 349–370, Amsterdam, The Netherlands, July 2007. IOS Press. ISBN: 978-1-58603-767-3.

[43] Peter H. Welch and Paul D. Austin. *Communicating Sequential Processes for Java (JCSP) Home Page*. Systems Research Group, University of Kent, 2010.

[44] Maarten M. Bezemer, Robert J. W. Wilterdink, and Jan F. Broenink. LUNA: Hard Real-Time, Multi-Threaded, CSP-Capable Execution Framework. In Peter H. Welch, Adam T. Sampson, Jan B. Pedersen, Jon M. Kerridge, Jan F. Broenink, and Frederick R. M. Barnes, editors, *Communicating Process Architectures 2011, Limmerick*, volume 68 of *Concurrent System Engineering Series*, pages 157–175, Amsterdam, November 2011. IOS Press BV. ISBN 978-1-60750-773-4.

[45] Sarwar Alam, Humaira Kamal, and Alan Wagner. Service-Oriented Programming in MPI. In Peter H. Welch and Jan B. Pedersen and Kevin Chalmers and Adam T. Sampson and Frederick R. M. Barnes, editor, *Communicating Process Architectures 2013*. Open Channel Publishing, August 2013.

[46] Jan B. Pedersen and Matthew Sowders. Static Scoping and Name Resolution for Mobile Processes with Polymorphic Interfaces. In *The thirty-third Communicating Process Architectures Conference, CPA 2011, organised under the auspices of WoTUG, Limerick, Ireland, June 19-22 2011*, pages 71–85, 2011.

[47] Matthew Sowders and Jan B. Pedersen. Mobile Process Resumption In Java Without Bytecode Rewriting. In *Proceedings of Parallel and Distributed Processing Techniques and Applications (PDPTA'11)*, July 2011.

[48] Matthew Sowders. ProcessJ: A Process Oriented Programming Language. Master's thesis, Department of Computer Science, University of Nevada Las Vegas, 2011.

[49] Frederick R. M. Barnes. Towards a New Language for Concurrent Programming. *CPA 2011 Fringe:* slides at http://www.wotug.org/papers/CPA-2011/Barnes11/Barnes11-slides.pdf, August 2011. Guppy Home Page: http://frmb.org/guppy.html.

[50] Peter H. Welch. Life of occam-π. In *Communicating Process Architectures 2013*. Open Channel Publishing, August 2013.

[51] Neil C. C. Brown and Marc L. Smith. Representation and Implementation of CSP and VCR Traces. In *Communicating Process Architectures 2008*, volume 66 of *Concurrent Systems Engineering*, pages 329–

345, Amsterdam, The Netherlands, September 2008. WoTUG, IOS Press.

[52] Neil C. C. Brown and Marc L. Smith. Relating and Visualising CSP, VCR and Structural Traces. In Peter H. Welch, Herman W. Roebbers, Jan F. Broenink, Frederick R. M. Barnes, Carl G. Ritson, Adam T. Sampson, Gardiner G. Styles, and Brian Vinter, editors, *Communicating Process Architectures 2009*, volume 67 of *Concurrent Systems Engineering*, pages 89–103, Amsterdam, The Netherlands, November 2009. WoTUG, IOS Press.

[53] Jan B. Pedersen and Peter H. Welch. Concurrency, Intuition and Formal Verification: Yes, We Can! In *Workshop on Curricula for Concurrency and Parallelism, SPLASH 2010*, October 2010. Position paper and slides: `ftp://ftp.cs.ukc.ac.uk/pub/phw/sei-cmu/occam/papers/README.txt` [Online; accessed 01-August-2013].

[54] Peter H. Welch and Jan B. Pedersen. Santa Claus - with Mobile Reindeer and Elves. In *Fringe Presentation at Communicating Process Architectures conference*, September 2008.

[55] Frederick R. M. Barnes, Peter H. Welch, Jim Moores, and David C. Wood. *The KRoC Home Page*. Programming Languages and Systems Research Group, University of Kent, `http://www.cs.kent.ac.uk/projects/ofa/kroc/`, 2010.

[56] Neil C. C. Brown. Communicating Haskell Processes: Composable Explicit Concurrency Using Monads. In Peter H. Welch, Susan Stepney, Fiona A. C. Polack, Frederick R. M. Barnes, Alistair A. McEwan, Gardner S. Stiles, Jan F. Broenink, and Adam T. Sampson, editors, *Communicating Process Architectures 2008*, volume 66 of *Concurrent Systems Engineering*, pages 67–83, Amsterdam, The Netherlands, September 2008. WoTUG, IOS Press.

[57] Verum Software Technologies B.V. The ASD:Suite. `http://www.verum.com/`, 2013. [Online; accessed 01-August-2013].

[58] Jack Dongarra. MPI: A Message Passing Interface Standard. *The International Journal of Supercomputers and High Performance Computing*, 8:165–184, 1994.

[59] Chris Lattner and Vikram Adve. LLVM: A Compilation Framework for Lifelong Program Analysis & Transformation. In *Proceedings of the 2004 International Symposium on Code Generation and Optimization (CGO'04)*, Palo Alto, California, March 2004.

[60] Dropbox. Online Cloud Storage. `http://www.dropbox.com/`. [Online; accessed 01-August-2013].

[61] Google. Google Drive SDK – Google Developers. `https://developers.google.com/drive/`. [Online; accessed 01-August-2013].

[62] GitHub - Build Better Software Together. Technical report. `http://github.com`, [Online; accessed 01-August-2013].

[63] ACM/IEEE-CS Joint Task Force on Computing Curricula. ACM/IEEE Computing Curricula 2001 Final Report. Technical report, 2001. `http://www.acm.org/sigcse/cc2001` [online; accessed 01-August-2013].

[64] ACM/IEEE-CS Joint Task Force for Computing Curricula. Computing Curricula 2005: An Overview Report. Technical report, 2005. `http://www.acm.org/education/curric_vols/CC2005-March06Final.pdf`, [Online; accessed 01-August-2013].

[65] ACM/IEEE-CS Joint Interim Review Task Force. Computer Science Curriculum 2008: An Interim Revision of CS 2001, Report from the Interim Review Task Force. Technical report, 2008. `http://www.acm.org/education/curricula/ComputerScience2008.pdf`, [Online; accessed 01-August-2013].

[66] Gene M. Amdahl. Validity of the Single Processor Approach to Achieving Large Scale Computing Capabilities. In *Proceedings of the AFIPS Spring Joint Computer Conference*, volume 30, pages 483–485, 1967.

[67] Herb Sutter. The Free Lunch is Over: a Fundamental Turn Toward Concurrency in Software. *Dr. Dobb's Journal*, 30(3), 2005.

[68] Herb Sutter and James Larus. Software and the Concurrency Revolution. *Queue*, 3(7), September 2005.

[69] Michael Kölling, Bruce Quig, Andrew Patterson, and John Rosenberg. The BlueJ system and its pedagogy. *Journal of Computer Science Education, Special issue on Learning and Teaching Object Technology*, 13(4), December 2003.

[70] Hans Muller and Kathy Walrath. Threads and Swing, September 2000. `http://java.sun.com/products/jfc/tsc/articles/threads/threads1.html` [Online; accessed 01-August-2013].

[71] The Go Community. The Go Language. `http://golang.org/`, 2013. [Online; accessed 01-August-2013].

[72] Zach Sims and Ryan Bubinski. Codecademy. `http://www.codecademy.com/`, 2013. [Online; accessed 01-August-2013].

[73] Nick Parlante. CodingBat Code Practice. `http://codingbat.com/`, 2013. [Online; accessed 01-August-2013].

A. The ProcessJ Home Page

We currently have a Web page in place at `http://www.processj.org`, the main page of which can be seen in Figure 6.

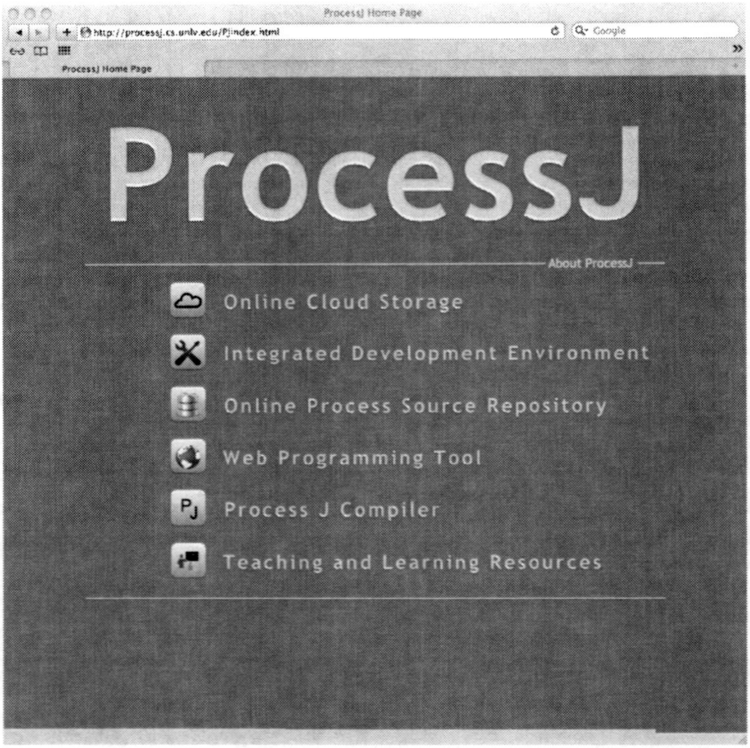

Figure 6. The ProcessJ home page.

Communicating Process Architectures 2013
P.H. Welch et al. (Eds.)
Open Channel Publishing Ltd., 2013

Specifying and Analysing Networks of Processes in CSP$_T$ (or In Search of Associativity)

Paul HOWELLS [a,1] and Mark D'INVERNO [b]
[a] *Department of Computer Science & Software Engineering, University of Westminster, UK*
[b] *Department of Computing, Goldsmiths, University of London, UK*

Abstract. In proposing theories of how we should design and specify networks of processes it is necessary to show that the semantics of any language we use to write down the intended behaviours of a system has several qualities. First in that the meaning of what is written on the page reflects the intention of the designer; second that there are no unexpected behaviours that might arise in a specified system that are hidden from the unsuspecting specifier; and third that the intention for the design of the behaviour of a network of processes can be communicated clearly and intuitively to others. In order to achieve this we have developed a variant of CSP, called CSP$_T$, designed to solve the problems of termination of parallel processes present in the original formulation of CSP. In CSP$_T$ we introduced three parallel operators, each with a different kind of termination semantics, which we call synchronous, asynchronous and race. These operators provide specifiers with an expressive and flexible tool kit to define the intended behaviour of a system in such a way that unexpected or unwanted behaviours are guaranteed not to take place. In this paper we extend out analysis of CSP$_T$ and introduce the notion of an *alphabet diagram* that illustrates the different categories of events that can arise in the parallel composition of processes. These alphabet diagrams are then used to analyse networks of three processes in parallel with the aim of identifying sufficient constraints to ensure associativity of their parallel composition. Having achieved this we then proceed to prove associativity laws for the three parallel operators of CSP$_T$. Next, we illustrate how to design and construct a network of three processes that satisfy the associativity law, using the associativity theorem and alphabet diagrams. Finally, we outline how this could be achieved for more general networks of processes.

Keywords. concurrency, CSP, CSP$_T$, parallel operators, associativity, process networks

Introduction

In the original *failure-divergence* semantic model for *Communicating Sequential Processes* (CSP) developed by Hoare, Brookes and Rosoce [1,2,3,4] the incomplete treatment of successful process termination, and in particular parallel termination, permitted intuitively contradictory processes to be defined. For example, it was possible to define a parallel process that appeared to have terminated several times before actually doing so. In many cases this meant that the behaviour of the system did not match the intention of the specifier when it was specified formally using CSP.

[1]Corresponding Author: *Paul Howells, Department of Computer Science & Software Engineering, University of Westminster, 115 New Cavendish St., London, W1W 6RU, UK.* `E-mail: P.Howells@westminster.ac.uk`.

In response to this problem we developed a variant of CSP as defined in the original work of Hoare and its development by Roscoe [5,6], called CSP_T that has been introduced elsewhere [7,8]. In this work we demonstrated how CSP_T, solves the original termination problems by introducing three distinct but related parallel operators that between them provide a transparent and intuitive means for specifying the desired termination of networks of processes. These three parallel operators each have a different form of termination semantics based on different ways of combining the termination of its component processes. The three forms are *synchronous* termination, *asynchronous* termination and *race* termination (where the first component process to terminate, terminates the composite process).

The design of these three parallel operators, was motivated by the desire to provide a system designer with both the *precision* and *flexibility* to be able to select as wide a range of different kinds of termination and interaction for networks of parallel processes. In doing so we had two overarching principles: (i) to design a semantic model as consistent as possible with the original model defined for CSP; and (ii) to design a model that provided any system designer with the confidence that the system would behave as intended.

By providing these three operators, we claim (and indeed have proved) that any system specifier can now make precise choices about how parallel networks of processes are required to successfully terminate and interact by means of synchronised events.

The need for such parallel operators can be seen when we consider the range of concurrent systems, with different termination behaviour, that we would wish to specify and design. For example, (distributed) client/server systems usually require the asynchronous termination of the client and server processes. In contrast, for a system of hardware components, synchronous termination would normally be a fundamental requirement. For parallel searching, race termination would be desirable, since, it is only necessary to wait for the first (and possibly only) successful process to terminate. By providing parallel operators with different termination semantics, a more natural way of modelling and designing modern parallel systems in different environments becomes possible.

In [8], we focused on the specification of how parallel processes should terminate using CSP_T's three parallel operators. In this paper we continue the investigation of CSP_T's parallel operators, by focusing on the specification of parallel process interaction. In particular, the different types of synchronous and asynchronous events that can occur in a parallel composition. A key aspect of this investigation is the discovery of constraints for and proof of associativity laws for the three parallel operators. These laws provide specifiers and designers with an essential law for developing, analysing and reasoning about their specifications and designs of networks of processes. For example, as we will demonstrate, they allow us to construct a process network using the parallel operators based on the types of interactions required. Further, the associativity constraints can be used to determine if the associativity law applies for a given process network and determine if a network with specific types of interactions is constructable.

Aims and Motivations

The main goal of the paper is to prove as general an associativity law as possible for the generalised parallel operator of CSP and of our three operators - race, synchronous and asynchronous of CSP_T. As Roscoe [6] acknowledges the generalised parallel operator for CSP has become the most widely used and so there is a need for as strong an associativity law as possible to enable CSP practitioners to design and analyse parallel networks of processes. Roscoe has introduced a weak notion of associativity and we have improved the strength of the associativity law. The weak law, for example, provides little practical use when designing and analysing networks of parallel processes that are constructed using it. Therefore, there is a clear need for a strengthened version of the associativity law. Of course Roscoe does provide a strong associativity law for the alphabetised parallel operator (see Section 3.2). However, this law only deals with the case when the synchronising alphabets of the operator are the same as the alphabets of the process operands. Clearly this is limiting for any specifier.

Furthermore, one of our aims is to make the analysis and design of CSP process networks more tractable for system developers. To date there appears to be no methodology specifically aimed at designing and analysing the interactions in these networks. In this paper we attempt to address this by presenting a detailed analysis and categorisation of the event types that can occur in these types of networks. When this is combined with the use of alphabet diagrams, it provides a simple methodology to help developers design and understand the interactions between processes in these networks.

Outline of the Rest of the Paper

In Section 1, we provide a brief introduction to CSP_T. *Alphabet diagrams* are introduced in Section 2. These illustrate the different categories of events, and thus interactions, that can arise in the parallel composition of processes. In Section 3, we use alphabet diagrams to identifying constraints sufficient to ensure associativity laws for the three parallel operators of CSP_T. We then illustrate, in Section 4, how to use the associativity law and alphabet diagrams to design and construct a simple network of process that satisfy the associativity law. We suggest further work, in Section 5. Finally, we present our conclusions in Section 6. (Appendix A contains a summary of the notation, definitions and functions used in the paper. Appendix B details the proof of the main Associativity result presented in Section 3.5.)

1. An Introduction to CSP_T[1]

In this section[2],we provide an overview of CSP_T, for a complete description please see [7, 8,10]. (For a summary of the CSP and CSP_T notation and definitions used throughout the paper, see Appendix A.)

In Hoare, Brookes and Roscoe's original CSP [1,2,3,4] there was a well known problem concerning the incomplete treatment of parallel termination by two of original parallel operators: alphabetised $({}_A\|_B)$[3] and asynchronous/interleaving ($|||$). In particular, the type of termination that occurred using these operators was inconsistent and could vary depending on the termination of the processes being combined.

We shall now give an example to illustrate the type of problems that could arise. First recall, that the *successful termination* of a process in CSP (and CSP_T) is modelled by the event *tick* ($\checkmark$). Now consider the following process equivalence derivable within the original CSP:

$$\begin{aligned}(a \to SKIP)|||(b \to SKIP) \equiv \quad & (a \to ((\checkmark \to b \to SKIP) \,\Box\, (b \to \checkmark \to SKIP))) \\ & \Box\, (b \to ((a \to \checkmark \to SKIP) \,\Box\, (\checkmark \to a \to SKIP)))\end{aligned}$$

Clearly the $\checkmark$s on the right hand side cannot be interpreted as the successful termination of the process $(a \to SKIP)|||(b \to SKIP)$, since it continues to perform a, b and $\checkmark$ events. This illustrates how the original semantics for CSP could gave rise to these obviously undesirable and intuitively contradictory processes.

[1]This section contains material extracted from Sections 3.1, 5.2 and 5.6 of [7] and Sections 3.2, 3.3 and 3.4 of [8] by permission of Springer and Elsevier, respectively.

[2]This section is identical to Section 1 of [9], a companion paper in these same Proceedings (CPA 2013). This duplication, by permission of the Editors, is to let both papers be self-contained.

[3]Here and henceforth, A and B represent the alphabets of the processes being composed. For example, as in $P_A\|_B Q$, where $A = \alpha(P)$ and $B = \alpha(Q)$ respectively.

Several solutions have been proposed to this problem, including Tej and Wolf [11], Roscoe [5,6], Hoare and He [12] and the authors's own solution CSP_T [7]. For a detailed comparison of these solutions the interested reader is referred to [7].

Our starting point in defining CSP_T was the original failure-divergence model developed by Hoare, Brookes and Roscoe [3]. Our aim in modifying this model was to provide a more robust treatment of termination through the consistent and special handling of ✓ by the language (processes and operators) and semantics (failures and divergences).

For CSP_T, this was achieved by defining a new process axiom that captured our view of termination:

$$t \neq \langle\rangle \ \wedge \ (s^\frown\langle\checkmark\rangle^\frown t, \varnothing) \in F \ \Rightarrow \ s \in D \tag{T1}$$

where s and t are traces, F and D are the failure and divergence sets respectively of a process. This axiom means that if a process indicates that it has terminated (by means of the ✓) but continues to perform events (t), then it must have started diverging before it performed the ✓ (i.e. $s \in D$). For the rational behind this axiom, see our companion paper [9].

This new termination axiom resolves the termination issues of the original semantics, and it is added to the existing CSP process axioms (D1) to (N5), see Appendix A, to define CSP_T. Note that our view of tick (✓) is consistent with Hoare's, i.e. that it is a normal event, and have not adopted Roscoe's view that it is a special *signal* event. In doing this, we have defined a sub-model of the original failure-divergence model N_T such that, within this sub-model, all processes are well-behaved with respect to termination. In addition, three new forms of parallel operators were defined for CSP_T, each with a different form of termination semantics, as replacements for the original ones. We now introduce the language and model for CSP_T; we begin by introducing the three new parallel operators.

1.1. Parallel Operators of CSP_T

Our three new parallel operators are defined to be used as replacements for the original synchronous ($\|$), interleaving ($|||$) and alphabetised (${}_A\|_B$) parallel operators. It is necessary to define replacements for $|||$ and ${}_A\|_B$, as they do not satisfy (T1). These new parallel operators are *generalised* (or *interface*) style parallel operators, i.e. are parameterised by the set of events the processes are required to synchronise on. Each of these three operators has a distinct type of parallel termination semantics, and thus are distinct operators, see [8] for details. We call them *synchronous*, *asynchronous* and *race*, that we define here.

Synchronous: requires the successful termination of both P and Q; and the synchronisation of their termination, that is, ✓.

Asynchronous: requires the successful termination of both P and Q; and P and Q terminated asynchronously, i.e. they do not synchronise on ✓. (Roscoe [5,6] refers to this type of parallel termination semantics as *distributed termination*.)

Race: requires the successful termination of either P or Q asynchronously. Successful termination fails to occur only if both P and Q fail to terminate. Unlike synchronous and asynchronous termination where the environment observes a single ✓, under *race* termination semantics it can observe any ✓ that is performed by P or by Q. Consequently, the first ✓ the environment observes is taken as representing the termination of the parallel composition and whichever of P or Q did not terminate, i.e. did not preform the ✓, is aborted. Hence, with this type of termination semantics termination occurs as soon as either P or Q does so.

Note that in [5,6] Roscoe chooses to reject race termination semantics, due to the problems of dealing with the non-terminated process, e.g. the need for some powerful mechanism to manage its termination. We believe, however, that it is preferable at least to offer system specifiers the choice of using this type of termination semantics and let them resolve these issues, rather than ban it outright. We believe that this is not against the spirit of CSP, since there are other features available in CSP that also raise similar implementation issues, e.g. the various interrupt style operators.

The CSP_T parallel operators with these different forms of parallel termination semantics, are given in Table 1. Each operator is parametrised by a *synchronisation set* (Ω, Δ, Θ), that is the set of events on which the combined processes are required to synchronise. Events which are not in the synchronisation set but that can be performed by either P or Q or both are asynchronous events.

Table 1. CSP_T parallel operators

Termination Semantics	Operator	Synchronisation Set	Notes
Generalised	$P\|_{\Omega}Q$	$\varnothing \subseteq \Omega \subseteq \Sigma$	
Synchronous	$P\|_{\Delta}Q$	$\{\checkmark\} \subseteq \Delta \subseteq \Sigma$	$\checkmark \in \Delta$
Asynchronous	$P\|\|_{\Theta}Q$	$\varnothing \subseteq \Theta \subseteq \Sigma - \{\checkmark\}$	$\checkmark \notin \Theta$
Race	$P\|_{\Theta}Q$	$\varnothing \subseteq \Theta \subseteq \Sigma - \{\checkmark\}$	$\checkmark \notin \Theta$

We define a generalised parallel operator (denoted $P\|_{\Omega}Q$) that is used to define the operators with synchronous and race termination semantics. To distinguish this from the synchronous termination operator we denote the synchronisation set by Ω rather than Δ. The synchronous parallel operator $\|_{\Delta}$ is simply the generalised one $\|_{\Omega}$, with the constraint that $\checkmark \in \Delta$, thus ensuring synchronous termination. The race termination operator $|_{\Theta}$ can also be defined using $\|_{\Omega}$ and *SKIP* as follows:

$$P|_{\Theta}Q \mathrel{\widehat{=}} (P\|_{\Theta}Q); SKIP \qquad [\varnothing \subseteq \Theta \subseteq \Sigma - \{\checkmark\}]$$

where Σ is the set of all events and, for CSP_T, includes $\checkmark$. Note that CSP_T does not have the law ($P; SKIP = P$), otherwise the above would mean the race and generalised operators were the same. To illustrate how this definition behaves, consider the process $(P\|_{\Theta}Q); R$, with $\Theta \subseteq \Sigma - \{\checkmark\}$. Whichever of the two $\checkmark$s R (the environment) observes first it takes as representing the termination of $P\|_{\Theta}Q$ and, hence, R proceeds to execute; whichever of P or Q did not terminate is aborted.

The asynchronous parallel operator $|||_{\Theta}$ is defined independently, as its form of termination semantics is not compatible with $\|_{\Omega}$, i.e. it cannot be used to define $|||_{\Theta}$. Both process operands must terminate for $|||_{\Theta}$ to terminate, so it is not the same as $\|_{\Theta}$ (see previous paragraph). Both processes must terminate for $\|_{\Delta}$, if $\Delta = \Theta \cup \{\checkmark\}$, but they *synchronise* on their $\checkmark$s (i.e. a non-terminating process can prevent the other operand from terminating, which is *not* the case for $|||_{\Theta}$).

Here are examples of the three parallel operators of CSP_T, that illustrate the difference between $|_{\Theta}$ and the two others (which for these examples, with empty synchronisation sets, are the same):

$$\begin{aligned} a \to SKIP\|_{\varnothing}b \to SKIP &\equiv a \to SKIP|||_{\varnothing}b \to SKIP \\ &\equiv (a \to b \to SKIP) \mathrel{\Box} (b \to a \to SKIP) \\ a \to SKIP|_{\varnothing}b \to SKIP &\equiv (a \to (SKIP \sqcap (SKIP \mathrel{\Box} b \to SKIP))) \\ &\quad \mathrel{\Box} (b \to (SKIP \sqcap (SKIP \mathrel{\Box} a \to SKIP))) \end{aligned}$$

It is important to note that we do not include the generalised parallel operator $\|_{\Omega}$ in the language of CSP_T, since this would defeat the purpose of the whole exercise. Since, it would again result in inconsistent processes, similar to the one given above in Section 1, with multiple ticks in its traces. However, by restricting its use to defining $\|_{\Delta}$ and $|_{\Theta}$ we ensure these forms of processes do not occur. The semantic functions for the generalised operator are given in Appendix A. Full details for the other processes and operators, including the operational semantics and all semantic functions, can be found in [7,8].

1.2. The Model and Language of CSP_T

The model for CSP_T, N_T is defined by adding the Termination axiom (T1) to the process axioms (D1) to (N5) of the original CSP model N, given in Appendix A. Hence, in N_T the failure and divergence sets of a process are defined as for N, except that they also satisfy the process axiom (T1) as well as (D1) to (N5). The semantic functions $\mathcal{F}$ and $\mathcal{D}$ for N_T are the same as for N, and are given in [7].

The language for CSP_T is the same as that of the original CSP, but uses the more recent form of relational renaming instead of functional renaming and uses the three new parallel operators as replacements for $\|$, $|||$ and ${}_A\|_B$, see Table 2. It is defined as follows:

$$P ::= \bot \mid STOP \mid SKIP \mid a \rightarrow P \mid P \sqcap P \mid P \,\Box\, P \mid P;P \mid P\backslash a$$
$$\mid P[\![R]\!] \mid \mu p.F(p) \mid p \mid P\|_{\Delta}P \mid P|||_{\Theta}P \mid P|_{\Theta}P$$

where $a \in \Sigma - \{\checkmark\}$. $\bot$ is the divergent process (can be defined as $\mu p.p$). *STOP* is the deadlocked process and *SKIP* is the successfully terminating process. $a \rightarrow P$ is action prefix. $P \sqcap Q$ is nondeterministic choice and $P \,\Box\, Q$ is deterministic choice. $P; Q$ is sequential composition. $P\backslash a$ is event hiding. $P[\![R]\!]$ is action (relational) renaming, with the usual constraint [5,6] applying to the renaming relation R with respect to $\checkmark$, i.e. that no other event is mapped to it or that it is mapped to another event. p is a process variable, $\mu p.F(p)$ is recursion and in the definition of $F(p)$, only the above processes and operators can be used. $P\|_{\Delta}Q$, $P|||_{\Theta}Q$ and $P|_{\Theta}Q$ are the generalised synchronous, asynchronous and race parallel operators respectively. The processes and operators of CSP_T are *well-defined* and *well-behaved* in N_T.

Table 2. CSP_T replacements for $\|$, $|||$ and ${}_A\|_B$

Termination Semantics	$\vert\vert\vert$	${}_A\Vert_B$	$\Vert$
Synchronous ($\Vert_{\Delta}$)	$\Delta = \{\checkmark\}$	$\Delta = (A \cap B) \cup \{\checkmark\}$	$\Delta = (A \cup B) \cup \{\checkmark\}$
Asynchronous ($\vert\vert\vert_{\Theta}$)	$\Theta = \varnothing$	$\Theta = (A \cap B) - \{\checkmark\}$	$\Theta = (A \cup B) - \{\checkmark\}$
Race ($\vert_{\Theta}$)	$\Theta = \varnothing$	$\Theta = (A \cap B) - \{\checkmark\}$	$\Theta = (A \cup B) - \{\checkmark\}$

We can now build on the account given so far in order to begin to investigate associativity of networks of CSP_T processes in the following sections.

2. Alphabet Diagrams

We introduce the notion of an *alphabet diagram*, which is a method of analysing parallel composition by means of the types of events that could occur or could not occur in its execution. An *alphabet diagram* is simply a Venn diagram that allows us to consider the different types of events that are involved in a parallel process throughout its lifetime. These events are subdivided into *synchronous* and *asynchronous* events. This subdivision allows us to characterise a parallel process depending on the presence or absence of these types of events. In [13], Schneider uses a related form of event diagram for parallel composition, but only to illustrate what events are initially possible.

As an example, consider the generalised parallel composition $P\|_{\Omega}Q$ of two processes P and Q, with alphabets A and B respectively, that are required to synchronise on a particular set of events Ω. The semantics of $P\|_{\Omega}Q$ is that P and Q are composed in parallel and are required to synchronise on every event in Ω, irrespective of their alphabets. Events which are not in Ω but which can be performed by either P or Q or both are asynchronous events. The event types for $P\|_{\Omega}Q$ are illustrated in the alphabet diagram of Figure 1 and defined below.

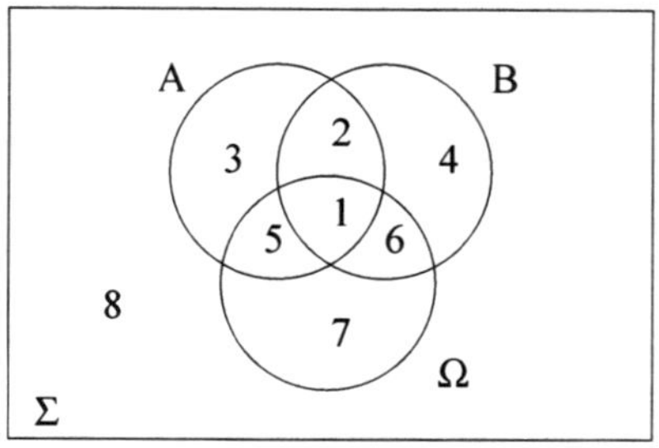

Figure 1. Alphabet diagram for $P\|_\Omega Q$.

1. *Possible synchronous* events $(A \cap B \cap \Omega)$: P and Q could synchronise on these since they are possible for both of them. Events of this type can only occur when both processes agree. They are called *possible* synchronous events because P and Q may never be willing to perform them at the same time, hence, they may never be performed.
2. *Common asynchronous* events $(A \cap B \cap \overline{\Omega})$: P and Q do not synchronise on these, even though they are possible for both of them. This type of event can be performed by either P or Q.
3. *P's private asynchronous* events $(A \cap \overline{B} \cap \overline{\Omega})$: only possible for P and since they are not in the synchronisation set they are performed independently of Q.
4. *Q's private* asynchronous events $(\overline{A} \cap B \cap \overline{\Omega})$: as for *P's*.
5. *P's inhibited synchronous* events $(A \cap \overline{B} \cap \Omega)$: only possible for P but they are also in the synchronisation set. Therefore, they require the participation of Q, which is impossible as they are not in Q's alphabet, hence, these events cannot be performed.
6. *Q's inhibited synchronous* events $(\overline{A} \cap B \cap \Omega)$: as for *P's*.
7. *Irrelevant synchronous* events $(\overline{A} \cap \overline{B} \cap \Omega)$: not in the alphabets of either process, and consequently cannot be performed.
8. *Irrelevant* events $(\overline{A} \cap \overline{B} \cap \overline{\Omega})$: not in the alphabets of either of the two processes nor are they required to be synchronised on.

Clearly *Irrelevant synchronous* events and *Irrelevant* events can never occur in the trace of any parallel process composed of P and Q because they are not in their alphabets. We therefore, only need to consider events which are actually possible for the two processes, i.e. within the union of their alphabets $A \cup B$, regions 1 to 6.

The eight types of events are the most which can occur for any given P, Q and Ω using this operator. However, for a particular instance of $P\|_\Omega Q$ only a subset of these types of events may be present. The presence or absence of these event types in a given instance of $P\|_\Omega Q$ can be used to classify different types of parallel composition.

3. Associativity of the Parallel Operators

In our search for an associativity law for each of the CSP_T's parallel operators we concentrate on finding an associativity law for the generalised parallel operator – $\|_\Omega$. The reason for this is that it is used to define two of the three operators: synchronous ($\|_\Delta$) and race ($|_\Theta$). Further more, any associativity law for this operator will form the basis for the associativity law for the asynchronous operator ($|||_\Theta$). In addition, for the purposes of this present investigation we treat tick (✓) as any other event that a process could perform.

3.1. Associativity of the Generalised Operator

The most important property required of $P\|_\Omega Q$ after it has been shown to be a well-defined and well-behaved process is *associativity*. That is, for what values of Λ_1, Λ_2, Π_1, Π_2, Γ_1 and Γ_2 does the following hold?

$$P\|_{\Lambda_1}(Q\|_{\Lambda_2}R) \equiv Q\|_{\Pi_1}(P\|_{\Pi_2}R) \equiv (P\|_{\Gamma_1}Q)\|_{\Gamma_2}R$$

We will refer to each of these processes as the (Λ), (Π) and (Γ) versions respectively.

It is obvious that constraints must be placed on the synchronisation sets, since combining processes using the existing parallel operators either in a different order or with different operators are rarely equivalent; that is, in general none of the following hold:

$$\begin{aligned} P\|(Q|||R) &\equiv (P\|Q)|||R \\ P|||(Q\|R) &\equiv (P|||Q)\|R \\ P|||(Q_B\|_C R) &\equiv (P|||Q)_{A\cup B}\|_C R \\ P|||(Q_B\|_C R) &\equiv (P\|Q)\|R \end{aligned}$$

3.2. Related Work

Roscoe [6] page 60, gives the following "weak (in that both interfaces are the same)" associativity law for his generalised parallel operator (that has asynchronous termination semantics).

$$P\|_X(Q\|_X R) = (P\|_X Q)\|_X R \qquad \langle\|_X\text{-assoc}\rangle$$

He states that it is difficult to "... construct a universally applicable and elegant associativity law.", for this type of operator due to the various types of events that can occur. In this context "universal" means that the synchronisation set X is independent of the alphabets of the three processes. He gives as an example, the process $P\|_X(Q\|_Y R)$ and an event that could occur in X but not in Y that both Q and R can perform. In what follows we shall denote this type of event as a *synchronous common* event, and denote it by *(QRa)Ps*, for processes similar to Roscoe's example. In addition, we shall show that if an event of this type is present in these types of processes then they cannot be transformed using a "universally applicable" associativity law. Schneider [13] gives a similar universal associativity law, for his "interface parallel" operator which has synchronous termination semantics.

Roscoe [6] gives an associativity law for his alphabetised parallel operator (with asynchronous termination) and gives an equivalence relating it to his generalised one as follows:

$$\begin{aligned} (P_A\|_B Q)_{A\cup B}\|_C R &= P_A\|_{B\cup C}(Q_B\|_C R) \qquad \langle_A\|_B\text{-assoc}\rangle \\ (P_A\|_B Q) &= P\|_{A\cap B}Q \end{aligned}$$

where A, B and C are $\alpha(P)$, $\alpha(Q)$ and $\alpha(R)$ respectively. By combining these two, it is possible to produce a less universal but stronger associativity law for the generalised operator than $\langle\|_X\text{-assoc}\rangle$. This new law would be applicable when all events in the intersections of the processes' alphabets are synchronised. However, it does not cover the cases when any common asynchronous events are present, for example, when an event in $A\cap B$ is required to be asynchronous, because by definition of $_A\|_B$ it must be synchronous.

Our aim in Section 3 is, therefore, to discover as general an associativity law as possible for our generalised parallel operator. Such laws are relevant to standard CSP because they capture intrinsic properties about parallel associativity and thus can be translated directly into associativity laws for the parallel operators of standard CSP.

3.3. Event Types for Three Processes

To identify the constraints on the synchronisation sets that are required to make the operator associative, we shall consider all of the possible event types that can occur when three processes are combined.

We have seen that when two processes are combined there are eight types of events but only six of these are relevant. When three are combined there are thirty two, of which twenty eight are relevant. Some of these are new types of events, others are just a natural extension of the existing types when three processes are combined. Below are listed all of the types of events that can occur, with a definition of the new types of events.

Private asynchronous events: are only performed by a single process, denoted for each process by – *Pa, Qa, Ra.*

Possible binary synchronous events: pairs of processes can synchronise on these. This excludes other forms of synchronous events such as ternary synchronous events etc. Denoted for each pair of processes by – *PQs, PRs, QRs.*

Common binary asynchronous events: are restricted to only two processes. Denoted for each pair of processes by – *PQa, PRa, QRa.*

Inhibited events: are due to the first synchronisation set that has effect on the process. Denoted for each process by – *Pi, Qi, Ri.* For example, in $P\|_{\Lambda_1}(Q\|_{\Lambda_2}R)$, *Pi* events are due to Λ_1, *Qi* and *Ri* events are due to Λ_2.

Inhibited private events: are private asynchronous events under the first synchronisation set but are then inhibited by the second synchronisation set which has effect on the process. Denoted for each process by – *(Pa)i, (Qa)i, (Ra)i.* For example, in $P\|_{\Lambda_1}(Q\|_{\Lambda_2}R)$ only *(Qa)i* and *(Ra)i* events are present, they are not in Λ_2 but are in Λ_1. No *(Pa)i* events are present since only one synchronisation set affects *P*.

Possible ternary synchronous events: all three processes are required to synchronise on these. Therefore, they are included in both synchronisation sets. Denoted by – *PQRs.*

Common ternary asynchronous events: all three processes can perform these asynchronously, hence, they are not included in either synchronisation set. Denoted by – *PQRa.*

Common synchronous events: are possible synchronous events because of the first synchronisation set but then become common asynchronous events with the third process. Denoted by – *(PQs)Ra, (PRs)Qa, (QRs)Pa.* Note that only one of these types of events can occur in any one of the three processes, for example, in $P\|_{\Lambda_1}(Q\|_{\Lambda_2}R)$ only *(QRs)Pa* events can occur.

Synchronous common events: are common asynchronous events under the first synchronisation set but then become possible synchronous events when combined with the third process. Denoted by – *(PQa)Rs, (PRa)Qs, (QRa)Ps.* Again only one of these types of event can be present in either of the three processes, for example, in $Q\|_{\Pi_1}(P\|_{\Pi_2}R)$ only *(PRa)Qs* events can occur.

Inhibited synchronous events: are possible synchronous events because of the first synchronisation set but are then required to be possible synchronous events with the third process which cannot perform them and are therefore, inhibited. Denoted by – *(PQs)i, (PRs)i, (QRs)i.* Only one of these event types can be present in each of the three processes, for example, *(QRs)i* in $P\|_{\Lambda_1}(Q\|_{\Lambda_2}R)$.

Inhibited common events: are common asynchronous events under the first synchronisation set but are then required to be possible synchronous events with the third process which cannot perform them and are therefore, inhibited. Denoted by – *(PQa)i, (PRa)i, (QRa)i.* Only one of these event types can be present, for example, *(PRa)i* in $Q\|_{\Pi_1}(P\|_{\Pi_2}R)$.

Irrelevant synchronous events: are not in the alphabets of the two processes being combined, and consequently have no effect on the resultant process. Denoted by – *PQis*, *PRis*, *QRis*.

Irrelevant events: are outside the alphabets of all three processes and the synchronisation sets involved, denoted by – *PQRi*.

3.3.1. Alphabet Diagram for Three Processes

Only certain combinations of these events can occur in one of the ways of combining P, Q and R. To see what part of the three processes' alphabets form each event type for each of the different combinations we give one alphabet diagram, see Figure 2, and use it to represent each of the three processes one at a time. That is S_1 and S_2 represent Λ_1, Λ_2, Π_1, Π_2, Γ_1 and Γ_2 respectively.

Note that even though a particular area is (visually) the same in the diagram under all three interpretations, this is because the same two groups of areas are used to represent the three pairs of synchronisation sets. The areas are not intended to be equal, as would be seen if four other groups of areas were introduced to represent the other pairs of synchronisation sets.

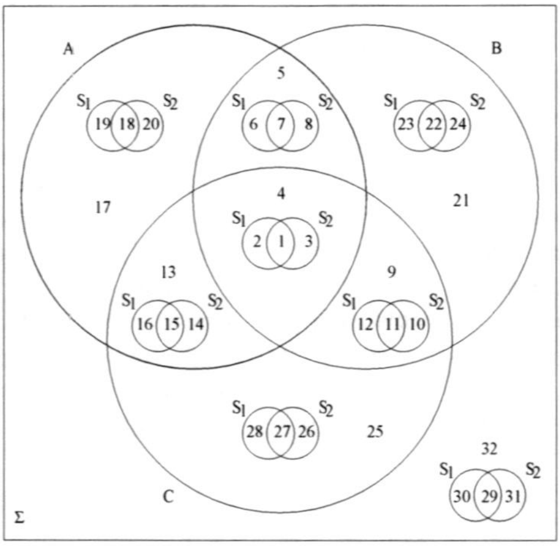

Figure 2. Alphabet diagram for three processes.

3.3.2. Tabulation of Event Types

Below in Tables 3 – 9 are tabulated the areas of the alphabet diagram of Figure 2, which comprise each of the event types for each of the three processes. Each area corresponds to a distinct region of the alphabet diagram, indicated by the number associated with it. If an event type is not possible then its area is given as the empty set. Due to limitations of space we have not included the tables for the various types of inhibited and irrelevant events[4]. We use the additional notation of Γ, Λ and Π as superscripts of the event types to indicate which of the three processes are under consideration, e.g. Pa^{Γ} represents the private asynchronous events of P in the Γ process.

[4]An extended version of this paper is in preparation and the full details of the inhibited and irrelevant events will be included there.

Table 3. Private asynchronous events: *Pa*, *Qa*, *Ra*

Event	Set of Events	*Area*
Pa^{Γ}	$(A \cap \overline{B} \cap \overline{C} \cap \overline{\Gamma_1} \cap \overline{\Gamma_2})$	17
Pa^{Λ}	$(A \cap \overline{B} \cap \overline{C} \cap \overline{\Lambda_1} \cap \Lambda_2) \cup (A \cap \overline{B} \cap \overline{C} \cap \overline{\Lambda_1} \cap \overline{\Lambda_2})$	20, 17,
	$\cup (A \cap B \cap \overline{C} \cap \overline{\Lambda_1} \cap \Lambda_2) \cup (A \cap \overline{B} \cap C \cap \overline{\Lambda_1} \cap \Lambda_2)$	8, 14
Pa^{Π}	$(A \cap \overline{B} \cap \overline{C} \cap \overline{\Pi_1} \cap \overline{\Pi_2})$	17
Qa^{Γ}	$(\overline{A} \cap B \cap \overline{C} \cap \overline{\Gamma_1} \cap \overline{\Gamma_2})$	21
Qa^{Λ}	$(\overline{A} \cap B \cap \overline{C} \cap \overline{\Lambda_1} \cap \overline{\Lambda_2})$	21
Qa^{Π}	$(A \cap B \cap \overline{C} \cap \overline{\Pi_1} \cap \Pi_2) \cup (\overline{A} \cap B \cap C \cap \overline{\Pi_1} \cap \Pi_2)$	8, 10,
	$\cup (\overline{A} \cap B \cap \overline{C} \cap \overline{\Pi_1} \cap \Pi_2) \cup (\overline{A} \cap B \cap \overline{C} \cap \overline{\Pi_1} \cap \overline{\Pi_2})$	24, 21
Ra^{Γ}	$(\overline{A} \cap B \cap C \cap \Gamma_1 \cap \overline{\Gamma_2}) \cup (\overline{A} \cap \overline{B} \cap C \cap \overline{\Gamma_1} \cap \overline{\Gamma_2})$	12, 25
	$\cup (\overline{A} \cap \overline{B} \cap C \cap \Gamma_1 \cap \overline{\Gamma_2}) \cup (A \cap \overline{B} \cap C \cap \Gamma_1 \cap \overline{\Gamma_2})$	28, 16
Ra^{Λ}	$(\overline{A} \cap \overline{B} \cap C \cap \overline{\Lambda_1} \cap \overline{\Lambda_2})$	25
Ra^{Π}	$(\overline{A} \cap \overline{B} \cap C \cap \overline{\Pi_1} \cap \overline{\Pi_2})$	25

Table 4. Common binary asynchronous events: *PQa*, *PRa*, *QRa*

Event	Set of Events	*Area*
PQa^{Γ}	$(A \cap B \cap \overline{C} \cap \overline{\Gamma_1} \cap \overline{\Gamma_2})$	5
PQa^{Λ}	$(A \cap B \cap \overline{C} \cap \overline{\Lambda_1} \cap \overline{\Lambda_2})$	5
PQa^{Π}	$(A \cap B \cap \overline{C} \cap \overline{\Pi_1} \cap \overline{\Pi_2})$	5
PRa^{Γ}	$(A \cap \overline{B} \cap C \cap \overline{\Gamma_1} \cap \overline{\Gamma_2})$	13
PRa^{Λ}	$(A \cap \overline{B} \cap C \cap \overline{\Lambda_1} \cap \overline{\Lambda_2})$	13
PRa^{Π}	$(A \cap \overline{B} \cap C \cap \overline{\Pi_1} \cap \overline{\Pi_2})$	13
QRa^{Γ}	$(\overline{A} \cap B \cap C \cap \overline{\Gamma_1} \cap \overline{\Gamma_2})$	9
QRa^{Λ}	$(\overline{A} \cap B \cap C \cap \overline{\Lambda_1} \cap \overline{\Lambda_2})$	9
QRa^{Π}	$(\overline{A} \cap B \cap C \cap \overline{\Pi_1} \cap \overline{\Pi_2})$	9

Table 5. Common ternary asynchronous events: *PQRa*

Event	Set of Events	*Area*
$PQRa^{\Gamma}$	$(A \cap B \cap C \cap \overline{\Gamma_1} \cap \overline{\Gamma_2})$	4
$PQRa^{\Lambda}$	$(A \cap B \cap C \cap \overline{\Lambda_1} \cap \overline{\Lambda_2})$	4
$PQRa^{\Pi}$	$(A \cap B \cap C \cap \overline{\Pi_1} \cap \overline{\Pi_2})$	4

3.4. Defining Associative Synchronisation Sets

For associativity we require the three parallel alternatives to be equivalent. For this to be the case a necessary condition is firstly that all three processes have the same event types present and secondly that each event type contains the same set of events in each process. We shall apply this constraint only to the event types that could actually occur and not to any particular type of inhibited events, because in general it is unimportant how an event is inhibited.

So from Table 3 it is clear that the *Pa*, *Qa* and *Ra* must be restricted in some way. Consider *Pa*, it contains events which are present in the (Λ) process which are not of the same type in the other two processes, i.e. areas 8, 14 and 20. Therefore, as a first step we require the following to hold:

Table 6. Possible binary synchronous events: *PQs*, *PRs*, *QRs*

Event	Set of Events	*Area*
PQs^{Γ}	$(A \cap B \cap \overline{C} \cap \Gamma_1 \cap \overline{\Gamma_2})$	6
PQs^{Λ}	$(A \cap B \cap \overline{C} \cap \Lambda_1 \cap \overline{\Lambda_2})$	6
PQs^{Π}	$(A \cap B \cap \overline{C} \cap \Pi_1 \cap \overline{\Pi_2})$	6
PRs^{Γ}	$(A \cap \overline{B} \cap C \cap \overline{\Gamma_1} \cap \Gamma_2)$	14
PRs^{Λ}	$(A \cap \overline{B} \cap C \cap \Lambda_1 \cap \overline{\Lambda_2})$	16
PRs^{Π}	$(A \cap \overline{B} \cap C \cap \overline{\Pi_1} \cap \Pi_2)$	14
QRs^{Γ}	$(\overline{A} \cap B \cap C \cap \overline{\Gamma_1} \cap \Gamma_2)$	10
QRs^{Λ}	$(\overline{A} \cap B \cap C \cap \overline{\Lambda_1} \cap \Lambda_2)$	10
QRs^{Π}	$(\overline{A} \cap B \cap C \cap \Pi_1 \cap \overline{\Pi_2})$	12

Table 7. Possible ternary synchronous events: *PQRs*

Event	Set of Events	*Area*
$PQRs^{\Gamma}$	$(A \cap B \cap C \cap \Gamma_1 \cap \Gamma_2)$	1
$PQRs^{\Lambda}$	$(A \cap B \cap C \cap \Lambda_1 \cap \Lambda_2)$	1
$PQRs^{\Pi}$	$(A \cap B \cap C \cap \Pi_1 \cap \Pi_2)$	1

Table 8. Common synchronous events: *(PQs)Ra*, *(PRs)Qa*, *(QRs)Pa*

Event	Set of Events	*Area*
$(PQa)Rs^{\Gamma}$	$(A \cap B \cap C \cap \overline{\Gamma_1} \cap \Gamma_2)$	3
$(PQa)Rs^{\Lambda}$	$\varnothing$	
$(PQa)Rs^{\Pi}$	$\varnothing$	
$(PRa)Qs^{\Gamma}$	$\varnothing$	
$(PRa)Qs^{\Lambda}$	$\varnothing$	
$(PRa)Qs^{\Pi}$	$(A \cap B \cap C \cap \Pi_1 \cap \overline{\Pi_2})$	2
$(QRa)Ps^{\Gamma}$	$\varnothing$	
$(QRa)Ps^{\Lambda}$	$(A \cap B \cap C \cap \Lambda_1 \cap \overline{\Lambda_2})$	2
$(QRa)Ps^{\Pi}$	$\varnothing$	

$$
\begin{aligned}
Pa &= (A \cap \overline{B} \cap \overline{C} \cap \overline{\Gamma_1} \cap \overline{\Gamma_2}) \\
&= (A \cap \overline{B} \cap \overline{C} \cap \overline{\Lambda_1} \cap \Lambda_2) \cup (A \cap \overline{B} \cap \overline{C} \cap \overline{\Lambda_1} \cap \overline{\Lambda_2}) \\
&\quad \cup (A \cap B \cap \overline{C} \cap \overline{\Lambda_1} \cap \Lambda_2) \cup (A \cap \overline{B} \cap C \cap \overline{\Lambda_1} \cap \overline{\Lambda_2}) \\
&= (A \cap \overline{B} \cap \overline{C} \cap \overline{\Pi_1} \cap \overline{\Pi_2})
\end{aligned}
$$

As a second step it is necessary to eliminate those events which are present in the (Λ) process which do not form part of *Pa* in the others. Therefore, the following equalities must hold:

$$
\begin{aligned}
Pa &= (A \cap \overline{B} \cap \overline{C} \cap \overline{\Gamma_1} \cap \overline{\Gamma_2}) \\
&= (A \cap \overline{B} \cap \overline{C} \cap \overline{\Pi_1} \cap \overline{\Pi_2}) \\
&= (A \cap \overline{B} \cap \overline{C} \cap \overline{\Lambda_1} \cap \Lambda_2) \cup (A \cap \overline{B} \cap \overline{C} \cap \overline{\Lambda_1} \cap \overline{\Lambda_2})
\end{aligned}
$$

Table 9. Synchronous common events: *(PQa)Rs*, *(PRa)Qs*, *(QRa)Ps*

Event	Set of Events	*Area*
$(PQs)Ra^{\Gamma}$	$(A \cap B \cap C \cap \Gamma_1 \cap \overline{\Gamma_2})$	2
$(PQs)Ra^{\Lambda}$	$\varnothing$	
$(PQs)Ra^{\Pi}$	$\varnothing$	
$(PRs)Qa^{\Gamma}$	$\varnothing$	
$(PRs)Qa^{\Lambda}$	$\varnothing$	
$(PRs)Qa^{\Pi}$	$(A \cap B \cap C \cap \overline{\Pi_1} \cap \Pi_2)$	3
$(QRs)Pa^{\Gamma}$	$\varnothing$	
$(QRs)Pa^{\Lambda}$	$(A \cap B \cap C \cap \overline{\Lambda_1} \cap \Lambda_2)$	3
$(QRs)Pa^{\Pi}$	$\varnothing$	

In addition, the following equality must hold, because there is no way that the *Pa* events of the (Γ) and (Π) versions could contain any events from the $A \cap B \cap \overline{C}$ and $A \cap \overline{B} \cap C$ areas.

$$(A \cap B \cap \overline{C} \cap \overline{\Lambda_1} \cap \Lambda_2) \cup (A \cap \overline{B} \cap C \cap \overline{\Lambda_1} \cap \Lambda_2) = \varnothing$$

Two possible restrictions that satisfy these equalities are the following:

$$A \cap \Lambda_2 = \varnothing \qquad \text{and} \qquad A \cap \overline{\Lambda_1} \cap \Lambda_2 = \varnothing$$

The problem with the first is that it would also eliminate the possibility of *PQRs* events, for this reason we choose the second. This leads to similar constraints for Γ_1, Γ_2, Π_1 and Π_2 derived from *Qa* and *Ra* as follows:

$$B \cap \overline{\Pi_1} \cap \Pi_2 = \varnothing \qquad\qquad C \cap \Gamma_1 \cap \overline{\Gamma_2} = \varnothing$$

From Table 8 it is clear that none of the events: *(PQa)Rs*, *(PRa)Qs* and *(QRa)Ps*, should occur in either of the three alternatives. Since there is no way in which they can occur in all of them. Therefore, constraints must be placed on Γ_1, Γ_2, Λ_1, Λ_2, Π_1 and Π_2 so that they are eliminated. If these events are eliminated then so are the corresponding inhibited ones: *(PQa)i*, *(PRa)i* and *(QRa)i*.

Similarly, from Table 9 we must eliminate: *(PQs)Ra*, *(PRs)Qa* and *(QRs)Pa*. However, the constraints that were introduced to eliminate the problems involving *Pa*, *Qa* and *Ra* also eliminate these events, as can be seen from Table 9.

Therefore, we must only find constraints to eliminate *(PQa)Rs*, *(PRa)Qs* and *(QRa)Ps*. So we require the following equalities to hold:

$$(QRa)Ps = (A \cap B \cap C \cap \Lambda_1 \cap \overline{\Lambda_2}) = \varnothing$$
$$(PQa)Rs = (A \cap B \cap C \cap \overline{\Gamma_1} \cap \Gamma_2) = \varnothing$$
$$(PRa)Qs = (A \cap B \cap C \cap \Pi_1 \cap \overline{\Pi_2}) = \varnothing$$

Just considering *(QRa)Ps* three possible restrictions which satisfy this equality are the following:

$$B \cap \Lambda_1 \cap \overline{\Lambda_2} = \varnothing \qquad \text{or} \qquad C \cap \Lambda_1 \cap \overline{\Lambda_2} = \varnothing \qquad \text{or} \qquad B \cap C \cap \Lambda_1 \cap \overline{\Lambda_2} = \varnothing$$

The problem with the first and second is that they would also eliminate the possibility of *PQs* and *PRs* events respectively (see Table 6), for this reason we choose the third. This leads to similar constraints for Γ_1, Γ_2, Π_1 and Π_2 derived from *(PQa)Rs* and *(PRa)Qs* as follows:

$$A \cap C \cap \Pi_1 \cap \overline{\Pi_2} = \varnothing \qquad A \cap B \cap \overline{\Gamma_1} \cap \Gamma_2 = \varnothing$$

So summarising, if Γ_1, Γ_2, Λ_1, Λ_2, Π_1 and Π_2 satisfy the above constraints then parts of event types and complete event types are eliminated. These events must be eliminated because if one of these categories of events were present in one of the three processes then it could not in general be equivalent to either of the other two irrespective of what values were chosen for the synchronisation sets. It is also the case that adopting these constraints reduces all of the equalities on the event types which can occur to equalities of just one area in all three processes.

3.5. Associativity Law

Using the constraints and equalities derived in the previous section as well as those from the tabulation of the event types we arrive at the following associativity law for the $\|_\Omega$ operator:

$$P\|_{W\cup X\cup Y}(Q\|_{W\cup Z}R) \;\equiv\; Q\|_{W\cup X\cup Z}(P\|_{W\cup Y}R) \;\equiv\; R\|_{W\cup Y\cup Z}(P\|_{W\cup X}Q) \tag{1}$$

where $W \subseteq \Sigma$, $A \cap Z = \varnothing$, $B \cap Y = \varnothing$, $C \cap X = \varnothing$ and A, B, C are the alphabets of P, Q and R respectively. Intuitively W represents the events that all three processes are required to synchronise, X represents the events that only P and Q are required to synchronise, Y represents the events that only P and R are required to synchronise and Z represents the events that only Q and R are required to synchronise. For an outline of a proof of this law see Appendix B.

From this, the appropriate values for the original synchronisation sets are as follows:

$$\begin{array}{ll} \Lambda_1 = W \cup X \cup Y & \Lambda_2 = W \cup Z \\ \Pi_1 = W \cup X \cup Z & \Pi_2 = W \cup Y \\ \Gamma_1 = W \cup X & \Gamma_2 = W \cup Y \cup Z \end{array}$$

3.5.1. Tabulation and Alphabet Diagram of Event Types

The alphabet diagram for the associative case for the three processes is given in Figure 3. This diagram represents all three cases at once, i.e. the areas are the same for the three versions. This differs from the diagram for the general cases, Figure 2, which represented one of the three cases at a time. The event types, for events that are not inhibited in some way, are given in Table 10 and for the associative case the following event type equalities must hold:

$$\begin{array}{lllllll} (PQa)Rs & = & (PRa)Qs & = & (QRa)Ps & = & \varnothing \\ (PQs)Ra & = & (PRs)Qa & = & (QRs)Pa & = & \varnothing \\ (PQa)i & = & (PRa)i & = & (QRa)i & = & \varnothing \end{array}$$

Again due to space limitations we do not include the detailed tabulation of the inhibited and irrelevant events. The events which form each event type for each process can easily be derived from those in Tables 3 – 9, by substituting in the above values for the synchronisation sets.

3.6. Associativity Laws for CSP_T Parallel Operators

Based on the associativity law for the generalised parallel operator $\|_\Omega$, we can now state (proofs similar) the associativity laws for the synchronous ($\|_\Delta$), asynchronous ($|||_\Theta$) and race ($|_\Theta$) parallel operators as follows:

$$P\|_{W\cup X\cup Y}(Q\|_{W\cup Z}R) \;\equiv\; Q\|_{W\cup X\cup Z}(P\|_{W\cup Y}R) \;\equiv\; R\|_{W\cup Y\cup Z}(P\|_{W\cup X}Q) \tag{2}$$

$$P|||_{W\cup X\cup Y}(Q|||_{W\cup Z}R) \;\equiv\; Q|||_{W\cup X\cup Z}(P|||_{W\cup Y}R) \;\equiv\; R|||_{W\cup Y\cup Z}(P|||_{W\cup X}Q) \tag{3}$$

$$P|_{W\cup X\cup Y}(Q|_{W\cup Z}R) \;\equiv\; Q|_{W\cup X\cup Z}(P|_{W\cup Y}R) \;\equiv\; R|_{W\cup Y\cup Z}(P|_{W\cup X}Q) \tag{4}$$

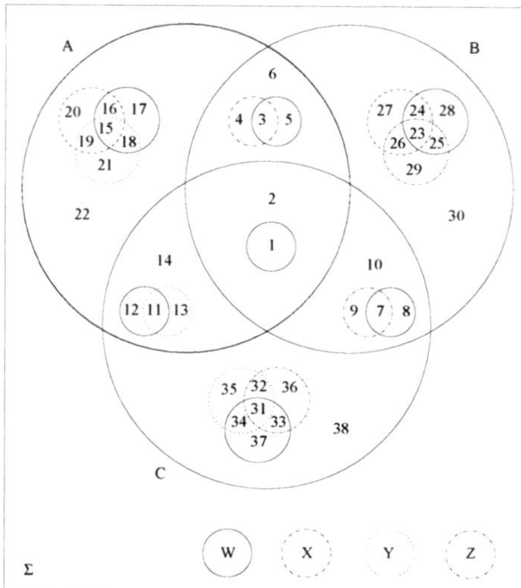

Figure 3. Alphabet diagram for associative case.

Table 10. Associative case: possible events.

Event	Set of Events	*Area*
Pa	$(A \cap \overline{B} \cap \overline{C} \cap \overline{W} \cap \overline{X} \cap \overline{Y})$	22
Qa	$(\overline{A} \cap B \cap \overline{C} \cap \overline{W} \cap \overline{X} \cap \overline{Z})$	30
Ra	$(\overline{A} \cap \overline{B} \cap C \cap \overline{W} \cap \overline{Y} \cap \overline{Z})$	38
PQa	$(A \cap B \cap \overline{C} \cap \overline{W} \cap \overline{X})$	6
PRa	$(A \cap \overline{B} \cap C \cap \overline{W} \cap \overline{Y})$	14
QRa	$(\overline{A} \cap B \cap C \cap \overline{W} \cap \overline{Z})$	10
PQRa	$(A \cap B \cap C \cap \overline{W})$	2
PQs	$(A \cap B \cap \overline{C} \cap \overline{W} \cap X)$	4
PRs	$(A \cap \overline{B} \cap C \cap \overline{W} \cap Y)$	13
QRs	$(\overline{A} \cap B \cap C \cap \overline{W} \cap Z)$	9
PQRs	$(A \cap B \cap C \cap W)$	1

In these three laws the meanings of W, X, Y, Z and A, B, C are as for $\|_\Omega$; and the same constraints apply to all three laws: $W \subseteq \Sigma$, $A \cap Z = \varnothing$, $B \cap Y = \varnothing$ and $C \cap X = \varnothing$.

However, due to the termination semantics of each operator there are additional constraints with respect to tick (✓). For the synchronous operator we require synchronous termination, therefore, $\checkmark \in W$. In the case of the asynchronous and race operators termination is not synchronised, i.e. ✓ is an asynchronous event, therefore, $\checkmark \notin W, X, Y, Z$.

3.7. *Testing for Associativity*

Now that an associativity law has been proved for the generalised operator, we can use it to determine whether an instance of either $P\|_{\Lambda_1}(Q\|_{\Lambda_2}R)$ or $Q\|_{\Pi_1}(P\|_{\Pi_2}R)$ or $(P\|_{\Gamma_1}Q)\|_{\Gamma_2}R$ can be transformed into an equivalent instance of the other two.

This equivalence preserving transformation can be achieved if the two synchronisation sets used to combine the three processes satisfy two conditions. The two conditions have been derived from the way in which the pairs of synchronisation sets in the associativity law are defined in terms of the sets W, X, Y, Z and the constraints placed on each of them. This question can also be answered using the laborious technique of working out the event types which occur and checking that none of them prohibit the transformation, i.e. occur only in that particular alternative.

The two conditions for the pair of synchronisation sets for a process of the form $P\|_{\Lambda_1}(Q\|_{\Lambda_2}R)$ are the following:

$$\begin{array}{ll}(1) & A\cap\overline{\Lambda_1}\cap\Lambda_2=\varnothing\\ (2) & B\cap C\cap\Lambda_1\cap\overline{\Lambda_2}=\varnothing\end{array}$$

If Λ_1 and Λ_2 satisfy these conditions then the process can be re-written as either of the other two forms, by using Λ_1 and Λ_2 to define W, X, Y and Z. It is possible to assign different values to these sets while still ensuring equivalence, the differences only affect how events are inhibited. One set of values for these variables is the following:

$$W=\Lambda_1\cap\Lambda_2 \qquad X=\overline{C}\cap\Lambda_1\cap\overline{\Lambda_2} \qquad Y=\overline{B}\cap\Lambda_1\cap\overline{\Lambda_2} \qquad Z=\overline{\Lambda_1}\cap\Lambda_2$$

These in turn are used to define the synchronisation sets for the other two processes as specified in the associativity law.

The conditions on the pairs of synchronisation sets and the definitions of W, X, Y and Z for the (Π) and (Γ) versions are as follows:

$$\begin{array}{ll}(1) & B\cap\overline{\Pi_1}\cap\Pi_2=\varnothing\\ (2) & A\cap C\cap\Pi_1\cap\overline{\Pi_2}=\varnothing\end{array}$$

$$W=\Pi_1\cap\Pi_2 \qquad X=\overline{C}\cap\Pi_1\cap\overline{\Pi_2} \qquad Y=\overline{\Pi_1}\cap\Pi_2 \qquad Z=\overline{A}\cap\Pi_1\cap\overline{\Pi_2}$$

$$\begin{array}{ll}(1) & C\cap\Gamma_1\cap\overline{\Gamma_2}=\varnothing\\ (2) & A\cap B\cap\overline{\Gamma_1}\cap\Gamma_2=\varnothing\end{array}$$

$$W=\Gamma_1\cap\Gamma_2 \qquad X=\Gamma_1\cap\overline{\Gamma_2} \qquad Y=\overline{B}\cap\overline{\Gamma_1}\cap\Gamma_2 \qquad Z=\overline{A}\cap\overline{\Gamma_1}\cap\Gamma_2$$

So for example, assuming Λ_1 and Λ_2 satisfy both conditions then the following holds:

$$P\|_{\Lambda_1}(Q\|_{\Lambda_2}R) \;\equiv\; Q\|_{\Pi_1}(P\|_{\Pi_2}R) \;\equiv\; (P\|_{\Gamma_1}Q)\|_{\Gamma_2}R$$

And the definition of the other synchronisation sets are as follows:

$$\begin{aligned}
\Pi_1 &= (\Lambda_1\cap\Lambda_2)\cup(\overline{C}\cap\Lambda_1\cap\overline{\Lambda_2})\cup(\overline{\Lambda_1}\cap\Lambda_2) &&= \Lambda_2\cup(\Lambda_1\cap\overline{C})\\
\Pi_2 &= (\Lambda_1\cap\Lambda_2)\cup(\overline{B}\cap\Lambda_1\cap\overline{\Lambda_2}) &&= (\Lambda_1\cap\Lambda_2)\cup(\Lambda_1\cap\overline{B})\\
\Gamma_1 &= (\Lambda_1\cap\Lambda_2)\cup(\overline{C}\cap\Lambda_1\cap\overline{\Lambda_2}) &&= (\Lambda_1\cap\Lambda_2)\cup(\Lambda_1\cap\overline{C})\\
\Gamma_2 &= (\Lambda_1\cap\Lambda_2)\cup(\overline{B}\cap\Lambda_1\cap\overline{\Lambda_2})\cup(\overline{\Lambda_1}\cap\Lambda_2) &&= \Lambda_2\cup(\Lambda_1\cap\overline{B})
\end{aligned}$$

4. Specifying and Implementing the Parallel Composition of Three Processes

In this section we show how to specify a parallel composition of three processes in terms of event types, and implement it using $\|_\Omega$ by defining the two synchronisation sets required.

The specification of the parallel composition of processes in terms of event types can be viewed as the specification of the inter-connections/interface between a network of three processes. Here we only consider a network of three processes, but we believe the present treatment could be extended to deal with a network consisting of an arbitrary number of processes.

This reverses the type of analysis performed during the search for the associativity law, where given the processes and two synchronisation sets, it was defined what type each event would be. The present approach is more useful when using $\|_\Omega$ to implement networks of processes, whereas the other is more useful when analysing existing ones.

The way in which we specify the parallel composition and interface of three processes in terms of event types is done in three stages:

1. Define the alphabets A, B and C of the three processes P, Q and R.
2. Determine how the events in each of the following seven regions $A \cap \overline{B} \cap \overline{C}$, $\overline{A} \cap B \cap \overline{C}$, $\overline{A} \cap \overline{B} \cap C$, $A \cap B \cap \overline{C}$, $A \cap \overline{B} \cap C$, $\overline{A} \cap B \cap C$ and $A \cap B \cap C$ should be partitioned into event types
3. Define the two synchronisation sets using these partitions.

We shall now give an example of how to apply these three stages for the associative case.

4.1. Example for the Associative Case

Our aim is to illustrate how to choose two interfaces (synchronisation sets) I_1 and I_2, such that we can apply the associativity law to the following simple network:

$$P\|_{I_1}(Q\|_{I_2}R)$$

Thus, we need to ensure that the event types that cause the associativity law to fail, for this form of network, cannot arise with our chosen interfaces.

Stage one is to define the alphabets for the three processes. For this example, we chose the arbitrary alphabets A, B and C respectively. Hence, based on our previous investigation, we must ensure that certain categories of *Pa* events and both *(QRa)Ps* and *(QRs)Pa* events do not occur. Therefore, we need to define the sets W, X, Y and Z as used in law (1) to define the two interfaces as follows:

$$I_1 = W \cup X \cup Y \qquad \text{and} \qquad I_2 = W \cup Z$$

where the W, X, Y and Z have the same intuitive meanings and satisfy the same constraints as given previously. By defining these four sets in this way, we are ensuring that the associativity law is applicable for our two interfaces.

For stage two, we use Table 11 to tabulate the events and components of events which fall into the seven areas. Note that we have ignored *irrelevant* synchronisation events and *impossible* events. As an example of how to read the table consider the region $A \cap \overline{B} \cap \overline{C}$; if the (Λ) form of process has been chosen the events in this area will be either *Pa* or *Pi* events, depending on how it is partitioned. A diagrammatic view of how the seven areas are partitioned for the (Λ) process is given in Figure 4.

There are a number of equalities which hold between the event types in each alternative and between the three alternatives. These additional equalities hold because we are dealing with the associative case. For simplicity we do not distinguishing between the different ways in which an event is inhibited, since for most cases this is irrelevant. Therefore, we group together the various categories of inhibited events using the following six equalities denoted by *Pw*, *Qw*, *Rw*, *PQw*, *PRw* and *QRw*. These inhibited events are then included in the synchronisation set W.

Table 11. Associate case: partitioned events

Area	(Λ)	(Π)	(Γ)
$A \cap \overline{B} \cap \overline{C}$	Pa, Pi_1	$Pa, Pi_1, (Pa)i$	$Pa, Pi_1, (Pa)i$
$\overline{A} \cap B \cap \overline{C}$	$Qa, Qi_1, (Qa)i$	Qa, Qi_1	$Qa, Qi_1, (Qa)i$
$\overline{A} \cap \overline{B} \cap C$	$Ra, Ri_1, (Ra)i$	$Ra, Ri_1, (Ra)i$	Ra, Ri_1
$A \cap B \cap \overline{C}$	PQa, PQs, Pi_2, Qi_2	PQa, PQs, Pi_2, Qi_2	$PQa, PQs, (PQs)i$
$A \cap \overline{B} \cap C$	PRa, PRs, Pi_3, Ri_2	$PRa, PRs, (PRs)i$	PRa, PRs, Pi_2, Ri_2
$\overline{A} \cap B \cap C$	$QRa, QRs, (QRs)i$	QRa, QRs, Qi_3, Ri_2	QRa, QRs, Qi_2, Ri_3
$A \cap B \cap C$	$PQRa, PQRs$	$PQRa, PQRs$	$PQRa, PQRs$

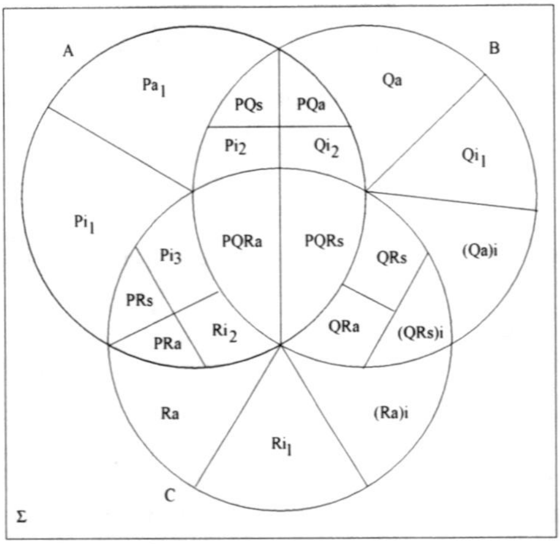

Figure 4. Partition of events for associative case of $P\|_{\Lambda_1}(Q\|_{\Lambda_2}R)$.

$$
\begin{array}{lrcl}
A \cap \overline{B} \cap \overline{C}: & Pw = Pi_1^{\Lambda} & = Pi_1^{\Pi} \cup (Pa)i^{\Pi} & = Pi_1^{\Gamma} \cup (Pa)i^{\Gamma} \\
\overline{A} \cap B \cap \overline{C}: & Qw = Qi_1^{\Lambda} \cup (Qa)i^{\Lambda} & = Qi_1^{\Pi} & = Qi_1^{\Gamma} \cup (Qa)i^{\Gamma} \\
\overline{A} \cap \overline{B} \cap C: & Rw = Ri_1^{\Lambda} \cup (Ra)i^{\Lambda} & = Ri_1^{\Pi} \cup (Ra)i^{\Pi} & = Ri_1^{\Gamma} \\
A \cap B \cap \overline{C}: & PQw = Pi_2^{\Lambda} = Qi_2^{\Lambda} & = Pi_2^{\Pi} = Qi_2^{\Pi} & = (PQs)i^{\Gamma} \\
A \cap \overline{B} \cap C: & PRw = Pi_3^{\Lambda} = Ri_2^{\Lambda} & = (PRs)i^{\Pi} & = Pi_2^{\Gamma} = Ri_2^{\Gamma} \\
\overline{A} \cap B \cap C: & QRw = (QRs)i^{\Lambda} & = Qi_3^{\Pi} = Ri_2^{\Pi} & = Qi_2^{\Gamma} = Ri_3^{\Gamma}
\end{array}
$$

Finally for stage three, we define the sets W, X, Y and Z to be used to define the interfaces I_1 and I_2 as follows:

$$W = PQRs \cup (Pw \cup Qw \cup Rw \cup PQw \cup PRw \cup QRw)$$

$$X = PQs \qquad Y = PRs \qquad Z = QRs$$

There are obviously other ways of combining the synchronous and inhibited events to achieve the desired result but we leave the definition of these to the reader.

5. Further Work

We have demonstrated how to use $\|_{\Omega}$ to implement a network of three processes with specific interfaces between the processes. These interfaces were defined in terms of the event types of the processes' alphabets. This approach entailed defining the processes' alphabets, then partitioning up the alphabets into the required event types and finally defining the two synchronisation sets using these event types. An obvious extension of this work is to develop a method, and probably more importantly some succinct notation, that would allow an arbitrary number of processes to be constructed into a network with the required event types present. For example, with n processes the network would have the form:

$$P_1 \parallel_{\Omega_1} (P_2 \parallel_{\Omega_2} (\ldots (P_{n-1} \parallel_{\Omega_{n-1}} P_n) \ldots)$$

The notation would need to be able to deal with the n alphabets of the processes, the $n-1$ synchronisation sets and the 2^{2n-1} different event types that could occur, although most of these event types would be classed as *inhibited*.

The process of constructing the required network would then be similar to that for three processes given in Section 4, and in outline would be as follows:

1. Define the alphabets $A_1, \ldots, A_n$ of the n processes, $P_1, \ldots, P_n$.
2. Determine how the events in each of the 2^n regions $A_1 \cap \overline{A}_2 \cap \ldots \cap \overline{A}_n$ to $\overline{A}_1 \cap \ldots \cap \overline{A}_{n-1} \cap A_n$ should be partitioned into event types.
3. Finally, define the $n-1$ synchronisation sets $\Omega_1, \ldots, \Omega_{n-1}$ using these partitions.

If the network was required to be "associative", in the sense that the order in which the processes were to be composed was irrelevant, then its construction would be greatly simplified. Since, the sets of synchronisation events between each pair of processes is one of the most important factors in ensuring associativity, we could use $X_{i,j}$ to denote the required set of synchronous events between P_i and P_j. Then for each $X_{i,j}$ require that it is disjoint with all other processes' alphabets in the network, i.e. the following condition should hold:

$$X_{i,j} \cap (\bigcup_{k \neq i,j} A_k) = \varnothing$$

We are grateful to one of the anonymous reviewers who raised the issue of whether it is possible to give an indication of the "order of magnitude" of the different types of events present. Currently, we are only able to provide an answer for the most important event types with respect to ensuring the associativity law holds, namely the *pure* synchronous and asynchronous events. In this context, pure means those types that are performed either synchronously by all participating processes, e.g. *PQRs*, or asynchronously by all processes, e.g. *PQRa*. For these we are able to state that: for n processes the total number of different (pure) synchronous event types is $2^n - (n+1)$ and for (pure) asynchronous event types it is $2^n - 1$. We leave the derivation of similar results for the more complex case for *mixed* event types, i.e. those that are performed synchronously by some processes and asynchronously by others, e.g. *(PQa)Rs*, as a topic for future research.

We are grateful to two of the anonymous reviewers who raised issues regarding the constraints on the two synchronisation sets that are required to ensure that the associativity law holds. In particular, are the conditions we give in Section 3.4 necessary? They are clearly sufficient, but currently we are unable to say with any confidence that they are also necessary. However, given that we consider a number of alternative conditions, and that one reviewer has also suggested:

$$\bar{C} \cap \Lambda_2 = \bar{B} \cap \Lambda_2 = \varnothing \qquad \text{(equivalently} \quad \Lambda_2 \subseteq B \cap C),$$

as an alternative constraint to eliminate the problem *Pa* events; we strongly suspect that they are not.

6. Summary and Conclusions

We have extended previous work describing our language and model CSP$_T$ for specifying the behaviour of networks of processes which is a variant (a variant specifically chosen to be as close as possible to the original formulation) of the original language of CSP developed by Hoare, Roscoe and others. Specifically, we have extended our analysis by the discovery and proof of associativity laws for the three parallel operators. By developing these laws we provide specifiers and designers with essential laws for specifying, designing, analysing and reasoning about networks of parallel processes.

In particular, our analysis of the event types which can occur in processes of the form $(P \|_{\Gamma_1} Q) \|_{\Gamma_2} R$, resulted in the discovery of constraints on the pair of synchronisation sets Γ_1 and Γ_2, which when satisfied allowed the definition of an associativity law. These constraints also provided us with a simple "test" for associativity. As a result of this analysis we were able to prove the general associativity law for $\|_\Omega$ and hence, for $\|_\Delta$, $|||_\Theta$ and $|_\Theta$. We believe these laws, although not "universally applicable" in the sense of Roscoe [5], that it is the most general and useful possible for these style of operators.

Furthermore, we demonstrated how to use $\|_\Omega$ to implement a network of three processes with specific interfaces (synchronisation sets) between the processes. These interfaces are defined in terms of the event types of the processes' alphabets. This approach entails defining the processes' alphabets, then partitioning up the alphabets into the required event types and finally defining the two interfaces using these event types. Clearly, this would be a useful technique when attempting to construct networks of processes using $\|_\Omega$. We believe this approach could be scaled up to deal with an arbitrary number of processes, with the aid of some well chosen notation, but we leave the details as a topic for further investigation.

Based on our investigations, we suggest the following general guideline for system designers, when choosing interfaces for these type of networks: if one wishes to be able to apply an associativity law, then one should avoid any interface that results in *mixed* event types, as these can only occur in networks that are constructed in particular ways.

Note that associativity is not always desired and that, in those cases, events can be (and need to be) of mixed type. For example, client-server systems have many client processes interleaving access to a pair of request and reply channels: first they send a request, then wait for a reply. Their use of those channels is asynchronous between them. The set of client processes (i.e. their parallel composition) now synchronises with the (single) server process on those request and reply channels: the server waits for a request, then sends a reply. The reply will be taken by the client that made the request since it will be the only client listening - the other clients are either blocked trying to make a request or doing something else. So, the request and reply channel interfaces are mixed: asynchronous between the clients and synchronous between the client group and the server. Associating the component processes in some other way (e.g. with some clients grouped with the server and some just with themselves) gives a system that is semantically *different* and incorrect[5].

However, when associativity matters, by identifying the associativity constraints and proving the set of associativity laws, we have provided a practical technique for designers to capture intuitively and flexibly the intended behaviour of the processes networks they wish to build.

Acknowledgements

We would like to acknowledge and thank the three anonymous referees for their feedback on the paper. We found this very helpful in improving the paper overall, but especially in

[5]This example was suggested by one of the anonymous reviewers.

improving the readability of the paper. We would also like to gratefully acknowledge the advice and assistance of the following individuals from whom at various stages we benefited a great deal over several very detailed discussions on our work: Richard Bornat, Jonathan Bowen, Michael Luck and Steve Schneider. Finally, we would like to acknowledge the help and assistance of the CPA 2013 Editors, especially Peter Welch, in preparing this paper.

References

[1] S. D. Brookes. *A Model for Communicating Sequential Processes*. PhD thesis, Oxford University, 1983.
[2] S. D. Brookes, C. A. R. Hoare, and A. W. Roscoe. A Theory of Communicating Sequential Processes. *Journal of the ACM*, 31(7), 1985.
[3] S. D. Brookes and A. W. Roscoe. An improved failures model for Communicating Sequential Processes. In *Proceedings of Pittsburgh Seminar on Concurrency*, LNCS 197, pages 281–305. Springer-Verlag, 1985.
[4] C. A. R. Hoare. *Communicating Sequential Processes*. Prentice-Hall International, 1985. ISBN: 0-131-53271-5.
[5] A. W. Roscoe. *The Theory and Practice of Concurrency*. Prentice-Hall International, 1997. ISBN: 0-13-674409-5.
[6] A.W. Roscoe. *Understanding Concurrent Systems*. Springer, 2010. ISBN: 978-1-84882-257-3.
[7] P. Howells and M. d'Inverno. A CSP model with flexible parallel termination semantics. *Formal Aspects of Computing*, 21(5):421–449, 2009.
[8] P. Howells and M. d'Inverno. Specifying Termination in CSP. *Theoretical Computer Science*, 2013. DOI: 10.1016/j.tcs.2013.05.008.
[9] P. Howells and Mark d'Inverno. Successful Termination in Timed CSP. In *Proceedings of Communicating Process Architectures 2013 (CPA13)*, 2013.
[10] P. Howells. *Communicating Sequential Processes with Flexible Parallel Termination Semantics*. PhD thesis, University of Westminster, 2005.
[11] H. Tej and B. Wolff. A Corrected Failure-Divergence Model for CSP in Isabelle/HOL. In *Proceedings of the FME '97 – Industrial Applications and Strengthened Foundations of Formal Methods*, LNCS 1313. Springer-Verlag, 1997.
[12] C. A. R. Hoare and He Jifeng. *Unifying Theories of Programming*. Prentice-Hall, 1998.
[13] S. Schneider. *Concurrent and Real-time Systems: The CSP Approach*. Wiley, 2000. ISBN: 0-471-62373-3.
[14] J. W. de Bakker. *Mathematical Theory of Program Correctness*. Prentice-Hall International, 1980.
[15] J. E. Stoy. *Denotational Semantics: The Scott-Strachey Approach to Programming Language Theory*. MIT Press, 1977.

A. (Appendix) CSP_T: Notation, Axioms and Semantic Definitions

This appendix contains a summary of the notation and definitions used in the paper and the proof of the associativity law for the generalised parallel operator $\|_\Omega$. For a more detailed description of CSP, see [4,5,6] and for CSP_T see [7,8].

A.1. Basic Notation

The notation used in CSP and CSP_T is the following. Σ is the set of all events, and is countable; denoted by a, b, c. $\mathbb{P}(\Sigma)$ is the power set of Σ; denoted by X, Y, Z. Σ^* is the set of finite sequences of events, i.e., *traces*; denoted by r, s, t, u. $\langle\,\rangle$ represents the empty sequence. $\langle a, b, c, d \rangle$ represents the sequence with members a, b, c, d. $s \frown t$ represents the concatenation of the two sequences s and t. $s \leq t$ is the sequence *prefix* relation and $s < t$ is the proper *prefix* relation. $\#s$ is the length of the sequence s. a in s is sequence membership. $\overline{X}$ represents the set complement of X, with respect to Σ, i.e., $\overline{X} \cap X = \varnothing$ and $\overline{X} \cup X = \Sigma$. $X - Y$ is the set difference of X and Y, i.e., $X - Y = X \cap \overline{Y}$. Finally, processes are denoted by P, Q, R and A, B, C denote their alphabets. The set of all process identifiers is denoted by Ide, and use $p, q, \ldots$ to range over Ide.

A.2. Axioms for Failures and Divergences

In this model a process P is denoted by an ordered pair $\langle F, D\rangle$, where F is the *failure set* of P, and D is the *divergence set* of P. A failure set F and divergence set D are any sets that satisfy the following conditions:

$$F \subseteq \Sigma^* \times \mathbb{P}(\Sigma) \qquad\qquad D \subseteq \Sigma^*$$

In addition, they must also satisfy the following *process axioms*, including the CSP_T *Termination* axiom (T1):

$$s \in D \Rightarrow s^\frown t \in D \tag{D1}$$

$$s \in D \Rightarrow (s^\frown t, X) \in F \tag{D2}$$

$$(\langle\rangle, \varnothing) \in F \tag{N1}$$

$$(s^\frown t, \varnothing) \in F \Rightarrow (s, \varnothing) \in F \tag{N2}$$

$$(s, X) \in F \wedge Y \subseteq X \Rightarrow (s, Y) \in F \tag{N3}$$

$$(s, X) \in F \wedge (\forall c \in Y : (s^\frown\langle c\rangle, \varnothing) \notin F) \Rightarrow (s, X \cup Y) \in F \tag{N4}$$

$$(\forall Y \in \mathbb{F}(X) : (s, Y) \in F) \Rightarrow (s, X) \in F \tag{N5}$$

$$t \neq \langle\rangle \wedge (s^\frown\langle\checkmark\rangle^\frown t, \varnothing) \in F \;\Rightarrow\; s \in D \tag{T1}$$

For a natural language interpretation of these axioms, see the Appendix of [9].

Definition A.1 The traces and alphabet of a process are now defined as follows:

$$\mathsf{traces}(P) = \{ s \mid \exists X \in \mathbb{P}(\Sigma) : (s, X) \in F \}$$
$$\alpha(P) = \{ a \mid \exists\, t \in \mathsf{traces}(P) : a\, in\, t \}$$

A.3. Nondeterministic (refinement) ordering on processes

The *refinement ordering* on CSP processes is defined in terms of the reverse subset ordering on the failure and divergence set components of processes, (the details of the underlying theory can be found in [14,15]), and is defined as follows.

Definition A.2 The refinement ordering $\sqsubseteq$ is defined as follows:

$$\langle F_1, D_1\rangle \sqsubseteq \langle F_2, D_2\rangle \;\;\widehat{=}\;\; F_1 \supseteq F_2 \wedge D_1 \supseteq D_2$$

where $\langle F_1, D_1\rangle$ and $\langle F_2, D_2\rangle$ represent the processes P_1 and P_2 respectively.

$P_1 \sqsubseteq P_2$, (*P_1 is refined by P_2*) means that P_1 is *more nondeterministic* than P_2 if it can diverge whenever P_2 can diverge and fail whenever P_2 can fail, i.e. any behaviour which is possible for P_2 is also possible for P_1 and that P_1 may also exhibit behaviour which is not possible for P_2.

Definition A.3 The ordering $\sqsubseteq$ induces an equivalence relation $\equiv$ on processes defined as follows:

$$P \equiv Q \;\;\widehat{=}\;\; P \sqsubseteq Q \wedge Q \sqsubseteq P$$

Hence two processes are equivalent in the model N_T when their failure and divergent set components are equal. This space of processes together with the ordering, i.e., ($PROC_T$, $\sqsubseteq$) is a *partially ordered set*, with least element $\bot$.

A.4. Semantic functions for CSP_T

This section contains the semantics functions $\mathcal{F}$ and $\mathcal{D}$ for the generalised parallel operator $P\|_\Omega Q$. The CSP_T semantic functions for the other "standard" processes and operators, the two new parallel operators: $P|||_\Theta Q$ and $P|_\Theta Q$, can be found in [7] and for CSP in [4,5,6].

First, we define an environment that is used to define process identifiers and the failure and divergence semantics functions as follows.

Definition A.4 An *environment e* is a function which maps process identifiers (Ide) to failure and divergence pairs $\langle F, D\rangle$ as follows.

$$e : \text{Ide} \rightarrow \mathbb{P}(\Sigma^* \times \mathbb{P}(\Sigma)) \times \mathbb{P}(\Sigma^*)$$

The set of all such environments is denoted by NEnv.

Definition A.5 The semantic function $\mathcal{N}$ for CSP_T is defined as follows:

$$\mathcal{N} : \mathbf{CSP}_T \rightarrow [\text{NEnv} \rightarrow N_T]$$
$$\mathcal{N}[\![P]\!]e = \langle \mathcal{F}[\![P]\!]e, \mathcal{D}[\![P]\!]e\rangle$$

where the two functions $\mathcal{F}$ and $\mathcal{D}$ are defined as follows:

$$\mathcal{F} : \mathbf{CSP}_T \rightarrow [\text{NEnv} \rightarrow \mathbb{P}(\Sigma^* \times \mathbb{P}(\Sigma))]$$
$$\mathcal{D} : \mathbf{CSP}_T \rightarrow [\text{NEnv} \rightarrow \mathbb{P}(\Sigma^*)]$$

We now give the definition of the semantic functions $\mathcal{F}$ and $\mathcal{D}$ for $\|_\Omega$ (and hence $\|_\Delta$). The semantic functions for $P\|_\Omega Q$ use the merge function to construct the *legal* traces for the parallel operator. merge takes three arguments, two traces and a synchronisation set, and maps them to the (possibly empty) set of traces. If the result of merging two traces, *s* and *t*, is the empty set, then the two traces are *incompatible*, because of synchronous events.

Definition A.6 The function $\mathsf{merge}(s, t, \Omega)$ is defined as follows:

$$\mathsf{merge} : \Sigma^* \times \Sigma^* \times \mathbb{P}(\Sigma) \rightarrow \mathbb{P}(\Sigma^*)$$

In the following $a \neq b$.

$$\mathsf{merge}(\langle a\rangle^\frown s, \langle b\rangle^\frown t, \Omega) = \mathsf{merge}(\langle b\rangle^\frown t, \langle a\rangle^\frown s, \Omega) \qquad [a, b \notin \Omega]$$
$$= \{\langle a\rangle^\frown u \mid u \in \mathsf{merge}(s, \langle b\rangle^\frown t, \Omega)\}$$
$$\cup \{\langle b\rangle^\frown u \mid u \in \mathsf{merge}(\langle a\rangle^\frown s, t, \Omega)\}$$
$$\mathsf{merge}(\langle a\rangle^\frown s, \langle a\rangle^\frown t, \Omega) = \{\langle a\rangle^\frown u \mid u \in \mathsf{merge}(s, t, \Omega)\} \qquad [a \in \Omega]$$
$$\mathsf{merge}(\langle a\rangle^\frown s, \langle b\rangle^\frown t, \Omega) = \mathsf{merge}(\langle b\rangle^\frown t, \langle a\rangle^\frown s, \Omega) \qquad [a \notin \Omega, b \in \Omega]$$
$$= \{\langle a\rangle^\frown u \mid u \in \mathsf{merge}(s, \langle b\rangle^\frown t, \Omega)\}$$
$$\mathsf{merge}(\langle a\rangle^\frown s, \langle b\rangle^\frown t, \Omega) = \varnothing \qquad [a, b \in \Omega]$$
$$\mathsf{merge}(\langle\rangle, \langle a\rangle^\frown s, \Omega) = \mathsf{merge}(\langle a\rangle^\frown s, \langle\rangle, \Omega) \qquad [a \notin \Omega]$$
$$= \{\langle a\rangle^\frown u \mid u \in \mathsf{merge}(\langle\rangle, s, \Omega)\}$$
$$\mathsf{merge}(\langle\rangle, \langle a\rangle^\frown s, \Omega) = \mathsf{merge}(\langle a\rangle^\frown s, \langle\rangle, \Omega) = \varnothing \qquad [a \in \Omega]$$
$$\mathsf{merge}(\langle\rangle, \langle\rangle, \Omega) = \{\langle\rangle\}$$

We now define the failure and divergence set semantic functions $\mathcal{F}$ and $\mathcal{D}$ for $P\|_\Omega Q$.

$$\begin{aligned}\mathcal{D}[\![P\|_\Omega Q]\!]e =& \{\, u^\frown v \mid \exists\, s,t : u \in \mathsf{merge}(s,t,\Omega) \wedge ((s \in \mathcal{D}[\![P]\!]e \wedge t \in \mathsf{traces}(Q)) \\ & \qquad \vee (s \in \mathsf{traces}(P) \wedge t \in \mathcal{D}[\![Q]\!]e))\,\} \\ \mathcal{F}[\![P\|_\Omega Q]\!]e =& \{\, (u, X \cup Y \cup Z) \mid X \subseteq \overline{\Omega} \wedge Y, Z \subseteq \Omega \wedge \exists\, s,t : u \in \mathsf{merge}(s,t,\Omega) \\ & \qquad \wedge (s, X \cup Y) \in \mathcal{F}[\![P]\!]e \wedge (t, X \cup Z) \in \mathcal{F}[\![Q]\!]e \,\} \\ & \cup \{\, (u, X) \mid u \in \mathcal{D}[\![P\|_\Omega Q]\!]e \wedge X \in \mathbb{P}(\Sigma) \,\}\end{aligned}$$

B. (Appendix) Proof of the Associativity Law

In this section we present an outline of just the following part of the proof of:

$$P\|_{W\cup X\cup Y}(Q\|_{W\cup Z}R) \;\equiv\; R\|_{W\cup Y\cup Z}(P\|_{W\cup X}Q) \tag{1}$$

The proof of equivalence of the other process with either of these two is similar. Equation (1) is proved[6] by showing that the divergence and failure sets of the two sides are equal. First, we need some lemmas.

B.1. Merge Lemmas

To prove the associativity law we require the following merge lemmas: symmetry, trace prefix and associativity.

$$\mathsf{merge}(s,t,\Omega) = \mathsf{merge}(t,s,\Omega) \qquad \text{(merge-Sym)}$$

$$s^\frown t \in \mathsf{merge}(u,w,\Omega) \Rightarrow \exists\, u', w' : u' \le u \wedge w' \le w \wedge s \in \mathsf{merge}(u', w', \Omega) \qquad \text{(merge-Prefix)}$$

[6]One of the anonymous reviewers suggested that our law (1) was a special case of Roscoe's $\langle{}_A\|_B\text{−assoc}\rangle$ law, see Section 3.2, and suggested the following proof:

Let D, E, F be the private asynchronous events of P, Q, R. Then the left-hand side equals:

$$P\;{}_{D\cup W\cup X\cup Y}\|_{E\cup F\cup W\cup X\cup Y\cup Z}\;(Q\;{}_{E\cup W\cup X\cup Z}\|_{F\cup W\cup Y\cup Z}\;R)$$

using the law relating alphabetised and generalised parallel (see Section 3.2). Then by Law $\langle{}_A\|_B\text{−assoc}\rangle$, the above equals:

$$(P\;{}_{D\cup W\cup X\cup Y}\|_{E\cup W\cup X\cup Z}\;Q)\;{}_{D\cup E\cup W\cup X\cup Y\cup Z}\|_{F\cup W\cup Y\cup Z}\;R$$

Which equals the right-hand side, again by the law relating alphabetised and generalised parallel. In particular, the above shows that the law you produce is just a particular instance of the law for alphabetised parallel.

We agree that this equality above is correct. However, we believe that the sets used do not include any of the shared asynchronous events between the three processes, i.e. *PQa*, *PRa*, *QRa* or *PQRa*. Furthermore, if these were included in the above example, then due to the definition of this operator they would automatically become synchronised events, since they are all within the intersections of the sets used. In particular, (assuming that the W, X, Y are as defined in our law) then $D \cup W \cup X \cup Y$ does not necessarily define the alphabet of P. Since even with the inclusion of P's private asynchronous events D, this set does not include P's shared asynchronous events with either Q (*PQa*) or R (*PRa*) or both (*PQRa*). Therefore, we believe that Roscoe's law $\langle\|_X\text{−assoc}\rangle$ is a special case of our associativity law, since ours also deals with this more general case when shared asynchronous events are present, which Roscoe's law does not.

$$\begin{aligned} &\{\, u \mid u \in \mathsf{merge}(p, x, W \cup X \cup Y) \wedge x \in \mathsf{merge}(q, r, W \cup Z)\,\} && \text{(merge-Ass)} \\ =&\{\, u \mid u \in \mathsf{merge}(q, y, W \cup X \cup Z) \wedge y \in \mathsf{merge}(p, r, W \cup Y)\,\} \\ =&\{\, u \mid u \in \mathsf{merge}(r, z, W \cup Y \cup Z) \wedge z \in \mathsf{merge}(p, q, W \cup X)\,\} \end{aligned}$$

where W, X, Y, Z, A, B and C are as before and $p \in A^*$, $q \in B^*$ and $r \in C^*$.

We omit the poofs of these lemmas as they are straightforward (by structural induction) and lengthy. The details can be found in [10]. □

B.2. Proof of Divergence Set Equality

We need to prove the following:

$$\mathcal{D}[\![P\|_{W\cup X\cup Y}(Q\|_{W\cup Z}R)]\!]e \;=\; \mathcal{D}[\![R\|_{W\cup Y\cup Z}(P\|_{W\cup X}Q)]\!]e$$

This proof is split into the following two parts:

(a) $s \in \mathcal{D}[\![P\|_{W\cup X\cup Y}(Q\|_{W\cup Z}R)]\!]e \;\Rightarrow\; s \in \mathcal{D}[\![R\|_{W\cup Y\cup Z}(P\|_{W\cup X}Q)]\!]e$
(b) $s \in \mathcal{D}[\![R\|_{W\cup Y\cup Z}(P\|_{W\cup X}Q)]\!]e \;\Rightarrow\; s \in \mathcal{D}[\![P\|_{W\cup X\cup Y}(Q\|_{W\cup Z}R)]\!]e$

To prove (a) we start by expanding the *lhs* using the definition of $\mathcal{D}[\![P\|_{\Omega}Q]\!]e$.

$$\begin{aligned} &\mathcal{D}[\![P\|_{W\cup X\cup Y}(Q\|_{W\cup Z}R)]\!]e \\ =\;& \{u^\frown v \mid \exists p, q, r, x : u \in \mathsf{merge}(p, x, W \cup X \cup Y) \wedge p \in \mathcal{D}[\![P]\!]e \\ &\qquad \wedge x \in \mathsf{merge}(q, r, W \cup Z) \wedge q \in \mathsf{traces}(Q) \wedge r \in \mathsf{traces}(R)\} \\ \cup\,& \{u^\frown v \mid \exists p, q, r, x, y : u \in \mathsf{merge}(p, x^\frown y, W \cup X \cup Y) \wedge p \in \mathsf{traces}(P) \\ &\qquad \wedge x \in \mathsf{merge}(q, r, W \cup Z) \wedge q \in \mathcal{D}[\![Q]\!]e \wedge r \in \mathsf{traces}(R)\} \\ \cup\,& \{u^\frown v \mid \exists p, q, r, x, y : u \in \mathsf{merge}(p, x^\frown y, W \cup X \cup Y) \wedge p \in \mathsf{traces}(P) \\ &\qquad \wedge x \in \mathsf{merge}(q, r, W \cup Z) \wedge q \in \mathsf{traces}(Q) \wedge r \in \mathcal{D}[\![R]\!]e\} \end{aligned}$$

We shall refer to these three sets as (*lhs-P*), (*lhs-Q*) and (*lhs-R*) respectively. The *rhs* can be expanded similarly as follows:

$$\begin{aligned} &\mathcal{D}[\![R\|_{W\cup Y\cup Z}(P\|_{W\cup X}Q)]\!]e \\ =\;& \{u^\frown v \mid \exists p, q, r, z : u \in \mathsf{merge}(r, z, W \cup Y \cup Z) \wedge r \in \mathcal{D}[\![R]\!]e \\ &\qquad \wedge z \in \mathsf{merge}(p, q, W \cup X) \wedge p \in \mathsf{traces}(P) \wedge q \in \mathsf{traces}(Q)\} \\ \cup\,& \{u^\frown v \mid \exists p, q, r, z, y : u \in \mathsf{merge}(r, z^\frown y, W \cup Y \cup Z) \wedge r \in \mathsf{traces}(R) \\ &\qquad \wedge z \in \mathsf{merge}(p, q, W \cup X) \wedge p \in \mathcal{D}[\![P]\!]e \wedge q \in \mathsf{traces}(Q)\} \\ \cup\,& \{u^\frown v \mid \exists p, q, r, z, y : u \in \mathsf{merge}(r, z^\frown y, W \cup Y \cup Z) \wedge r \in \mathsf{traces}(R) \\ &\qquad \wedge z \in \mathsf{merge}(p, q, W \cup X) \wedge p \in \mathsf{traces}(P) \wedge q \in \mathcal{D}[\![Q]\!]e\} \end{aligned}$$

We shall refer to these three sets as (*rhs-R*), (*rhs-P*) and (*rhs-Q*) respectively. Note that both the *rhs* and the *lhs* divergent sets are the union of three sets and that each one represents the case where one of the processes is definitely diverging. To prove part (a) it is sufficient to show that each of the *rhs* sets is a subset of the corresponding *lhs* sets, so we must prove the following:

(i) $s \in$ (*lhs-P*) $\Rightarrow$ $s \in$ (*rhs-P*)
(ii) $s \in$ (*lhs-Q*) $\Rightarrow$ $s \in$ (*rhs-Q*)
(iii) $s \in$ (*lhs-R*) $\Rightarrow$ $s \in$ (*rhs-R*)

The proof of (i) is as follows:

(1)	$s \in (lhs\text{-}P)$	Assumption
(2)	$\exists u, p, q, r, x : u \leq s \wedge u \in \mathsf{merge}(p, x, W \cup X \cup Y)$ $\wedge\, x \in \mathsf{merge}(q, r, W \cup Z) \wedge p \in \mathcal{D}[\![P]\!]e$ $\wedge\, q \in \mathsf{traces}(Q) \wedge r \in \mathsf{traces}(R)$	
(3)	$\exists u, p, q, r, z : u \leq s \wedge u \in \mathsf{merge}(r, z, W \cup Y \cup Z)$ $\wedge\, z \in \mathsf{merge}(p, q, W \cup X) \wedge p \in \mathcal{D}[\![P]\!]e$ $\wedge\, q \in \mathsf{traces}(Q) \wedge r \in \mathsf{traces}(R)$	(merge-Sym), (merge-Ass)
(4)	$s \in (rhs\text{-}P)$	$y = \langle\,\rangle$, $(rhs\text{-}P)$

This completes the proof of part (i). □

The proof of (ii) is as follows:

(1)	$s \in (lhs\text{-}Q)$	Assumption
(2)	$\exists u, p, q, r, x, y : u \leq s \wedge u \in \mathsf{merge}(p, x^\frown y, W \cup X \cup Y)$ $\wedge\, x \in \mathsf{merge}(q, r, W \cup Z) \wedge p \in \mathsf{traces}(P)$ $\wedge\, q \in \mathcal{D}[\![Q]\!]e \wedge r \in \mathsf{traces}(R)$	
(3)	$u \in \mathsf{merge}(p, x^\frown y, W \cup X \cup Y) \wedge p \in \mathsf{traces}(P)$	
(4)	$\exists u', p' : u' \leq u \wedge p' \leq p \wedge u' \in \mathsf{merge}(p', x, W \cup X \cup Y)$ $\wedge\, p' \in \mathsf{traces}(P)$	(merge-Prefix), N2(P)
(5)	$\exists u', p', q, r, x : u' \leq s \wedge u' \in \mathsf{merge}(p', x, W \cup X \cup Y)$ $\wedge\, x \in \mathsf{merge}(q, r, W \cup Z) \wedge p' \in \mathsf{traces}(P)$ $\wedge\, q \in \mathcal{D}[\![Q]\!]e \wedge r \in \mathsf{traces}(R)$	(2), (4)
(6)	$\exists u', p', q, r, z : u' \leq s \wedge u' \in \mathsf{merge}(r, z, W \cup Y \cup Z)$ $\wedge\, z \in \mathsf{merge}(p', q, W \cup X) \wedge r \in \mathsf{traces}(R)$ $\wedge\, p' \in \mathsf{traces}(P) \wedge q \in \mathcal{D}[\![Q]\!]e$	(merge-Ass)
(7)	$s \in (rhs\text{-}Q)$	$y = \langle\,\rangle$, $(rhs\text{-}Q)$

This completes the proof of part (ii). □

The proof of (iii) is similar and is omitted. Therefore, this completes the proof of part (a).

The proof of part (b) is similar and is omitted. Thus we have proved that the divergent sets of both sides are equal as required.

B.3. Proof of Failure Set Equality

We need to prove the following:

$$\mathcal{F}[\![P \|_{W \cup X \cup Y} (Q \|_{W \cup Z} R)]\!]e \;=\; \mathcal{F}[\![R \|_{W \cup Y \cup Z} (P \|_{W \cup X} Q)]\!]e$$

Again the proof is split into the following two parts:

(a) $(s, X) \in \mathcal{F}[\![P \|_{W \cup X \cup Y} (Q \|_{W \cup Z} R)]\!]e \;\Rightarrow\; (s, X) \in \mathcal{F}[\![R \|_{W \cup Y \cup Z} (P \|_{W \cup X} Q)]\!]e$;
(b) $(s, X) \in \mathcal{F}[\![R \|_{W \cup Y \cup Z} (P \|_{W \cup X} Q)]\!]e \;\Rightarrow\; (s, X) \in \mathcal{F}[\![P \|_{W \cup X \cup Y} (Q \|_{W \cup Z} R)]\!]e$.

Only the proof of (a) is given since that of (b) is similar. To prove (a) we start by expanding the *lhs* repeatedly using the definition of $\mathcal{F}[\![P \|_{\Omega} Q]\!]e$:

$$\begin{aligned}
&\mathcal{F}[\![P \|_{W \cup X \cup Y} (Q \|_{W \cup Z} R)]\!]e \\
=\;& \{\, (u, D \cup E \cup F) \mid \exists p, x : D \subseteq \overline{W} \cap \overline{X} \cap \overline{Y} \wedge E, F \subseteq W \cup X \cup Y \\
&\quad \wedge\, u \in \mathsf{merge}(p, x, W \cup X \cup Y) \wedge (p, D \cup E) \in \mathcal{F}[\![P]\!]e \\
&\quad \wedge\, (x, D \cup F) \in \mathcal{F}[\![Q \|_{W \cup Z} R]\!]e \,\} \\
&\cup \{\, (u, D) \mid u \in \mathcal{D}[\![P \|_{W \cup X \cup Y} (Q \|_{W \cup Z} R)]\!]e \,\}
\end{aligned}$$

Since the divergence sets were dealt with above we can ignore the second set.

$$\begin{aligned}
= \quad & \{ (u, D \cup E \cup F) \mid \exists p, q, r, x : D \subseteq \overline{W} \cap \overline{X} \cap \overline{Y} \wedge E, F \subseteq W \cup X \cup Y \\
& \quad \wedge D \cup F = U \cup V \cup M \wedge U \subseteq \overline{W} \cap \overline{Z} \\
& \quad \wedge V, M \subseteq W \cup Z \wedge u \in \mathsf{merge}(p, x, W \cup X \cup Y) \\
& \quad \wedge x \in \mathsf{merge}(q, r, W \cup Z) \wedge (p, D \cup E) \in \mathcal{F}[\![P]\!]e \\
& \quad \wedge (q, U \cup V) \in \mathcal{F}[\![Q]\!]e \wedge (r, U \cup M) \in \mathcal{F}[\![R]\!]e \} \\
& \cup \{ (u, D \cup E \cup F) \mid \exists p, x : D \subseteq \overline{W} \cap \overline{X} \cap \overline{Y} \wedge E, F \subseteq W \cup Z \\
& \quad \wedge u \in \mathsf{merge}(p, x, W \cup X \cup Y) \wedge (p, D \cup E) \in \mathcal{F}[\![P]\!]e \\
& \quad \wedge x \in \mathcal{D}[\![Q \|_{W \cup Z} R]\!]e \wedge D \cup F \in \mathbb{P}(\Sigma) \}
\end{aligned}$$

Again the second set can be ignored because it is included in the previously discarded failure set. The failure set for the *rhs* can be similarly expanded.

$$\begin{aligned}
& \mathcal{F}[\![R \|_{W \cup Y \cup Z} (P \|_{W \cup X} Q)]\!]e \\
= & \{ (u, D' \cup E' \cup F') \mid \exists p, q, r, z : D' \subseteq \overline{W} \cap \overline{Y} \cap \overline{Z} \wedge E', F' \subseteq W \cup Y \cup Z \\
& \quad \wedge D' \cup F' = U' \cup V' \cup M' \wedge U' \subseteq \overline{W} \cap \overline{X} \\
& \quad \wedge V', M' \subseteq W \cup X \wedge u \in \mathsf{merge}(p, z, W \cup Y \cup Z) \\
& \quad \wedge z \in \mathsf{merge}(q, r, W \cup X) \wedge (p, D' \cup E') \in \mathcal{F}[\![P]\!]e \\
& \quad \wedge (q, U' \cup V') \in \mathcal{F}[\![Q]\!]e \wedge (r, U' \cup M') \in \mathcal{F}[\![R]\!]e \}
\end{aligned}$$

Now both sides have been completely expanded we can proceed with the proof.

(1) $(s, G) \in \mathcal{F}[\![P \|_{W \cup X \cup Y} (Q \|_{W \cup Z} R)]\!]e$ — Assumption

(2) $\exists p, q, r, x, D, E, F, U, V, M : D \subseteq \overline{W} \cap \overline{X} \cap \overline{Y}$
$\wedge E, F \subseteq W \cup X \cup Y \wedge U \subseteq \overline{W} \cap \overline{Z}$
$\wedge V, M \subseteq W \cup Z \wedge D \cup F = U \cup V \cup M$
$\wedge G = D \cup E \cup F \wedge s \in \mathsf{merge}(p, x, W \cup X \cup Y)$
$\wedge x \in \mathsf{merge}(q, r, W \cup Z) \wedge (p, D \cup E) \in \mathcal{F}[\![P]\!]e$
$\wedge (q, U \cup V) \in \mathcal{F}[\![Q]\!]e \wedge (r, U \cup M) \in \mathcal{F}[\![R]\!]e$

(3) $\exists p, q, r, z, D, E, F, U, V, M : D \subseteq \overline{W} \cap \overline{X} \cap \overline{Y}$ — (merge-Ass)
$\wedge E, F \subseteq W \cup X \cup Y \wedge U \subseteq \overline{W} \cap \overline{Z}$
$\wedge V, M \subseteq W \cup Z \wedge D \cup F = U \cup V \cup M$
$\wedge G = D \cup E \cup F \wedge s \in \mathsf{merge}(r, z, W \cup Y \cup Z)$
$\wedge z \in \mathsf{merge}(p, q, W \cup X) \wedge (p, D \cup E) \in \mathcal{F}[\![P]\!]e$
$\wedge (q, U \cup V) \in \mathcal{F}[\![Q]\!]e \wedge (r, U \cup M) \in \mathcal{F}[\![R]\!]e$

Since we have shown the existence of the traces required to satisfy the *rhs*, traces can be ignored from now on. Therefore, all that remains to be shown is the existence of the failure sets needed to satisfy the *rhs*, i.e. D', E', F', U', V' and M'. These can be defined in terms of the failure sets of the *lhs* D, E, F, U, V and M since the traces are the same for both the *lhs* and the *rhs*. So we define the required failure sets as follows:

$$\begin{aligned}
U' &= ((U \cup V) \cap \overline{W} \cap \overline{X} \cap \overline{Y}) \cup (V \cap \overline{W} \cap \overline{X} \cap Y \cap Z) \cup (E \cap \overline{W} \cap \overline{X} \cap Y) \\
V' &= E \cap (W \cup X) \\
M' &= (U \cup V) \cap (W \cup X) \\
D' &= (U \cap \overline{W} \cap \overline{Y} \cap \overline{Z}) \cup (E \cap \overline{W} \cap \overline{Y} \cap \overline{Z} \cap X) \\
E' &= (U \cap \overline{W} \cap \overline{Z} \cap Y) \cup M \\
F' &= \ (U \cap X \cap Y) \cup (V \cap (W \cup (X \cap Y) \cup (X \cap Z))) \cup (V \cap \overline{W} \cap \overline{X} \cap Z) \\
& \quad \cup (E \cap (W \cup (X \cap Y) \cup (X \cap Z))) \cup (E \cap \overline{W} \cap \overline{X} \cap Y)
\end{aligned}$$

It is clear from the definitions of these sets that they satisfy the *rhs* requirements.

$$G = D' \cup E' \cup F' \;\wedge\; D' \subseteq \overline{W} \cap \overline{Y} \cap \overline{Z} \;\wedge\; E', F' \subseteq W \cup Y \cup Z$$

$$\wedge\; U' \subseteq \overline{W} \cap \overline{X} \;\wedge\; V', M' \subseteq W \cup X \;\wedge\; D' \cup F' = U' \cup V' \cup M'$$

Proof is trivial and is omitted. Now all that remains to be shown is that these sets are in fact refused by the relevant processes, that is prove each of the following:

$$(p, U' \cup V') \in \mathcal{F}[\![P]\!]e \qquad (q, U' \cup M') \in \mathcal{F}[\![P]\!]e \qquad (r, D' \cup E') \in \mathcal{F}[\![R]\!]e$$

The proof of $(p, U' \cup V') \in \mathcal{F}[\![P]\!]e$ is as follows:

(1)	$(p, D \cup E) \in \mathcal{F}[\![P]\!]e$	Assumption
(2)	$(U \cup V) \cap \overline{W} \cap \overline{X} \cap \overline{Y} \subseteq D$	def. D
(3)	$(E \cap \overline{W} \cap \overline{X} \cap Y) \cup (E \cap (W \cup X)) \subseteq E$	def. E
(4)	$(p, ((U \cup V) \cap \overline{W} \cap \overline{X} \cap \overline{Y}) \cup (E \cap \overline{W} \cap \overline{X} \cap Y)$ $\cup (E \cap (W \cup X))) \in \mathcal{F}[\![P]\!]e$	N3(P), (1), (2), (3)
(5)	$(V \cap \overline{W} \cap \overline{X} \cap Y \cap Z) \subseteq Z \subseteq \overline{A}$	def. Z
(6)	$\forall a \in (V \cap \overline{W} \cap \overline{X} \cap Y \cap Z) : (p^\frown\langle a \rangle, \varnothing) \notin \mathcal{F}[\![P]\!]e$	def. Z
(7)	$(p, U' \cup V') \in \mathcal{F}[\![P]\!]e$	N4(P), (4), (6)

Notes: (5) means that the events in $(V \cap \overline{W} \cap \overline{X} \cap Y \cap Z)$ are outside the alphabet A of P, i.e. are impossible for it. (6) states that in particular they are impossible after P has performed the trace p.

The proofs of $(q, U' \cup M') \in \mathcal{F}[\![Q]\!]e$ and $(r, D' \cup E') \in \mathcal{F}[\![R]\!]e$ are similar, the sets of impossible events for these are $(E \cap \overline{W} \cap \overline{X} \cap Y)$ and $(E \cap \overline{W} \cap \overline{Y} \cap \overline{Z} \cap X)$ respectively.

This concludes the proof of the failure sets equality. Therefore, we have proved that the failure sets and divergence sets of both sides are equal, and hence, the two processes are equivalent. This concludes the proof of the associativity law. □

Communicating Process Architectures 2013
P.H. Welch et al. (Eds.)
Open Channel Publishing Ltd., 2013

Efficient Simulation of CSP-Like Languages

Thomas GIBSON-ROBINSON
Department of Computer Science, University of Oxford, UK
thomas.gibson-robinson@cs.ox.ac.uk

Abstract. In *"On the Expressiveness of CSP"*, Roscoe provides a construction that, given the operational semantics rules of a *CSP-like* language and a process in that language, constructs a strongly bisimilar CSP process. Unfortunately, the construction provided is difficult to use and the scripts produced cannot be compiled by the CSP model-checker, FDR. In this paper, we adapt Roscoe's simulation to make it produce a process that can be checked relatively efficiently by FDR. Further, we extend Roscoe's simulation in order to allow recursively defined processes to be simulated in FDR, which was not supported by the original simulation. We also describe the construction of a tool that can automatically construct the simulation, given the operational semantics of the language and a script to simulate, both in an easy-to-use format.

Keywords. CSP, CSP-like, FDR, model checking

Introduction

Developing a model checker for a process algebra requires a significant investment of resources, particularly if the model checker is to be efficient. Therefore, rather than developing new model checkers for each new process algebra, it would make sense to reuse existing model checkers, where possible.

In papers on the *expressiveness* of CSP [1,2], Roscoe develops a simulation that, given a *CSP-like* language, constructs a strongly-bisimilar process in the process algebra CSP [3,4,5]. Thus, as CSP has an efficient, industrial-strength, model checker FDR [6], it should now be possible to check models written in other CSP-like languages without having to develop a new model checker.

Whilst the name CSP-like might appear to imply that only a small class of languages can be simulated using [1,2], most of the CSP-like conditions are merely well-formedness conditions and thus many other process algebras, such as CCS [7], the π-calculus [8] and Lowe's Readyness Testing CSP extension [9], are either CSP-like, or can be easily converted into equivalent, but CSP-like formulations. Further, many denotational models for CSP, such as Lowe's Availability model [10], have CSP-like derived operational semantics that enable the model to be extracted using the standard CSP models.

The above would suggest that a tool that could automatically construct a simulation of processes written in a CSP-like language, given the operational semantics rules as input, would be of value. Unfortunately, Roscoe's construction in [1,2] was never intended to be used as an actual simulation. In particular, only non-recursive processes can be simulated within FDR as the simulated recursive processes are infinite state. The simulation also imposes a very high computational overhead, making it impractical for any non-trivial problem. Further, the simulation is extremely complex to construct and the user has to construct the simulation manually.

Contributions In this paper we explain how to alter Roscoe's simulation to allow it to be encoded in machine-CSP (henceforth CSP_M). We also give a number of optimisations to Roscoe's simulation to ensure that it can efficiently simulate processes. We then prove that the resulting optimised simulation is still strongly bisimilar to the original process. We also extend Roscoe's simulation to enable recursive processes to be simulated correctly by FDR. We then prove that, whilst this alteration is no longer strongly bisimilar, it results in a process that is equivalent in all CSP models to the original process.

We also describe the construction of a tool, tyger, that takes as input the operational semantics rules of the language (in a easy-to-use format) and a list of process definitions to compile (again, in a natural format). The tool automatically produces the CSP_M simulation described above, and also automatically applies the necessary alternations to ensure that recursive processes can be successfully simulated. The output of this tool is thus a CSP file that can be passed to FDR, or any other CSP_M tool.

This has a number of advantages. Not only does it mean that model checkers can be easily constructed for other process algebras, but it allows a language designer to experiment with the language whilst developing it. For example, suppose a user wants to create a process algebra to enable easy verification of a certain problem. Previously, to experiment with the language, an entirely new model checker would have to be constructed, making it difficult for the language designer to experiment with the language's capabilities during design. Now, a simulation of the language can be constructed trivially, allowing experiments to be run whilst the language is being developed and therefore providing feedback into the design.

Outline In Section 1 we give a brief overview of CSP, before considering what languages we are able to simulate within CSP by defining Roscoe's CSP-like condition. In Section 2 we explain Roscoe's original construction before giving a number of modifications that enable it to run efficiently. In Section 3 we discuss the problem with recursion, explain the refactorings required to permit recursive definitions and prove that the refactorings produce semantically equivalent processes. Further, we discuss the limitations of the recursion refactorings. In Section 4 we describe the construction of the tool. In Section 5 we detail the results of several experiments that were run in order to verify the effectiveness and performance of the simulation. Lastly, in Section 6 we draw conclusions and discuss future work.

1. Background

In this section we provide a brief overview of the fragment of CSP necessary to understand the remainder of the paper. We also define what it means for an operational semantics to be *CSP-like*, which represents the class of operational semantic languages that we can simulate.

1.1. A Brief Overview of CSP

CSP [3,4,5] is a *process algebra* in which programs or *processes* that communicate events from a set Σ with an environment may be described. We sometimes structure events by sending them along a *channel* . For example, $c.3$ denotes the value 3 being sent along the channel c. Further, given a channel c the set $\{\!|c|\!\} \subseteq \Sigma$ contains those events of the form $c.x$.

The simplest CSP process is the process $STOP$, that can perform no events and thus represents a deadlocked process. The process $a \to P$ offers the environment the event $a \in \Sigma$ and then when it is performed, behaves like P. The process $P \mathrel{\Box} Q$ offers the environment the choice of the events offered by P and by Q. Alternatively, the process $P \sqcap Q$, nondeterministically chooses which of P or Q to behave like. Note that the environment cannot influence the choice, the process chooses *internally*. $P \rhd Q$ initially offers the choice of the events of P but can *timeout* and then behaves as Q.

Syntax	Meaning
$a \rightarrow P$	Performs the event a and then behaves like P.
$P \,\square\, Q$	Offers the choice between P and Q.
$P \sqcap Q$	Non-deterministically picks one of P and Q to behave as.
$P \setminus A$	Runs P but hides events from A (which become τ).
$P \mid\mid\mid Q$	Runs P and Q in parallel, enforcing no synchronisation.
$P \underset{A}{\parallel} Q$	Runs P and Q in parallel, enforcing synchronisation on events in A.
$P \;{}_A\!\parallel_B\; Q$	Runs P and Q in parallel, enforcing synchronisation on events in $A \cap B$.
$P[\![R]\!]$	Renames the events of P according to the relation R.
$P \,\Theta_A\, Q$	Runs P until an event in A occurs, at which point Q is run.
$P \triangle Q$	Runs P, but at any point Q may interrupt P, perform an action and discard P.

Table 1. A summary of the syntax of CSP.

TThe process $P \;{}_A\!\parallel_B\; Q$ allows P and Q to perform only events from A and B respectively and forces P and Q to synchronise on events in $A \cap B$. The process $P \underset{A}{\parallel} Q$ runs P and Q in parallel and forces them to synchronise on events in A. The *interleaving* of two processes, denoted $P \mid\mid\mid Q$, runs P and Q in parallel but enforces no synchronisation. The process $P \setminus A$ behaves as P but hides any events from A by transforming them into a special internal event, τ. This event does not synchronise with the environment and thus can always occur. The process $P[\![R]\!]$, behaves as P but renames the events according to the relation R. Hence, if P can perform a, then $P[\![R]\!]$ can perform each b such that $(a, b) \in R$. The process $P \triangle Q$ initially behaves like P but allows Q to *interrupt* at any point and perform an event, at which point P is discarded and the process behaves like Q. The process $P \,\Theta_A\, Q$ initially behaves like P, but if P ever performs an event from A, P is discarded and $P \,\Theta_A\, Q$ behaves like Q.

Recursive processes can be defined either equationally, or using the notation $\mu X \cdot P$. In the latter, every occurrence of X within P represents a recursive call to $\mu X \cdot P$.

There are a number of ways of giving meaning to CSP processes. The simplest approach is to give an operational semantics. The operational semantics of a CSP process naturally creates a labelled transition system (LTS) where the edges are labelled by events from $\Sigma \cup \{\tau\}$ and the nodes are process states. The usual way of defining the operational semantics of CSP processes is by presenting *Structured Operational Semantics* (SOS) style rules. For example, the operational semantics of $P \,\Theta_A\, Q$ can be defined by the following inductive rules:

$$\frac{P \xrightarrow{a} P'}{P \,\Theta_A\, Q \xrightarrow{a} Q}\; a \in A \qquad \frac{P \xrightarrow{b} P'}{P \,\Theta_A\, Q \xrightarrow{b} P' \,\Theta_A\, Q}\; b \notin A \qquad \frac{P \xrightarrow{\tau} P'}{P \,\Theta_A\, Q \xrightarrow{\tau} P' \,\Theta_A\, Q}$$

The interesting rule is the first, which specifies that if P performs an event $a \in A$, then $P \,\Theta_A\, Q$ can perform the event a and then behave like Q (i.e. the exception has been *thrown*). The last rule is known as a *tau-promotion* rule as it promotes any τ performed by a component (in this case P) into a τ performed by the operator. The justification for this rule is that τ is an unobservable event, and therefore the operator cannot prevent P from performing the event.

CSP also has a number of *congruent* denotational semantics, where congruence means that the denotational semantics may either be extracted from the operational semantics, or calculated directly given the process. For example, the *traces* model represents each process by the prefix-closed set of finite sequences of visible events that it can perform (i.e. a process is represented by a subset of Σ^*).

1.2. CSP-Like Languages

In [1] Roscoe proves that if an operator Op is *CSP-like*, then it is possible to construct a CSP process Op' such that the LTSs representing Op and Op' are strongly bisimilar. In this section we define what it means for an operator to be CSP-like.

Conventionally, operational semantics are represented by SOS rules, such as those given above for the CSP exception operator. However, given that the syntax of SOS rules is very general, defining what CSP-like means in terms of the rules would be rather difficult. For example, the following SOS rule is not a legal CSP-like rule (since no CSP operator allows an event to occur as the result of the absence of another event), but is a valid SOS rule:

$$\frac{\neg(P \xrightarrow{a} P')}{Op(P) \xrightarrow{a} Op(P')}.$$

Therefore, in [1], Roscoe instead uses a different presentation of operational semantics rules in which only CSP-like operators are definable. We now give a brief overview of this alternative presentation: more information can be found in [1,5].

Definition 1.1 (From [1,5]). The *combinator* operational semantics of an operator $F(\mathbf{P})$, where $\mathbf{P}$ is a vector of processes, is defined by:

- Each process of the vector $\mathbf{P}$ is defined as either ***on*** or ***off***. **on** arguments are indexed by positive numbers, whilst negative indices refer to **off** arguments. Intuitively, an **on** argument of a process can perform visible events, whilst an **off** argument cannot. For example, both arguments of $\Box$ are **on**, the first argument of Θ_A is **on**, whilst the second argument of Θ_A is **off**.
- A set of *combinator rules* that specify the operational semantics of the operator, except tau promotion rules which are implicitly defined for precisely the **on** arguments. A single combinator rule is of the form (ϕ, x, T) where:
  - $*$ ϕ is a partial function from the **on** indices to Σ. If $\phi(x) = y$ then the xth **on** process performs the event y. If $x \notin dom(\phi)$, then x does not perform an event.
  - $*$ $x \in \{\tau\} \cup \Sigma$ is the resulting event that the operator performs;
  - $*$ T is a piece of syntax that represents the state of the resulting process. T is generated from the following grammar:

$$T ::= -\mathbf{N} \mid \mathbf{N} \mid Op(T, \ldots, T)$$

  where: the first clause represents an **off** argument, the second clause a currently **on** argument, and the third clause represents the application of an operator to a number of process arguments. Further, there are a number of well-formedness conditions on T:

  1. **on** processes cannot be cloned, i.e. each **on** index **i** can appear at most once in T (there is no CSP operator that can clone an **on** argument);
  2. **on** processes cannot be suspended, i.e. if T contains $Op'(P_1, \ldots, P_N)$ where $P_i = \mathbf{j}$ (i.e. P_i is the jth currently **on** process of Op), then i must be an **on** argument of Op' (again, no CSP operator can suspend and then later reactivate an **on** argument).

  For example, if $T = \mathbf{1}$, then this represents the rule that discards the current operator and discards all process arguments except for $\mathbf{1}$. Alternatively, if $T = \mathbf{1} \Box -\mathbf{1}$, then the resulting state is the external choice of the 1st argument of the current operator and the newly started process $-\mathbf{1}$.

For example, the external choice operator has two arguments which are both **on** and the combinator rules consist of the set of all rules of the form $((a, \cdot), a, \mathbf{1})$ and $((\cdot, a), a, \mathbf{2})$, where a ranges over Σ. The exception operator has two arguments, the first of which is **on** whilst the second is **off**. The set of combinator rules for Θ_A consists of the set of all rules of the form $((a, \cdot), a, \mathbf{1}\ \Theta_A\ \mathbf{-1})$ where $a \in \Sigma \setminus A$, and rules of the form $((a, \cdot), a, \mathbf{-1})$ where $a \in A$.

A *CSP-like* operator is then defined as any operator whose operational semantics can be expressed by a combinator operational semantics. A CSP-like language is defined as a language in which every operator is CSP-like. Whilst this might appear to be restrictive, it turns out that many process algebras, including CCS[1] [7] and the π-calculus [8], are CSP-like. Further, many denotational models for CSP, such as Lowe's Availability Model [10], have a derived operational-semantics that is CSP-like, meaning that we are able to simulate these denotational models operationally.

2. Simulating CSP-Like Languages

In this section we define Roscoe's simulation and then define an optimised CSP$_M$ simulation, before proving the equivalence of the two. We start in Section 2.1 by describing Roscoe's simulation, including the internal format of the operational semantics and the CSP process $Operator$ that simulates a particular operator. In Section 2.2 we then explain the modifications that we made in order to construct a CSP$_M$ simulation. Further, we detail the changes that we made to enable the simulation to run efficiently and prove that the optimisation still result in a strong bisimulation.

Throughout this paper we will need to enlarge the set of events Σ that the environment offers. Thus, for the remainder of this paper we will use Σ to denote the set of events that the user uses in their script; $\Sigma_0 \supseteq \Sigma$ to include all semantic events (such as ✓ to indicate termination in the case of CSP); $\Sigma_1 \supseteq \Sigma_0$ to include all internal events used by the simulation. We also use a special event $tau \in \Sigma_0$ that will simulate a semantic τ.

2.1. Roscoe's Simulation

In order to simulate arbitrary CSP-like operators in [1,2], Roscoe defines a data structure in which the operational semantic rules of an operator are encoded. In this section, we start by defining this encoding before explaining the simulation.

Roscoe's encoding is an expansion of the combinator semantics, as presented in Definition 1.1, in order to make the combinators more easy to work with. As with the combinator semantics, operators such as $\cdot \setminus A$ are considered as a *family* of operators, one for each $A \subseteq \Sigma$. Thus, the only arguments of an operator are **on** and **off** processes. Given an operator α we define $n(\alpha)$ as the number of **on** arguments and $I(\alpha)$ as the number of **off** arguments. Further, we index the **on** arguments by $\{1 \ldots\}$ and **off** arguments by $\{\ldots - 1\}$ (i.e. a positive index i refers to the ith **on** process of α, whilst a negative index $-i$ refers to the ith **off** process of α). The operational semantics of an operator α is given by a set of tuples $(\phi, x, \beta, f, \psi, \chi, B)$ where:

- ϕ: specifies what event each **on** process must perform in order for this action to be performed. Formally, it is a partial map from $\{1 \ldots n(\alpha)\}$ to Σ_0, where $\phi(x) = b$ means the xth **on** process performs b. This corresponds precisely to ϕ in the combinator rules (cf. Definition 1.1).
- x: is the resulting event from Σ_0 that α performs.
- β: is the resulting operator that this process evolves into.

[1]CCS is not quite CSP-like as τ resolves choice. However, Roscoe shows how to simulate it in [1].

- f: denotes which **off** process of α is used for each newly turned **on** process. Formally, it is a map from $\{1\ldots k\}$ (where k is equal to the number of processes turned **on**) to $\{-I(\alpha)\ldots -1\}$ where $f(x) = y$ indicates the xth newly turned **on** process is a copy of the yth **off** process.
- ψ: specifies what the process arguments of β are, which can either be currently **on** processes of α, or newly turned **on** processes (according to f). Formally, it is a map from $\{1\ldots n(\beta)\}$ to $\{1\ldots n(\alpha)+k\}$ where $\psi(y) = x$ indicates that the yth argument of β is either, if $x \leq n(\alpha)$ then the xth **on** argument of α, or otherwise the $x - n(\alpha)^{th}$ newly turned **on** argument (i.e. the $f(x - n(\alpha))$ **off** process of α).
- χ: specifies what the **off** arguments of β are in terms of those of α. Formally, χ is a map from $\{-I(\beta)\ldots -1\}$ to $\{-I(\alpha)\ldots -1\}$ where $\chi(y) = x$ means that the yth **off** argument of β is the xth **off** argument of α.
- B: is the set of **on** processes that are discarded (this was not present in Roscoe's model but it is added for clarity).

In the following, we assume that there exists a function $Rules$ that gives the rules for each operator in the language that is being modelled. Thus, $Rules$ is a function from the set of all operators to the powerset of tuples $(\phi, x, \beta, f, \psi, \chi, B)$ of the above form. For example, assuming that the $a \rightarrow \cdot$ operator is denoted as $\mathit{Prefix}.a$, and that $\cdot \underset{A}{\|} \cdot$ is denoted as $\mathit{Parallel}.A$, a partial specification of the $Rules$ function for CSP can be encoded as follows (where the identity operator is denoted by $\mathit{Identity}$):

$$\begin{aligned}
Rules(\mathit{Identity}) &\mathrel{\widehat=} \{(\{(1,a)\}, a, \mathit{Identity}, \{\}, \{(1,1)\}, \{\}, \{\}) \mid a \in \Sigma_0\} \\
Rules(\mathit{Prefix}.a) &\mathrel{\widehat=} \{(\{\}, a, \mathit{Identity}, \{(1,-1)\}, \{(1,1)\}, \{\}, \{\})\} \\
Rules(\mathit{Parallel}.A) &\mathrel{\widehat=} \{(\{(1,a),(2,a)\}, a, \mathit{Parallel}.A, \{\}, \{(1,1),(2,2)\}, \{\}, \{\}) \mid a \in A\} \\
&\quad \cup \{(\{(1,a)\}, a, \mathit{Parallel}.A, \{\}, \{(1,1),(2,2)\}, \{\}, \{\}) \mid a \notin A\} \\
&\quad \cup \{(\{(2,a)\}, a, \mathit{Parallel}.A, \{\}, \{(1,1),(2,2)\}, \{\}, \{\}) \mid a \notin A\}
\end{aligned}$$

For the rest of this paper we assume that a function $Rules$ has been defined for the language that we are attempting to model.

Operator Simulation We now define the process $Operator(\alpha, \ldots)$ from [1,2] (with some minor alterations) that simulates the CSP-like operator α. The general form that the process will take is:

$$\begin{aligned}
&Operator(\alpha, \mathit{onProcs}, \mathit{offProcs}) \mathrel{\widehat=} \\
&\Bigg(\Big\|_{(n,P)\in \mathit{onProcs}} (Harness(P,n), AlphaProc(n)) \\
&\qquad \underset{\bigcup_{n\in\{1\ldots|\mathit{onProcs}|\}} AlphaProc(n)}{\|} \\
&\qquad Reg\left(\alpha, |\mathit{onProcs}|, id_{\{1\ldots|\mathit{onProcs}|\}}, id_{\{-|\mathit{offProcs}|..-1\}}\right)\Bigg)[\![Rename]\!] \setminus \{tau\}.
\end{aligned}$$

In the above, each **on** process is run in a harness that will both ensure that it is turned **off** when necessary and that renames its events to allow the simulation to function correctly. The regulator process is responsible for ensuring that only events according to the current operator are allowed to occur.

Observe that the left hand side of the parallel in the above definition does not depend on the current operator, only the **on** processes. Therefore, only Reg differs between, e.g. $Operator(ExternalChoice, \langle P, Q\rangle, \langle\rangle)$ and $Operator(Parallel.\Sigma, \langle P, Q\rangle, \langle\rangle)$. However, these have very different synchronisation patterns; in the former P and Q do not synchronise whereas in the latter, P and Q synchronise on Σ. Therefore, it follows that the regulator needs to allow all possible synchronisation patterns to occur.

In order to ensure that we can simulate the two different synchronisation patterns, we need to rename the events of each **on** process to allow the processes to synchronise together in any possible way. The regulator could then be defined to allow only those synchronisations that correspond to the current operator. Thus, the general event form that will be used in the operator simulation will be a tuple (ϕ, x, B) where:

- ϕ is a partial map from **on** process to Σ_0 (i.e. if $\phi(x) = y$ the xth process performs the event y);
- x is the resulting event;
- B is the set of **on** processes that are discarded.

As we have fixed the event format, we are now able to define $Rename$ as the function that renames each (ϕ, x, B) to x. Further, $AlphaProc(n)$ can be defined to be the set of all (ϕ, x, B) such that n is affected; i.e. $n \in dom(\phi)$ or $n \in B$.

We now consider the definition of $Harness(P, n)$. The harness needs to run the process P as if it were the nth **on** process, renaming every event that it performs to one of the above form. It also needs to turn this process **off** when an event of the form (ϕ, x, B) where $n \in B$ is performed. Hence, we define $Harness(P, n)$ as follows:

$$Harness(P, n) \mathrel{\widehat{=}} \Big((P[\![Prime]\!] \; \Theta_{\{x' \mid x \in \Sigma_0\}} \; STOP) \bigtriangleup \mathit{off} \to STOP\Big)[\![\mathit{HRename}]\!]$$

where *off* is a fresh event and $Prime$ maps every event $x \in \Sigma_0$ to both x and x'. ***HRename*** renames as follows:

- Each event $a \in \Sigma_0$ to every (ϕ, x, B) such that $\phi(n) = a$ and $n \notin B$ (i.e. where process n performs $\phi(n)$ and is not discarded);
- Each event $a' \in \{x' \mid x \in \Sigma_0\}$ to every (ϕ, x, B) such that $\phi(n) = a$ and $n \in B$ (i.e. process n performs a, but is discarded);
- off to every (ϕ, x, B) such that $n \notin dom(\phi)$ and $n \in B$ (i.e. process n is discarded by an event in which it does not partake).

We now define the regulator, which has three principle responsibilities: it must ensure that only events allowed by the current operator can be performed by the processes; it must track which process on the left hand side corresponds to which process of the current operator; and it must turn **on** processes according to the executed rule. Thus, the regulator takes the following parameters:

- λ: the current operator;
- m: the number of processes ever turned **on**;
- Ψ: a map from $\{1 \ldots n(\lambda)\}$ to $\{1 \ldots m\}$; in particular if $\Psi(x) = y$ it means that the the xth **on** argument of λ is the yth **on** argument;
- χ: a map from $\{1 \ldots I(\lambda)\}$ to $\{1 \ldots I(\alpha)\}$ where α is the starting operator (i.e. the operator for which $Operator(\alpha, \ldots)$ was called); in particular if $\chi(x) = y$ it means that the xth **off** argument of λ is the yth **off** argument of α.

We can then define the regulator, $Reg(\lambda, m, \Psi, \chi)$, as follows:

$$Reg(\lambda, m, \Psi, \chi) \mathrel{\widehat{=}}$$
$$\square_{(\phi,x,\beta,f,\psi,\chi',B)\in Rules(\lambda)} (\phi \circ \Psi^{-1}, x, \{\Psi(x) | x \in B\}) \rightarrow \left(\begin{array}{c} \|_{n\in\{m\ldots m+|f|\}} (Harness(offProcs(\chi(f(n-m)))), AlphaProc(n)) \\ \bigcup_{n\in\{1\ldots m+|f|\}} AlphaProc(n) \| \bigcup_{n\in\{m+|f|\ldots\}} AlphaProc(n) \\ Reg(\beta, m+|f|, (\Psi \cup id_{m+1,\ldots,m+|f|}) \circ \psi, \chi \circ \chi') \end{array} \right)$$

Note that the above regulator allows only those events that correspond to the current operator and, whenever such an event occurs, evolves into the state dictated by the rule that was chosen (with the state being updated according to the definitions of f etc.).

Roscoe then proves that the above simulation results in a CSP process that is strongly bisimilar to the original operator.

2.2. A CSP_M Simulation

In this section we outline the alterations to Roscoe's simulation that are required in order to allow the simulation to be successfully modelled in CSP_M. The first major issue that prevents the above solution being compiled into CSP_M is that the alphabets, such as $AlphaProc(n)$, are infinite because it is possible that an infinite number of processes can be turned **on** by an operator. Thus, since the **on** processes have to be able to be synchronised in any possible way, $AlphaProc(n)$ has to represent an infinite number of possible synchronisations patterns. In order to prevent this we disallow operators that could turn **on** an infinite number of arguments[2]. Hence, we can now statically bound the maximum number of processes turned **on** to some number N. Therefore, the set of all the (ϕ, x, B) events is given by:

$$\{(\phi, x, B) \mid \phi \in \{1 \ldots N\} \rightarrow \Sigma_0, x \in \Sigma_0, B \subseteq \{1 \ldots N\}\}.$$

We create a new CSP_M channel, $renamings$ that carries events from the above set. Using this, the remainder of the translation to CSP_M is largely mechanical and is therefore elided.

Efficiency Unfortunately, the naïve translation of the above into CSP_M results in scripts that can only support at most a handful of events. In order to make the simulation run efficiently a number of optimisations had to be made. Some of the simpler optimisations included using sequences rather than sets to represent partial functions and using integers to identify events of the form (ϕ, x, B), rather than tuples. However, in order to make the simulation really run efficiently, a number of more interesting optimisations were made, as we now describe.

One problem with $Operator$ is that the $renamings$ channel has $O(2^{N+|\Sigma_0|} \cdot |\Sigma_0| \cdot 2^N)$ members, meaning that computing the $Harness$ renaming or the set $AlphaProc(n)$ will be slow. However, most operators will never use the majority of the events. For example, CSP's parallel operator (over a set A), requires only events of the form $(\{(0, a), (1, a)\}, a, \{\})$, $(\{(0, b)\}, b, \{\})$, $(\{(1, b)\}, b, \{\})$, where $a \in A$, $b \notin A$. In fact, all standard CSP operators only use $O(|\Sigma_0|^2)$ events. Thus, when computing the renamings and the process alphabets we restrict ourself to the subset of $renamings$ that could be used either by the current operator, or any resulting operator. This requires us to compute the set of all (ϕ, x, B) such that (ϕ, x, B) is permitted either by the current regulator, or any regular we can evolve to.

To calculate the above set, we firstly calculate the set of regulator configurations that the regulator can evolve into. In particular, we define the function $configs(\lambda, m, \Psi, \chi)$, where the arguments are as per Reg, as the least fixed point of the following function h:

[2] Clearly no practical simulation could support such operators, so this is not a restriction in practice.

$$h(X) \mathrel{\widehat{=}} X \cup \{(\lambda, m, \Psi, \chi)\}$$
$$\cup \{(\beta, m + |f|, (\Psi \cup id_{m+1,\ldots,m+|f|}) \circ \psi, \chi \circ \chi')$$
$$\mid (\phi, x, \beta, f, \psi, \chi', B) \in Rules(\lambda), (\lambda, m, \Psi, \chi) \in X\}.$$

Thus, it follows that $configs(\lambda, m, \Psi, \chi)$ consists of the set of all states that $Reg(\lambda, m, \Psi, \chi)$ can evolve into.

Using the above, the set of all events of the form (ϕ, x, B) that $Operator(\alpha, \textit{onProcs}, \textit{offProcs})$ requires, denoted $closure(\alpha, \textit{onProcs}, \textit{offProcs})$, can be defined as:

$$\{(\phi \circ \Psi^{-1}, x, \{\Psi(x) | x \in B\}) \mid (\phi, x, \beta, f, \psi, \chi', B) \in Rules(\lambda),$$
$$(\lambda, m, \Psi, \chi) \in configs(\alpha, |\textit{onProcs}|, id_{\{1\ldots|\textit{onProcs}|\}}, id_{\{-|\textit{offProcs}|..-1\}})\}.$$

The correctness of the above is shown in the following lemmas.

Lemma 2.1. Suppose $Operator(\alpha, \textit{onProcs}, \textit{offProcs})$ evolves into state in which Reg is in state $Reg(\lambda, m, \Psi, \chi)$. Then, $(\lambda, m, \Psi, \chi) \in configs(\alpha, |\textit{onProcs}|, id_{\{1\ldots|\textit{onProcs}|\}}, id_{\{-|\textit{offProcs}|..-1\}})$.

Proof (Sketch). Observe that in the definition of $Reg(\lambda, m, \Psi, \chi)$, Reg evolves into states precisely the same as those on the left hand side of h in the definition of $configs$. □

Lemma 2.2. If $Operator(\alpha, \textit{onProcs}, \textit{offProcs})$ internally uses the event (ϕ, x, B) then $(\phi, x, B) \in closure(\alpha, \textit{onProcs}, \textit{offProcs})$.

Proof (Sketch). This follows from the observation that, internally, Reg participates in every event that $Operator$ performs, and thus the set of events that $Operator$ uses is bounded by the set of events that Reg allows, which, given Lemma 2.1, $closure$ extracts by definition. □

One other issue with the definition of $Operator$ is that it contains a large number of nested CSP operators, meaning that FDR has to deal with a large, complex, stack of operators. However, observe that if a particular component n can never be turned **off** (as is the case with CSP's parallel operator) then $Harness(P, n)$ is equivalent to $P[\![Rename]\!]$. Hence, if we can compute the set of arguments that can be turned **off**, it follows that we can easily decide when to compute the full harness. In order to compute the set of arguments that can be turned **off**, we define the function $turned\_off(\alpha, \textit{onProcs}, \textit{offProcs})$ as:

$$\{\Psi(b) \mid (\lambda, m, \Psi, \chi) \in configs(\alpha, |\textit{onProcs}|, id_{\{1\ldots|\textit{onProcs}|\}}, id_{\{-|\textit{offProcs}|..-1\}}),$$
$$(\phi, x, \beta, f, \psi, \chi', B) \in Rules(\lambda), b \in B\}.$$

The correctness of this optimisation can be proven as follows, noting that $Harness(P, n) = P[\![Rename]\!]$, if n can never be turned **off**. We elide the proof since it follows immediately from the definition of Reg and Lemma 2.1.

Lemma 2.3. If $Operator(\alpha, \textit{onProcs}, \textit{offProcs})$ evolves into a state in which $Harness(P, n)$ has turned P **off**, then $n \in turned\_off(\alpha, \textit{onProcs}, \textit{offProcs})$.

Having performed the optimisations we are now able to compile reasonably sized systems with little difficulty. Note that the above optimisations do not alter Roscoe's original result since, as proven above, the optimisations do not affect the semantics of the $Operator$ process.

3. Recursion

As discussed in the introduction, FDR cannot successfully compile simulations of recursive processes. For example, consider simulating CSP within CSP (which we do for ease of exposition only). According to the simulation given in Section 2, the process $P = a \to P$ would be simulated by the process $P = Operator(Prefix.a, \langle\rangle, \langle P\rangle)$. We calculate the definition of this as FDR would below, writing HR for the renaming done by the $Harness$ and R for the renaming done by $Operator$.

$$\begin{aligned}
P &= Operator(Prefix.a, \langle\rangle, \langle P\rangle) \\
&= Reg(Prefix, \ldots)[\![R]\!] \setminus \{tau\} \\
&= ((\{\}, a, \{\}) \to (Harness(P, 0) \underset{\ldots}{\|} Reg(Identity, \ldots)))[\![R]\!] \setminus \{tau\} \\
&= a \to (Harness(P, 0) \underset{\ldots}{\|} Reg(Identity, \ldots))[\![R]\!] \setminus \{tau\} \\
&= a \to (P[\![HR]\!] \underset{\ldots}{\|} Reg(Identity, \ldots))[\![R]\!] \setminus \{tau\}.
\end{aligned}$$

Thus, P_i, which is P unwrapped i times, is equivalent to:

$$\begin{aligned}
P_0 &= P \\
P_{i+1} &= a \to ((P_i[\![HR]\!] \underset{..}{\|} Reg(Identity, ...))[\![R]\!] \setminus \{tau\}).
\end{aligned}$$

Whilst it is certainly true that P_i is equivalent to any of the other P_i, FDR (unsurprisingly) does not detect this and instead loops, looking for a transition back to P. Therefore, this results in an unbounded number of copies of $Reg(Identity, \ldots)$ being put in parallel.

In this section, we start in Section 3.1 by outlining a simple solution to the recursion issue outline above. In Section 3.2 we then generalise this to allow a large class of recursive definitions to be compiled. We then formalise the transformation and prove it correct in Section 3.3. In Section 3.4 we expand the transformation to permit additional recursive processes to be compiled. Lastly, in Section 3.5 we discuss the potential limitations of the recursion refactorings presented in this section.

3.1. A Simple Solution

One solution to the problem above would be to periodically *throw away* the collection of identity operator regulators that have been spawned. For example, consider the following simulation of $P = a \to P$:

$$\begin{aligned}
PSim &\mathrel{\widehat{=}} Loop(Operator(Prefix.a, \langle\rangle, \langle callPSim \to STOP\rangle)) \\
Loop(Q) &\mathrel{\widehat{=}} Q \ominus_{\{callPSim\}} PSim.
\end{aligned}$$

When compiling $PSim$, according to the definition of $Loop$, whenever $Operator(Prefix.a \ldots)$ performs a $callPSim$ event the left argument of $\ominus_{\{callPSim\}}$ would be terminated and $PSim$ would be started again. In particular, if we calculate the definition of $PSim$ as FDR would, assuming that the identity operator allows $callPSim$ (which we discuss further below), we obtain the following:

$$\begin{aligned}
PSim &= Loop(Operator(Prefix.a, \langle\rangle, \langle callPSim \to STOP\rangle)) \\
&= Operator(Prefix.a, \langle\rangle, \langle callPSim \to STOP\rangle) \ominus_{\{callPSim\}} PSim
\end{aligned}$$

$$= (a \rightarrow ((callPSim \rightarrow STOP)[\![HR]\!] \underset{\dots}{\|} Reg(Identity, \dots))[\![R]\!] \setminus \{tau\})$$

$$\Theta_{\{callPSim\}}\ PSim$$

$$= a \rightarrow (((callPSim \rightarrow STOP)[\![HR]\!] \underset{\dots}{\|} Reg(Identity, \dots))[\![R]\!] \setminus \{tau\}$$

$$\Theta_{\{callPSim\}}\ PSim)$$

$$= a \rightarrow (callPSim \rightarrow ((STOP[\![HR]\!] \underset{\dots}{\|} Reg(Identity, \dots))[\![R]\!] \setminus \{tau\})$$

$$\Theta_{\{callPSim\}}\ PSim)$$

$$= a \rightarrow callPSim \rightarrow PSim.$$

Hence, it follows that $PSim \setminus \{callPSim\}$ simulates $PSim$ in a way that FDR can successfully compile.

Note that $PSim \setminus \{callPSim\}$ is no longer strongly bisimilar to P as there is a τ transition whenever $PSim$ recurses. However, as all CSP denotational models equate $\tau \rightarrow P$ and P, this is not of any consequence providing the processes are used in refinement checks (which is the intention, after all).

3.2. Generalising the Solution

In order to generalise the solution of Section 3.1 to arbitrary processes and operators we give a number of refactorings that must be performed on the simulated script. The first refactoring will be as above; thus, we declare a new channel $callProc$ of type $Proc$ (for ease we assume process names may be sent over channels). We then replace every call inside a recursive process to another recursive process P with a $callProc.P$ event. Using this we can now define a process $WrapThread$ that is analogous to $Loop$ above. Note that much of the complexity in the following comes from the fact that the exception operator does not pass which event from the exception set was performed to its right argument. We work around this by putting the exception operator in parallel with a regulator that remembers which $callProc$ event occurred.

$$WrapThread(P) =$$
$$((P\ \Theta_{\{|callProc|\}}\ startProc?Q \rightarrow WrapThread(Q)) \underset{\{|callProc, startProc|\}}{\|} \mu X \cdot callProc?p \rightarrow startProc!p \rightarrow X) \setminus \{|startProc, callProc|\}.$$

Note that no $callProc$ events can propagate out of a process of the form $WrapThread(P)$.

As $\{|callProc|\} \not\subseteq \Sigma_0$ it follows that we will have to add the following rule to the operational semantics of the $Identity$ operator in order to allow the $callProc$ events can propagate, as discussed above.

$$\frac{P \xrightarrow{callProc.p} P'}{Identity(P) \xrightarrow{callProc.p} Identity(P')}\ p \in Proc$$

We might ask if similar propagation rules should be added to other operators for **on** arguments. However, it is easy to see that in all but the case of the identity operator that if a $callProc$ event were to propagate then information would be lost. For example, suppose $P = a \rightarrow P \,\Box\, R$ and $R = b \rightarrow R$. As the call to R on the right of the external choice is replaced by a $callProc$ event, it follows that the recursion would resolve the external choice which is clearly incorrect.

3.3. Formalising the Solution

We now consider how to formalise the solution sketched above and, in particular, we prove that the process that results from the transformations is equivalent, in a sense we later formalise, to the original simulation. In order to prove this, we firstly formally define what we are actually simulating. In particular, we define a *syntactic process* which corresponds, essentially, to a process definition in a CSP_M file.

Definition 3.1. A *syntactic process* P is defined by the grammar $P ::= Op(P, \ldots, P) \mid N$, where N represents a *process name* and Op is a CSP-like operator. A *process environment* Γ is a function from process names to syntactic processes. For simplicity, we assume that $\Gamma(N)$ is never a process name, and is thus always of the form $Op(\ldots)$[3].

Given a syntactic process, it is possible to construct a simulation of it using the processes from Section 2.1. Thus, we define a function $sim^{\Gamma}(P)$ that, given a syntactic process P, returns the simulation of it. For ease, we assume that the *Operator* process defined in Section 2.1 can be given a single vector of processes, rather than two vectors of **on** and **off** processes, respectively.

Definition 3.2. The *simulation of P in Γ*, denoted $sim^{\Gamma}(P)$ is defined recursively by:

$$\begin{aligned} sim^{\Gamma}(Op(P_1, \ldots, P_N)) &\mathrel{\widehat{=}} Operator(Op, \langle sim^{\Gamma}(P_1), \ldots, sim^{\Gamma}(P_N)\rangle) \\ sim^{\Gamma}(N) &\mathrel{\widehat{=}} N \end{aligned}$$

Note that the above simulation cannot be compiled by FDR if it contains recursive processes, as discussed above. However, it is strongly bisimilar to the original process, given Roscoe's results.

We now consider how to formalise the recursion refactorings we sketched in the preceding section. In order to formalise them, we need to distinguish between the different ways that an operator can use the events performed by its arguments. For example, $\Box$ only looks at the initial visible events performed by each of its arguments, whilst $|||$ looks at all the events performed by its arguments. We formalise this difference in the following definition.

Definition 3.3. An **on** process argument P of a CSP-like operator Op is *1-required* iff P can perform a visible event, i.e. there exists an inductive rule of the form:

$$\frac{\ldots \wedge P \xrightarrow{a} P' \wedge \ldots}{Op(\ldots, P, \ldots) \xrightarrow{b} Op'(\ldots)}.$$

where $a \in \Sigma$. An argument P of a CSP-like operator Op is $(k+1)$*-required* iff there exists an inductive rule of the above form to a CSP-like operator $Op' \neq Identity$ such that P' is k-required for Op'. An argument P of an operator Op is *infinitely recursive* iff P is k-required for every $k \in \mathbb{N}$. An argument P of an operator Op is *only k-recursive* iff P is k-required, but not $k+1$-required.

For example, considering CSP: both arguments of $\|$ are infinitely recursive; both arguments of $\Box$ are only 1-required; the left argument of $\triangle$ is infinitely recursive but the right argument is only 1-required; the left argument of Θ_A is infinitely recursive but the right is not required as it is **off**. Note that, by definition of an **on** argument each **on** argument is at least 1-required.

[3]This is not a restriction in the tool, but is added only to simplify the presentation.

Using the above definitions, we can now formalise the recursion refactorings that we sketched in Section 3.2. Firstly, we change every recursive call[4] to a process N to be a $callProc$ event. Further, if an argument of an operator is infinitely recursive, we apply $WrapThread$, to ensure that no $callProc$ events propagate out. We formalise this as a function that returns a CSP process as follows.

Definition 3.4. Let Γ be a syntactic process environment and P a syntactic process. $rec\_sim^{\Gamma}(P)$ is defined inductively by:

$$
\begin{aligned}
rec\_sim^{\Gamma}(Op(P_1, \ldots, P_N)) &\mathrel{\widehat{=}} Operator(Op, \langle irec^{\Gamma}(Op, 1, P_1), \ldots, irec^{\Gamma}(Op, N, P_N)\rangle) \\
rec\_sim^{\Gamma}(N) &\mathrel{\widehat{=}} callProc.N \rightarrow STOP \\
irec^{\Gamma}(Op, i, N) &\mathrel{\widehat{=}} \begin{cases} WrapThread(rec\_sim^{\Gamma}(\Gamma(N))) & i \text{ is infinitely recursive} \\ rec\_sim^{\Gamma}(P) & \text{otherwise} \end{cases} \\
irec^{\Gamma}(Op, i, P) &\mathrel{\widehat{=}} rec\_sim^{\Gamma}(P)
\end{aligned}
$$

Note that the above simulation will no longer be strongly bisimilar to the original process (or indeed to $sim^{\Gamma}(P)$). This is because whenever a recursion occurs, there will be two extra τ's resulting from the hidden $callProc$ events (cf. $WrapThread$). Whilst this is of theoretical importance, this is not of any consequence to the tool, since all CSP denotational models equate P and $\tau \rightarrow P$. Hence, a particular refinement check would hold on the original process iff it holds on the above simulation.

Note that the above simulation is not sufficient for every CSP-like operator, nor every syntactic process. Thus, we now consider what syntactic processes the above simulation can handle. Firstly, observe that the process $P = Q \Box P$ cannot be simulated, since the available transitions of P directly depend on those of P. Prohibiting such processes is an entirely reasonable restriction: such a P is not well-defined and would be rejected by any CSP-based tool (including FDR). Also, consider the process $P = a \rightarrow P \mathbin{|||} STOP$. Again, the simulation will produce a process that cannot be compiled by FDR, since the left-hand argument is changed to $WrapThread(a \rightarrow callProc.P \rightarrow STOP)$. Hence, when FDR compiles P, after an a is performed another copy of P, itself containing another copy of $WrapThread$, will be spawned by the original $WrapThread$. We claim that prohibiting such processes is entirely reasonable, on the grounds that FDR cannot compile any such process either.

1-required arguments require extra restrictions to be placed since, by definition, no recursions can be permitted before a visible event has been performed. For example, the process $P = (P \sqcap STOP) \Box a \rightarrow STOP$ is not well-defined since the τ of the $\sqcap$ does not resolve the $\Box$, and thus the transitions of P again directly depend on those of P. We believe that such a restriction is reasonable, since FDR would also fail to compile such a process.

More problematically, any process of the form $P = Q \Box a \rightarrow STOP$ cannot be simulated since the Q will be converted to $callProc.Q \rightarrow STOP$, but $callProc$ events are not allowed by the simulation of $\Box$. Thus, it follows that the above simulation cannot deal with *immediately* recursive arguments. We extend the simulation to support such arguments in Section 3.4. We can formalise the set of processes for which the simulation is correct as follows. In the following definition, we only consider operators where each argument of the operator is either only 1-required or infinitely recursive, for simplicity. The definition and subsequent results can be generalised to arguments that are only k-required for $k > 1$.

[4]We don't actually need to change every recursive call; it actually suffices to change only recursive calls by non-recursive processes to recursive processes or by recursive processes to other recursive processes. We do not formalise this version since it adds extra complications.

Definition 3.5.

- Given a syntactic process P in a process environment Γ, P *recurses to* N iff either $P = N$, or $P = N'$ where $\Gamma(N')$ recurses to N, or $P = Op(P_1, \ldots, P_N)$ and one of the P_i recurses to N.
- A syntactic process P *recurses to* N *through only* ***on*** *arguments* iff $P = N$, or $P = N'$ and $\Gamma(N')$ recurses to N through only **on** arguments, or $P = Op(P_1, \ldots, P_N)$ and there exists an **on** P_i such that P_i recurses to N through only **on** arguments.
- A syntactic process P *has no named* ***on*** *arguments* iff P is not a name, and if $P = Op(P_1, \ldots, P_N)$ then each **on** P_i has no named **on** arguments.
- A syntactic process P *has 1-guarded recursion to* N iff P does not recurse to N, or $P = N'$ and $\Gamma(N')$ has 1-guarded recursion to N, or $P = Op(P_1, \ldots, P_N)$ and whenever $Op(P_1, \ldots, P_N) \xrightarrow{\tau} Op'(\mathbf{P})$, $Op'(\mathbf{P})$ has 1-guarded recursion to N.
- A process environment Γ *has correct simple recursion* iff for every $(N, P) \in \Gamma$, for every instance of $Op(P_1, \ldots, P_N)$ within P, and for each **on** P_i:
  - ∗ If P_i is infinitely recursive, then P_i does not recurse to N;
  - ∗ If P_i is only 1-required, then P_i does not recurse to N through only **on** arguments, P_i has no named **on** arguments and P_i has 1-guarded recursion to N.

Since the definition of *has correct simple recursion* only mentions operator arguments that are only 1-required and infinitely recursive, it follows that we can only allow operators that have only such arguments. With this observation in mind, we can state and prove our main result. Note in the following, as discussed after Definition 3.4, we weaken the equivalence from strong bisimulation to equivalence in any CSP denotational model.

Theorem 3.6. Let P be a syntactic process and Γ be a syntactic process environment with correct simple recursion. If every argument of every operator used in Γ or P is either infinitely recursive or only 1-required, then $WrapThread(rec\_sim^{\Gamma}(P)) = sim^{\Gamma}(P)$, where equality is interpreted in any CSP model.

Proof (Sketch). Let P and Γ be as per the lemma. Formally, this theorem can be proven by a structural induction over P. The correctness of this result follows primarily from the requirement that Γ has correct simple recursion. This means that, if an argument wishes to perform a $callProc$ event, then the operator must have evolved into the $Identity$ operator state, which allows the $callProc$ event to propagate, which can then be picked up by $WrapThread$. □

3.4. Immediate Recursions

The above solution enables many recursive process definitions to be compiled, but does not cover every possibility. As noted above, it does not correctly simulate processes that have *immediate* recursion, such as $P = a \to P \,\square\, R$ and $R = b \to R$. In particular, in $rec\_sim^{\Gamma}(P)$ the right hand side of the external choice of P would be replaced by a $callProc$ event. However, as these are not allowed to propagate through the external choice it follows that only the choice of the left hand side would be presented.

The simplest solution to the above would be to *inline* the definition of R within P to yield the process $P = a \to P \,\square\, b \to R$, which could be successfully simulated (in particular, it has correct simple recursion and thus Theorem 3.6 applies). However, this will not work on processes such as $P = a \to (P \,\square\, b \to P)$ since, after inlining, the process still contains an incorrect usage of external choice (in this case, the inlined version would be $P = a \to ((a \to (P \,\square\, b \to P) \,\square\, P))$. To solve the above we introduce a fresh name R for the process $P \,\square\, b \to P$. We then rewrite the definitions as:

$$P \mathrel{\widehat{=}} a \to R$$

$$R \mathrel{\widehat{=}} a \to R \,\square\, b \to P.$$

Note that the above processes have no disallowed immediate recursions.

In general, if an argument of a process is only k-required, we need to inline sufficiently to ensure that none of the first k events can ever be $callProc$ events. If we do this, then it follows that the resulting syntactic process will satisfy the preconditions of Theorem 3.6 and thus this can be applied to correctly simulate the process. In order to formalise this we firstly expand the definition of has correct simple recursion to allow immediate recursion. As before, we concentrate only on operators where each argument is either infinitely recursive, or only 1-required. Again, the definition could be generalised as required.

Definition 3.7. A process environment Γ *has correct recursion* iff for every $(N, P) \in \Gamma$, for every instance of $Op(P_1, \ldots, P_N)$ within P, and for each **on** P_i:

1. If P_i is infinitely recursive, then P_i does not recurse to N;
2. If P_i is only 1-required, then P_i does not recurse to N through only **on** arguments and P_i has 1-guarded recursion to N.

Note that the second clause prohibits syntactic processes where the events for a process depend recursively on the process itself. Clearly such processes are not well-defined. We now define how to inline process definitions, in order to support the refactoring that was sketched above. Note that the following transformation is only well-defined if the process environment has correct recursion.

Definition 3.8. Given a syntactic process environment Γ', the *inlined process environment* Γ' is defined on each syntactic process contained in any process in Γ. In particular, $dom(\Gamma')$ is the set of all $N_{P'}$ where P' is a subprocess of some $P \in img(\Gamma)$. Further, Γ' is defined by:

$$\Gamma'(N_{Op(P_1,\ldots,P_M)}) \mathrel{\widehat{=}} Op(inline(Op, 1, P_1), \ldots, inline(Op, N, P_N))$$

$$\Gamma'(N_{N'}) \mathrel{\widehat{=}} \Gamma'(\Gamma(N'))$$

$$inline(Op, i, P_i) \mathrel{\widehat{=}} \begin{cases} N_{P_i} & i \text{ is } \textbf{off} \\ \Gamma'(N_{N'}) & i \text{ is } \textbf{on} \text{ and } P_i = N' \\ \Gamma'(N_{Op'(\mathbf{P}')}) & i \text{ is } \textbf{on} \text{ and } P_i = Op'(\mathbf{P}') \end{cases}$$

Note that the second clause of Γ' is well-defined by the assumption in Definition 3.5 that $\Gamma(N) \neq N'$. Further, observe that the $inline$ is semantics-preserving since it is essentially just a renaming operation. Thus, if $N \in dom(\Gamma)$, then $sim^{\Gamma}(N)$ is strongly bisimilar to $sim^{\Gamma'}(\Gamma'(N))$.

Clearly, the inlining process will only terminate if no **on** argument recurses back to itself. This is implied by the definition of has correct recursion in Definition 3.7.

Theorem 3.9. Let Γ be a process environment with correct recursion and let Γ' be the flattened process environment of Γ. Then Γ' has correct simple recursion.

Proof. Let Γ and Γ' be as per the lemma. By definition of correct recursion and correct simple recursion it suffices to show: for each $(N, P) \in \Gamma'$, each $Op(P_1, \ldots, P_N)$ that is a subprocess of P, and each **on** argument P_i, that P_i has no named **on** arguments. This follows immediately from the definition of $inline$. □

For optimisation reasons, the tool does not implement the refactoring exactly like this, but instead detects when the recursion refactoring has to be applied and does so only to such processes. This is discussed further in Section 4.

3.5. Limitations

The recursion refactorings that we have discussed above allow a large class of recursive processes to be successfully simulated. However, there are fundamental limitations to the range of operators that can be simulated in such a way. One good example of an operator that cannot be simulated using the above construction is the new CSP operator *synchronising external choice*, $\Box_A$ [11]. This operator behaves likes a hybrid of external choice and generalised parallel. In particular, $P \Box_A Q$ is resolved by P or Q performing a visible event outside of A, whilst if P or Q wishes to perform an event from A, it must synchronise with the other process. Thus, $\Box_{\{\}}$ is equivalent to $\Box$, whilst $\Box_\Sigma$ is equivalent to $\underset{\Sigma}{\parallel}$.

This operator is particularly problematic for the recursion refactorings since it is the only CSP operator where an argument that is infinitely recursive can also be recursed through. For example, consider the process: $P = a \to b \to P \Box_{\{a\}} a \to STOP$. In this process, P recurses through the first argument of $\Box_{\{a\}}$, which is infinitely recursive[5], (cf. Definition 3.3). However, FDR has no problem compiling the above process, because when the b is performed, $\Box_{\{a\}}$ is discarded, meaning that a transition back to P is found, as required.

There is no obvious way of altering the simulation to support an operator that might sometimes recurse (and thus requires the $callProc$ event to not be consumed by a $WrapThread$ inside $\Box_{\{a\}}$), but sometimes never recurses (and thus requires that all $callProc$ events *must* be consumed by a $WrapThread$ inside $\Box_{\{a\}}$). In [1] Roscoe actually conjectures that it is impossible to simulate $\Box_A$ in such a way that FDR can successfully compile processes that use it. Thankfully, such operators are sufficiently uncommon that the simulation we provide is still applicable to a very wide variety of operational semantics.

4. Tool Support

As part of this work we have constructed a tool called tyger, written in Haskell, that automatically constructs the simulation and refactors the user's process definitions, as above. The input to tyger consists of two input files, the first of which specifies the operational semantics of the language. For example, the machine-readable version of the SOS rules for the CSP exception operator, as given in Section 1.1, can be defined as follows:

```
Operator Exception(P : InfRec, Q, A)
    Syntax Binary "[| $3 |>" 12 AssocNone
    Rule
        P =a=> P'
        ----------------------------- a <- diff(Sigma, A)
        P [| A |> Q =a=> P' [| A |> Q
    EndRule
    Rule
        P =a=> P'
        ------------------ a <- A
        P [| A |> Q =a=> Q
    EndRule
EndOperator
```

In the above, `Exception(P : InfRec, Q, A)` specifies that the exception operator takes 3 arguments, the first of which is a process that is infinitely recursive (cf. Definition 3.3). The `Syntax` line provides the information that tyger requires to parse the file. In particular, it specifies that the operator is a binary operator, with concrete syntax "`[| $3 |>`" where $3 refers to the third argument (i.e. `A`), is non-associative and has a precedence of 12 (which is used to disambiguate non-bracketed statements). Using this line tyger dynamically constructs a parser (using the Haskell Parsec library[6]) to parse this file. The two `Rule` constructs

[5] Hence, this violates the conditions of has correct recursion, since P recurses through an infinitely-recursive argument.

[6] `http://hackage.haskell.org/package/parsec`

are precisely the machine readable versions of the SOS rules from Section 1.1 (noting that `diff(Sigma, A)` is the machine readable form of $\Sigma \setminus A$). Note that tau promotion rules are omitted as they are automatically added.

After parsing the above file, tyger performs a sort of type-checking on the operational semantics and infers which arguments of the operators are **off** and **on**. The type-checking is necessary because the side conditions of the operational semantic conditions (e.g. the `a <- Sigma`) can be written in a simple functional language, consisting of a simple set comprehensions. It also checks that the operators are CSP-like by checking a number of conditions on each SOS rule. For example, it checks that no **on** arguments are cloned, no **on** arguments are suspended, amongst several other conditions. The operational semantics are then converted into the form specified in Section 2.1. At this point, the simulation script can be produced, which is done by concatenating some constant code that implements the $Operator$ process with some generated code for the $Rules$ function.

The second input file contains process definitions which, thanks to the dynamic parser construction, can be specified using natural syntax. For instance, when defining a CSP process one can simply write `P = a -> P [] b -> Q`. Further, the script may contain definitions using the functional portion of the CSP$_M$ language, such as `head(<x>^xs) = x`. The script is then type-checked (using an experimental CSP$_M$ type checker), and then has the recursion refactorings applied to it. In particular, the inferred types are used to identify all processes within a file (note that the processes here are essentially the syntactic processes of Definition 3.5). Then, a call graph of the processes is constructed and the strongly connected components (SCCs) are deduced. This enables the recursive processes to be identified, which then allows the recursion refactorings to be run.

If the recursion refactoring specified in Section 3.4 were applied exactly as specified, this would result in many more named processes than are strictly necessary, slowing down the simulation. Thus, instead the following recursion refactorings are applied instead:

- Every call from a recursive process to a recursive process is replaced by a $callProc$ event;
- Every call from a non-recursive process to a recursive process Q instead calls the wrapped version, $WrapThread(Q)$ (so that $callProc$'s are processed);
- Every infinitely recursive argument Q of an operator is replaced by $WrapThread(Q)$;
- Every argument that is only 1-required is inlined as per Section 3.4.

Further, in order to define the $WrapThread$ process (cf. Section 3.2), the set of values that each recursive process can take has to be specified. For example, the following defines a recursive process P that takes a single integer value that can either be 0 or 1:

```
P :: ({0, 1}) -> Proc
P(x) = if x == 0 then a -> P(1) else b -> P(0)
```

Lastly, the transformed processes are pretty printed to a file. This file can be loaded into FDR and any assertions that are contained can be checked in the usual way.

Availability tyger has been open-sourced and is available from `https://github.com/tomgr/tyger`. Included with the code are several examples, including a full simulation of CSP, a simulation of Lowe's availability models [10], a simulation of Lowe's Readyness Testing extension to CSP [9], and a simulation of a portion of CCS [7].

5. Experiments

In order to demonstrate the effectiveness of the tool and to establish how efficient the simulation is, we performed a number of experiments which we now describe. In this section, all

	Number of Philosophers							
Tool	3	4	5	6	7	8	9	10
Simulation	0.4	1.0	1.9	3.3	5.6	10.6	33.2	173.2
FDR	< 0.1	< 0.1	< 0.1	0.1	0.2	1.1	7.0	42.5

Table 2. Time in seconds to check if a CSP model of the Dining Philosophers is deadlock free for various numbers of philosophers.

	Number of Philosophers						
Tool	3	4	5	6	7	8	9
CWB	< 0.1	< 0.1	0.3	2.6	38.5	*	*
Simulation	0.6	0.8	1.8	3.0	8.9	53.9	384.0

Table 3. Time in seconds to check if a CCS model of the Dining Philosophers is deadlock free for various numbers of philosophers. The CWB was unable to falsify the property within 30 minutes with 8 or more philosophers.

experiments were performed on a 64-bit Linux virtual machine with access to 2GB of RAM and two 2.2GHz cores. The simulation was run using a pre-release version of FDR3, running in single-threaded mode. CSP model checking was performed using the same pre-release version of FDR3 in single-threaded mode. CCS model checking was performed using version 7.1 of The Edinburgh Concurrency Workbench [12].

In order to directly measure the overhead of the simulation, we simulated CSP itself. Clearly, there is no point in doing this in general, but if we compare the time taken to check if a CSP process P satisfies a property to the time taken to check if the same property is satisfies by the simulation of P, then it follows that any difference must be due to the overhead of the simulation. Table 2 gives the time taken to check if a simple CSP model of the classic Dining Philosophers problem is deadlock free (in particular, it gives the time taken to find the first counterexample). Whilst the simulation is slower, note that the magnitude of the difference becomes less as the number of philosophers is increased. Further, the simulation is still reasonably performant and scales in an almost identical way to the non-simulated version, indicating that there should be no reason why larger models cannot be checked.

In order to demonstrate the that tool can effectively model other CSP-like process algebras, we compared the performance of a simulation of CCS [7] with a native CCS tool, The Edinburgh Concurrency Workbench [12] (henceforth, CWB). As noted in Section 1.2, CCS is actually not quite CSP-like, since τ resolves $+$ (which is the CCS analog of $\Box$). Hence, in order to simulate CCS we instead introduce a new event, *ccs_tau*, which is used instead of *tau* in the operational semantics rules. Then, explicit *ccs_tau* promotion rules are added to each operator and *ccs_tau* is hidden at the top-level of the simulation. It follows that the resulting operational semantics is CSP-like, although it is not compositional. Once this change has been made, the translation of the operational semantics into the form required for tyger is routine.

Table 3 details how long it took to verify if the Dining Philosophers problem is deadlock free for an increasing number of philsophers. Whilst CWB outperforms the simulation for 6 philosophers or less, for 7 philosophers or more the simulation actually outperforms CWB. Further, CWB was unable to falsify the property within 30 minutes when there were 8 or more philosophers whilst the simulation took just 54s to find a counterexample. To a large extent, this data justifies the motivation behind the tool. Not only is the simulation able to effectively simulate other process algebras, but is actually able to improve on the performance of custom tools (largely thanks to the optimisation performed on FDR).

We have had success in simulating a number of other CSP-like languages using tyger. One interesting example is Lowe's readyness testing extension to CSP [9]. This adds a new operator, "if ready a then P else Q", that tests if the event a is being offered and behaves like

P if so and like Q otherwise. Lowe demonstrated its usefulness by giving a notably simple solution to the readers and writers problem [13] that was fair to the writers. We were able to successfully simulate this, with the simulation taking 60s to complete a safety check.

In the author's experience, the performance of the simulation is perfectly adequate providing the number of events is resaonble, but unsuitable when the set of events become large. This is primarily because the alphabets used by the simulation become prohibitively large, and FDR appears to have trouble manipulating them.

6. Conclusions

In this paper we presented a CSP_M adaptation of Roscoe's simulation that can be successfully used in FDR. Further, we gave a number of optimisations to Roscoe's simulation that ensure the resulting simulation runs reasonably efficiently. We then proved that these optimisations maintained the strong bisimulation between the original process and the simulation. We also provided some automated recursion refactorings that permit a large class of recursive definitions to be successfully compiled by FDR, even though the standard simulation could not be. We then sketched the construction of a tool that can produce the simulation automatically. Lastly, we described the results of several experiments that show the tool can effectively simulate other process algebras.

Future Work The overhead imposed by the simulation is essentially unavoidable and, therefore, there are unlikely to be ways of significantly improving the performance using the current approach. However, one option would be to add support directly within FDR for directly constructing the state machines for CSP-like languages. This would entirely avoid the overhead of the simulation and, further, would negate the need for the complicated recursion refactorings required.

There are several programs that produce CSP scripts as output, such as Roscoe's shared variable compiler [14] and Lowe's Casper Security Protocol Compiler [15]. The output of these tools can take a long time to compile, sometimes because it was not possible to express certain behaviours as a simple combination of CSP operators. Thus, it would be interesting to consider if domain-specific operators could be added to enable the above scripts to be expressed in a way that compiles more efficiently, using the new refinement checker mentioned in the previous paragraph.

Acknowledgements I would like to thank Gavin Lowe, who supervised a large part of this work as part of an Oxford undergraduate project [16]. Further, I would like to thank Gavin for reviewing an early draft of this paper and Bill Roscoe for many interesting discussions concerning this work. I would also like to thank the anonymous reviewers for providing many useful comments.

References

[1] A. W. Roscoe. On the Expressiveness of CSP. Draft of February 2011, available from `http://www.cs.ox.ac.uk/files/1383/complete(3).pdf`, 2011.

[2] A. W. Roscoe. CSP is Expressive Enough for Pi. In A.W. Roscoe C.B. Jones and K.R. Wood, editors, *Reflections on the work of C.A.R. Hoare*. Springer, 2010.

[3] C. A. R. Hoare. *Communicating Sequential Processes*. Prentice-Hall, Inc., Upper Saddle River, NJ, USA, 1985.

[4] A. W. Roscoe. *The Theory and Practice of Concurrency*. Prentice Hall, 1997.

[5] A. W. Roscoe. *Understanding Concurrent Systems*. Springer, 2010.

[6] Formal Systems (Europe) Ltd. *Failures-Divergence Refinement—FDR 2 User Manual*, 2011.

[7] R. Milner. *A Calculus of Communicating Systems*. Springer-Verlag New York, Inc., Secaucus, NJ, USA, 1982.

[8] R. Milner. *Communicating and Mobile Systems: the Pi-Calculus*. Cambridge University Press, June 1999.

[9] G. Lowe. Extending CSP with Tests for Availability. In *Communicating Process Architectures*, pages 325–347, 2009.

[10] G. Lowe. Models for CSP with Availability Information. In *EXPRESS'10*, pages 91–105, 2010. `http://dx.doi.org/10.4204/EPTCS.41.7`.

[11] P. Armstrong, G. Lowe, J Ouaknine, and A.W Roscoe. Model Checking Timed CSP. In *Proceedings of HOWARD (Festschrift for Howard Barringer)*. 2012.

[12] Faron Moller and Perdita Stevens. Edinburgh Concurrency Workbench user manual (version 7.1). Available from `http://homepages.inf.ed.ac.uk/perdita/cwb/`.

[13] P. J. Courtois, F. Heymans, and D. L. Parnas. Concurrent Control with "Readers" and "Writers". *Communications of the ACM*, 14(10):667–668, 1971.

[14] A. W. Roscoe and D. Hopkins. SVA, a Tool for Analysing Shared-variable Programs. In *Proceedings of AVoCS 2007*, pages 177–183, 2007.

[15] G. Lowe. Casper: A Compiler for the Analysis of Security Protocols. *Journal of Computer Security*, 6(1-2):53–84, 1998.

[16] Thomas Gibson-Robinson. Tyger: A Tool For Automatically Simulating CSP-Like Languages In CSP. MCompSci Thesis, University of Oxford, 2010.

Communicating Process Architectures 2013
P.H. Welch et al. (Eds.)
Open Channel Publishing Ltd., 2013

Selective Choice 'Feathering' with XCHANs

Øyvind TEIG[1]
Autronica Fire and Security AS[2], Trondheim, Norway

Abstract. This paper suggests an additional semantics to **XCHAN**s, where a sender to a synchronous channel that ends up as a component in a receiver's selective choice (like **ALT**) may (if wanted) become signaled whenever the **ALT** has been (or is being) set up with the actual channel *not* in the active channel set. Information about this is either received as the standard return on **XCHAN**'s attempted sending or on the built-in feedback channel (called **x**-channel) if initial sending failed. This semantics may be used to avoid having to send (and receive) messages that have been seen as *uninteresting*. We call this scheme *feathering*, a kind of low level *implicit* subscriber mechanism. The mechanism may be useful for systems where channels that were not listened to while listening on some *other* set of channels, will not cause a later including of those channels to carry already declared uninteresting messages. It is like not having to treat earlier bus-stop arrival messages for the wrong direction after you sit on the first arrived bus for the correct direction. The paper discusses the idea as far as possible, since modeling or implementation has not been possible. This paper's main purpose is to present the idea.

Keywords. channels, synchronous, asynchronous, buffers, overflow, flow control, CSP, modeling, semantics, feathering

Introduction

The idea of the suggested semantics appeared after reading Tony Hoare's lecture "Concurrent programs wait faster" from 2003 [1]. After some pondering of a situation described there, we saw that it might be viable to use **XCHAN** [2] as a vehicle for a secondary problem not mentioned in Hoare's lecture. The problem is how to avoid having to relate to information of busses that would potentially stop at a bus stop but heading in the wrong destination. It was believed that this could map to a new software pattern: *feathering*.

The word *feathering* is used like "turning an oar parallel to the water between pulls" [3]. Metaphorically a pull is like an **ALT** selective choice. The oar is pushed into the water at one place: only one channel is taken. But we can hear the oar whip the top of the small waves on its way saying "was there, but not interested". So we take the step to name barely touching the small waves as *feathering*.

The author is not aware of **XCHAN** having been implemented in any language or runtime system. The time when a "channel was a channel" seems to be over; the plethora of channel types and channel usage seems to increase. Worth mentioning here is Go's channel usage and semantics [4], which are quite different from what we are used to in the **occam** tradition. The **XCHAN** is here perceived as being an expansion of the traditional channel (**CHAN**) type.

This *feathering* pattern has not been formally modeled or proven. The paper informally introduces the ideas. No literature search has been done to find similar or equal ideas.

[1] The author works with concurrent software for fire detection systems, but this "industrial paper" does not necessarily reflect views taken by the company. See: http://www.teigfam.net/oyvind/work/work.html
[2] A UTC Fire & Security company. http://www.autronicafire.com

The **occam** language [5] is used as a vehicle for the reasoning here because of its clean design. However, Listing 1 will show ANSI C macro usage, close to how we might have implemented it at Autronica.

1. Original XCHAN

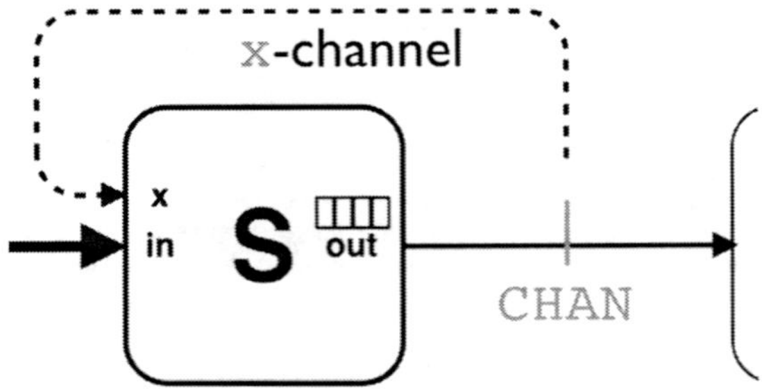

Figure 1. **XCHAN** is **CHAN** plus **x**-channel.

We repeat the basics from [2]. The whole idea with **XCHAN** is to merge asynchronous and synchronous traditions and allow application level (a sender or server) to handle overflow while it always is allowed to continue (like accept new input). **XCHAN** may be buffered or not; in the latter case a buffer could instead be inside the sender, to allow more control of overflow handling like flushing (see Figure 1). Also shown by Figure 1, an **XCHAN** is like a standard **CHAN** with a bundled **x**-channel feedback channel. This feedback originates in the scheduler[3]. The sending on **XCHAN** always returns with a status telling about *success* or *not success*; so sending never blocks. In the *success* case the message passes through like for a standard **CHAN**, and all is done. *Not success* means that there was no reader or buffer available. In this case the sender is required to listen on the **x**-channel and not send again. When **x**-channel arrives the sender must *commit to send* immediately - old, new or even data ignorantly uninteresting to the receiver. The receiver side does not know whether data arrived on an **XCHAN** or a **CHAN**; and the **ALT** as seen by the application in the receiver is not changed from a standard **ALT**. The discussion uses standard single channel output, not from a selective choice with output guards. The receiver side is a selective choice with input guards only, or a single channel input.

The semantics in summary:

1. **XCHAN** may be used in any type of input or output construct except sending from a selective choice with output guards (not in **occam** anyhow).
2. The sending call never blocks but returns a status value.
3. If the sending call returns *success* then the message has been taken by the channel (if buffered) or by a receiver (if non-buffered). This may be repeated any number of times, without use of the **x**-channel.
4. If the sending call returns *await_commit* then the sender is committed to listen on the **x**-channel amongst its other work. This may be **PRI ALT**ing on the **x**-channel

[3] We will mention the runtime *scheduler* several times in this paper, even if it would be invisible if **XCHAN** and *feathering* were burnt into a language. Still *scheduler* is seen as having some value for the reasoning.

along with other input sources or it may mean regular polling of the **x**-channel. When the **x**-channel is triggered, the receiver is committed to receive (if the **XCHAN** is unbuffered) or the channel has space (if buffered) and the sender must send (old or new data) as soon as possible.

To synchronise the sender and receiver when they are both committed, the receiver's input will be two phased. The first phase will accept the request and the second will take the data. This is an implementation detail invisible to the programmer.

To prepare **XCHAN** for *feathering* (discussed in the following sections):

a. We will extend the sending return value set to also include a *feathered* value, in addition to *success* and *not success*, the latter called *await_commit*.
b. We will let **x**-channel carry data with ***x-committed***, ***x-feathered*** or ***x-unfeathered*** values (naming convention is to prefix with '*x_*' all that comes in on **x**-channel).

The **x**-channel usage may alternatively be implemented to provide the sender with a precondition and then become involved on every sending [6]. We have not used these semantics here.

2. XCHAN with Feathering Semantics

2.1 Problem: Busses in the Wrong Direction

Hoare's lecture [1] starts by describing the average waiting time for a bus that will take a passenger from A to B. Instead of taking a particular route and wait for it to arrive, Hoare shows that the standard practice in concurrent programming to instead take the *first* bus which could take the passenger to the same place (e.g. bus routes 0 and 2 in Figure 2, Section 2.2) would make the system "wait faster" in Hoare's wording. The waiting time may become shorter than one would assume at first thought, Hoare shows.

This author's first whim to this was to think of each bus route as a software process, and the bus stop as a multiplexer, containing a selective choice. In **occam** this is an **ALT**; in most modern embodiments this is a variant of **select**. Being used to thinking that any unnecessary communication should be avoided; we saw that messages of wrong busses arriving at the bus stop might be a problem. Entering just any first bus could be wrong, as having to rise from the bus stop's bench every time to inspect the line number on the bus perhaps would be unnecessary. Being near-sighted easily causes more work than necessary - and besides, entering a wrong bus could cause a meeting to be lost. So we would not want a paradigm where we would be fed and have to inspect every bus passage or message.

A process could enter an **ALT** and then switch off the set of non-interesting bus arrival messages, by having the Boolean expression in those guards evaluate to **FALSE**. This way the process could wait for only the correct set of buses. This is a standard *channel based programming* pattern.

What happens after the first correct bus has arrived is not of interest. However, if in fact uninteresting buses had arrived during the waiting time, the bus stop process would have to treat those messages of older arrivals *afterwards*, even with the pattern described. The bus processes may be blocking to send - or have sent and forgotten the message in a queue or buffered channel. There is no way to avoid having to flush these messages. *But we could have avoided sending them.*

2.2 Suggestion

It appeared that an **XCHAN** as described in [2] would not help; but might it be a possible vehicle on which to build a mechanism to avoid having to send the old bus arrival messages? We think so.

Unlike the traditional **CHAN**, the **XCHAN** sender already has a means to treat messages that were not sent immediately. As we have seen we suggest this will mostly be interesting when:

1. the receiver end is a pre-conditioned guard in an **ALT**
2. the **XCHAN** is not buffered (see above)

Events participating in an **ALT** are "enabled". When or if one is ready and selected, both it and all the enabled channels are "disabled"; the **ALT** is "torn down". So, with **CHAN** or **XCHAN** there already is a way to handle the *interesting* channels that did not win the **ALT**.

However, in the present implementation of an **ALT**, a guard's Boolean pre-condition (if present) is evaluated and if the result is **FALSE**, then the channel is not touched. To the channel it looks like the input side is not present, like no 'call' to the input[4].

We propose marking **XCHAN** guards that have **FALSE** pre-conditions in a blocked receiving **ALT** as *feathered* - i.e. that this process is not interested in them[5]. A sender attempting to send to an **XCHAN** in that state (or has already attempted to send and been told to wait) is now signalled (on the **x**-channel) about that state. The sender should abandon its message and should not send anything more until the **XCHAN** is *unfeathered*, which happens at the end of the **ALT** (and signalled on the **x**-channel).

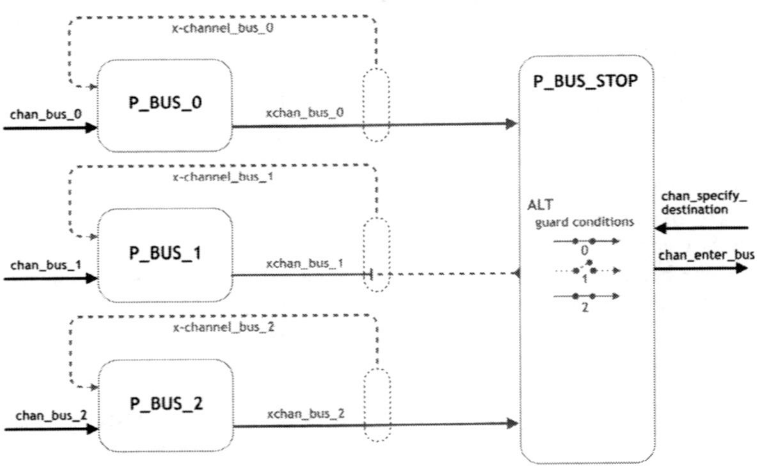

Figure 2. XCHAN (array of 3) and feathering, with only buses #0 and #2 possible to ride.

Figure 2 shows an example of this. The *bus stop* process is waiting on its **ALT** for either buses 0 or 2 (the channels from those buses have **TRUE** pre-conditions) – but is, on this occasion, not interested in bus 1 (whose channel is pre-conditioned **FALSE**).

---

[4] This author is not aware of any implementation where it is any other way.
[5] This is information that is 'lost' in the traditional semantics of a channel.

2.3 Feathering Semantics (Discussion)

With *feathered* semantics we are given a means not to have to send that unnecessary data. This section is a discussion. Section 2.4 tries to summarise.

A sending is done with an output attempt in the **XCHAN** the usual way. However, we need to declare that we want *feathering* semantics.

2.3.1 Successful immediately

Sending attempt will go through as successful if the receiver was able to receive. As mentioned, the **x**-channel is not involved.

2.3.2 Unsuccessful

If sending were unsuccessful there are two outcomes: either there was no receiver or the channel had been *feathered*, meaning that the receiver is already in an **ALT** waiting on other events and explicitly excluding the sender's **XCHAN**. In both cases the sender receives this status (*success*, *feathered* or *await_commit*) after the **XCHAN** output. It never blocks.

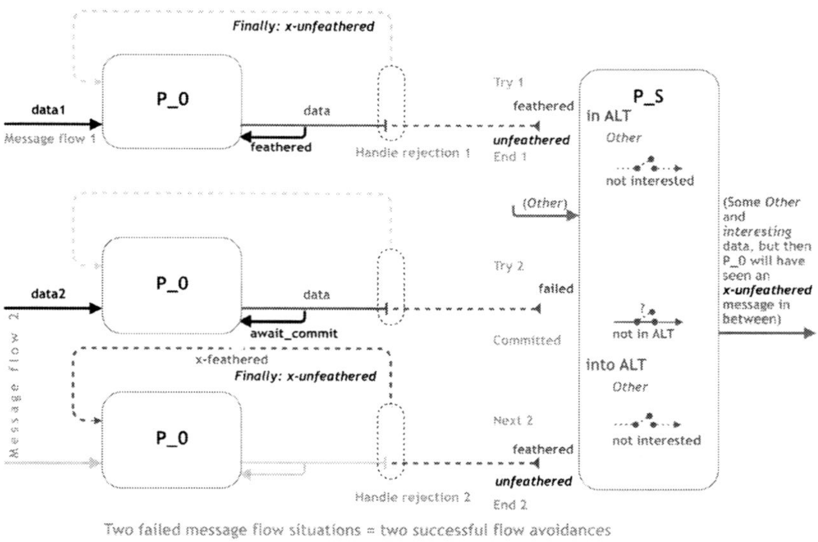

Figure 3. Two mind map scenarios that show message avoidance.

Figure 3 resembles Figure 2, but in addition to topology (process, data flow) it also attempts to overlay sequences. The diagram may be seen as an informal mind map showing message avoidance only. The first scenario (top) shows an initial *feathered* which is terminated with an ***x-<u>un</u>feathered***. The second scenario shows first an initial *await_commit* (centre) followed by an ***x-feathered*** and final ***x-<u>un</u>feathered*** (bottom).

The sender is finished if it receives *success*. It will handle *feathered* at application level, so it is per design what it would do with the message that seemed to be uninteresting. We shall see below that *feathered* is not a state in which the application has finished with its attempted send.

2.3.3 Towards feathering

If the sender gets an *await_commit* return then it is obliged to listen on the **x**-channel. If the receiver commits to the read (or buffer space in the **XCHAN** becomes available), then *x-committed* will be sent and the sender must send as soon as possible. If it receives *x-feathered* then it means that the receiver has already *feathered* the channel, and in this way told any potential sender (of which it is not aware) that it is not interested. The sender should treat this as it will with the *feathered* status return.

2.3.4 Ready for next with x-unfeathered

Observe that when a message has been "warned away" with *feathering* we propose that there should be no undo; that it shall not be possible not send *anyhow*. If the receiver end of the channel later should becomes ready that should in the default semantics be for a next message.

Feathered status is never returned when the receiver is not in the act of entering or being in an **ALT**, so it must be removed when the **ALT** is torn down.

It is important that there should be no divergence (livelock) between the sender and the receiving **ALT**. The allowable action for a sender after seeing a *feathered* receiver end should probably be the same for a *feathered* status return or an *x-feathered* value. In the first case the receiver is already in an **ALT** with that **XCHAN** guard pre-conditioned **FALSE** (we have already mentioned that we would never return *feathered* if the receiver is anywhere else in its code). Retrying to send would in that case cause divergence, as the same *feathered* status return would be repeated infinitively. The same would happen with an *x-feathered* value. Trying to send immediately would therefore also cause a repeated *feathered* status return. Therefore, after any *feathered* status we need a channel 'link level' intermediate state to avoid this. During this state it is not allowed to resend. To terminate this state we propose to wait for an *x-unfeathered* value.

This is sent when the **ALT** has been taken by one of the other inputs after having waited; on 'tearing down' the **ALT**. It will send this to all the *feathered* **x**-channels.

2.3.5 Observation

The described scheme certainly motivates a need to model *feathering* semantics and to see if there are any unseen pitfalls. Also, this pattern may be asserted with a proper rule for the compiler to (usage) check.

Section 2.4 summarises the proposed semantics of *feathering*. Section 2.5 gives a coding template for a process sending on an **XCHAN** with *feathering*. Further discussion is in Section 3.

2.4 Feathering Semantics (Summarised)

This is a summary of the suggested semantics, not a formal model.

1. *Feathering* semantics inherits **XCHAN** semantics (see Section 1)
    a. However, output and input constructs are limited by point 2 (here)
    b. This may not include buffered **XCHAN**; we are uncertain about the usability of *feathering* semantics in that case.
2. *Feathering* is possible if the receiver end of **XCHAN** being processed by an **ALT**, not a single channel input. However, the sending side is a single channel output, not part of an **ALT** with output guards (not possible in **occam**).
3. *Feathering* semantics is specified with a parameter in the **XCHAN** send call.
4. The *feathered* status is returned to a sender that is trying to send when a receiver is in an **ALT** and the requested channel has been tagged by the receiver as *not-interesting* (i.e. its pre-condition is **FALSE**).
5. The *feathered* status is sent to a sender on **x**-channel if it has been trying to send but got *await_commit* reply; when the receiver enters an **ALT** and the requested channel is being tagged as *not-interesting*.
    a. The **ALT** in the receiver sends these *synchronous* messages to all pre-condition excluded feathered **XCHAN** senders at the end of the **ALT** setup if the **ALT** blocks - i.e. it is not immediately taken by another guard (channel, timeout or **SKIP**).
    b. None of these will block indefinitely, as all the receivers will already have committed to listen on **x**-channel.
6. Whenever a sender knows that a channel is *feathered* it shall obey the rule not to resend before an *unfeathered* message has been received on **x**-channel.
7. The *unfeathered* status is delivered to a *feathered* **x**-channel when the **ALT** is later on taken (by another guard) and 'torn down', in the same synchronous scheme as described above (5.a-b)
8. The **x**-channel will only carry an **x**-*unfeathered* after a *feathered* situation.
9. The **x**-channel will only carry **x**-*feathered* or *x-committed* after an *await_commit* status return on the initial sending call.
10. A receiver could possibly do a system call to learn if a message in fact did get rejected. This information could alternatively be delivered on an "**n**-channel"[6] that could be "parallel" with the **XCHAN**'s input on the receiver side. This probably is a complicating matter since type of channel is transparent on the receiver side. We will not discuss this here.

[6] Like "**x**-channel" on the sender side of **XCHAN** and "**n**-channel" on the receiver side of **XCHAN**.

2.5 Code example

```
01 CP->Tag = READY; // READY, SUCCESS, AWAIT_READY, FEATHERED
02 while (true) {
03   PRIALT ();
04     ALT_CHAN_IN (X_CHANNEL, X_Tag); // X_COMMITTED,
05                                     // X_FEATHERED, X_UNFEATHERED
06     ALT_CHAN_IN (CHAN_DATA_IN, Value);
07   ALT_END (); // delivers ThisChannelId
08
09   switch (ThisChannelId) {
10     case X_CHANNEL: { // After CHAN_OUT ret AWAIT_READY or FEATHERED
11       if (X_Tag == X_FEATHERED) {
12         ... handle not interested
13         CP->Tag = FEATHERED;                    // stop
14       } else if (X_Tag == X_COMMITTED){
15         CHAN_OUT (XCHAN_DATA_OUT,Value,NIL);    // will succeed
16         CP->Tag = READY;                        // finished
17       } else { // (X_Tag == X_UNFEATHERED)
18         CP->Tag = READY;                        // finished
19       }
20     } break;
21     case CHAN_DATA_IN: {
22       if ((CP->Tag == AWAIT_READY) or (CP->Tag == FEATHERED)) {
23         ... handle overflow (decide what value(s) to discard)
24       } else { // (CP->Tag == READY)
25         CP->Tag = CHAN_OUT (XCHAN_DATA_OUT,Value,ALLOW_FEATHERING);
26         if (CP->Tag == SUCCESS) {
27           CP->Tag = READY;                      // finished
28         } else if (CP->Tag == FEATHERED) {
29           ... handle not interested
30         } else { // (CP->Tag == AWAIT_READY)
31         }
32       }
33     } break;
34   }
35 }
```

Listing 1. Overflow and 'feathered' handling on an **XCHAN** (ANSI C and macros).

The listing above basically follows listing 1 in [2]. **CP** is the Context Pointer to process data that has been allocated on the heap. It shows an example of a server that always can input new data since output is non-blocking. Data arriving on **CHAN_DATA_IN** is always accepted (line 21) within a bounded time (since no code in listing blocks and an **X_CHANNEL** signal can only happen at most twice by a failed send on **XCHAN_DATA_OUT**, in response to the previous input from **CHAN_DATA_IN**). If there is no overflow situation then line 25 attempts to send. If this succeeds immediately with **SUCCESS**, it is finished. If the result is **FEATHERED** then handle the fact that the receiver was "not interested" immediately (e.g. abandon processing of this input). If the result is **AWAIT_READY** or **FEATHERED** then loop around to wait for an **X_CHANNEL** signal; or more from **CHAN_DATA_IN**. Observe that in the example we always include **X_CHANNEL** in the **ALT**; this is ok since it will not be signaled out of order (though it might be more efficient to exclude it, by pre-conditioning, when not expected). If the **X_CHANNEL** signal is **X_FEATHERED**, then handle this "not interested" state (responses in lines 12 and 29 should be the same). If it is **X_COMMITTED**, send the latest received value (line 15). Once **FEATHERED**, it cannot retry sending before **X_UNFEATHERED** is received. Observe that we now use a **PRIALT** in 03, so **CHAN_DATA_IN** cannot jam **X_CHANNEL**. The code has no explicit error handling and there are no compiler usage checks.

3. Discussion

3.1 Feathering is "Low Level"

The receiver is able to tell to the sender at a potential synchronisation point that it is '*not interested*'. However, it is not able to give any *reason*. It is a basic, low-level type of rejection. This way the **ALT** construct need not be changed. It may also be discussed whether alternatively 'sending' some reason across is counter-productive. One must probably take care that this does not become an opposite direction channel hidden in the channel. We therefore propose only the lowest level: the implicit "not interested". (However, a higher level channel with these properties could potentially be part of some other software pattern.)

There is no other type of state exchange either. The sender may not know the receiver's state; like a priori information whether a channel would have become *feathered* or if the receiver is just doing something else; like being in a session with a third party and that it certainly would want to receive the pending message later. Coupling between processes should be kept at a minimum, and *feathering* does not seem to increase it – rather the opposite – with its implicit subscriber mechanism and, after all, less point-to-point messages between the parties.

Instead of sending on **XCHAN** with a parameter, we could have defined a new type of channel called **FXCHAN**, ("**F**eathering **XCHAN**") with *feathering* as standard semantics. In that case the receiver would know about its *feathering* capability. However, a generic type would make the semantics selectable at run-time (although with some cost: **FALSE** pre-conditioned **XCHAN** guards always need processing). Usage rules would be a little more complex with **XCHAN** plus parameter, certainly also depending on whether this parameter is a variable or an invariable in the actual scope.

With the suggested scheme the receiver will not know if it will implicitly stop a message by *feathering*, or if *feathering* has any impact. It would only know that this new **XCHAN** or **FXCHAN** has a potential to be *feathered.*

Sending outdated messages across would also need a design criterion as to decide whether the message is outdated. This also implies some degree of coupling between the processes (like a common clock).

Like any language primitive there may be uses for *feathering* and cases where it is not applicable or plainly wrong.

As mentioned we have not studied the alternative precondition semantics of **x**-channel [6].

3.2 Anti-Deadlock

We argue that the anti-deadlocking property of **XCHAN** is kept by this additional semantics. We have shown that the extra synchronisation needed to feed **x**-*feathered* or **x**-*unfeathered* back on the **x**-channels is not blocking, asserted by rules and contracts. However, the associated scheduling may cause the sender in a 'new round' to block on other channel outputs, but this will be part of another communication graph. Therefore **XCHAN** in itself still breaks any cycle (and 'removes' deadlock).

Setting of the *feathered* status in the **ALT** construct is done by calls to scheduler code. It has full control of all parties (no pre-emption), so there should be no race between the two sides of **XCHAN**. However, this needs to be further studied for multicore or distributed systems.

3.3 Run-Time Penalty

As we see, there would have to be some reporting back to the sender on every channel that is *feathered*. One might argue that the goal not to send unnecessary data is compromised when some other (data or not) will be sent instead. Not having to send lots of, for example, 1500 bytes long messages would probably save bandwidth. However, the language design to offer the receiver a pattern of an implicit subscriber mechanism is perhaps even more interesting. It is this idea that has driven the writing of this paper.

Some communication and associated scheduling will be removed. Since nothing is free, there is a cost associated with supplying the information enabling this removal. The extra two communications for **x**-*feathered* and **x**-*unfeathered* on the **x**-channels will after all interfere with state in the sender, which is then allowed to take proper action, like reporting back to some other component.

There is no busy-polling in any of the algorithms.

3.4 Feathering Requires "Visible" Pre-Conditions on **ALT** Guards

An **occam** **ALT** with conditional guards may be transformed to an **IF** structure:

```
01 ALT                                 -- feathering semantics hidden
02   condition.0 & in.0 ? x.0
03     ...  response 0
04   condition.1 & in.1 ? x.1
05     ...  response 1

10 IF                                  -- no feathering
11   condition.0 AND condition.1
12     ALT
13       in.0 ? x.0
14         ...  response 0
15       in.1 ? x.1
16         ...  response 1
17   condition.0                       -- condition.1 must be FALSE
18     SEQ
19       in.0 ? x.0
20       ...  response 0
21   condition.1                       -- condition.0 must be FALSE
22     SEQ
23       in.1 ? x.1
24       ...  response 1
```

Listing 2. *Feathering* loss of semantic equivalence (**occam**).

The programmer must be aware that, for *feathering* semantics, the pre-condition on the guard must be "visible" in the code. Without *feathering*, the two blocks of code (lines 1-5 and 10-24) above are equal. With *feathering*, the **ALT** in line 12 will never take part in any *feathering*; neither will the two **SEQ** blocks starting at lines 18 and 22. Since *feathering* means to have a "tap to turn off", all should be obvious: there is no "tap" or boolean expression. The sender does not know which of the methods are used; but the block starting in line 10 will return neither *feathered* status nor **x**-*feathered* nor **x**-*unfeathered*.

The receiver may use this loss in semantic equivalence to ensure that *feathering* is off (although the **IF**-version gets exponentially more complex to program as the number of guards in the **ALT** increases). Expressions evaluated to **FALSE** are not so easily spotted. The code on the sender side should probably not be implemented with this knowledge: an

XCHAN's receiver may at any time decide to engage in *feathering*. The compiler or linker could inform about this. This would probably also go for visible invariant conditionals. But there is nothing wrong with coding that at a certain time *all* components are hard coded as *interesting*.

3.5 Feathering Adds Non-Determinism

Whether a message gets sent or not depends on when the attempt is made. This adds non-determinism. However, *feathering* should only be used when this non-determinism is deliberate. Just as using **XCHAN** opens for application control of message handling (like data loss) on the sender side of the **XCHAN**, *feathering* takes the receiver's face value for throwing a message. "Not listening on that ear" really means "please, drop it" – by design.

By hiding this non-determinism inside a block that will equally allow different behaviour, *feathering* should not hinder any formal proofs.

4. Alternative

We have not discussed whether *feathering* solves real problems where programmers have longed for a solution.

An alternative to this mechanism could be an *explicit* subscriber mechanism. Mixing real life and software we could say that the bus-stop subscribes to the busses, on behalf of the passenger, which of the busses that are interesting. When the correct bus has been taken, all busses are unsubscribed to. This way there are no arrival messages to flush, provided there are no possible race conditions in the algorithm used. Some of us may have missed a bus if it were so close to the bus in front that we did not see the badge.

Feathering may also be seen as an implicit embodiment of the *publish-subscribe pattern* [7]. *Feathering* status would be an a *not publish* status handled to the publisher (sender) when the subscriber (receiver) implicitly has decided not to *subscribe*. However, differently from the *publish-subscribe pattern* there is no explicit message or calls done by the subscriber to subscribe to publishments or to unsubscribe. Examples for the usability of *feathering* may probably be collected from the *publish-subscribe pattern* examples.

Since *feathering* may be considered a low-level implicit subscriber mechanism, it has no messages to flush. The algorithm we have described here should be without race condition, but as mentioned – this is not formally proven.

5. Summary

The proposed **XCHAN** [2] already provides actionable feedback to the sender on the state of the receiver. This proposal extends that information (beyond success/fail/ready) to include *feathered* (i.e. not interested, do not resend) and *unfeathered* (which ends a period of feathering). *Feathered* signals are sent (to relevant potential senders) if the receiver's **ALT** blocks (and the relevant **XCHAN** guard is excluded by a **FALSE** pre-condition). Unfeathered signals are sent upon exit from the **ALT**.

6. Conclusion

This paper takes the rather drastic attempt to change the state of an **XCHAN** when, in the standard **occam** semantics of **CHAN**, there would be no state change when *not* touching a guard of an **ALT**. The suggested *feathering* semantics exploits information in the running code that a channel's receiving end has been excluded from an **ALT** – and propagates this information to a potential sender. This was our starting point. We have suggested a usage of this idea that could be of interest in some systems.

The sum of the benefits of **XCHAN** and *feathering* may justify adding these features to a concurrent language.

This paper is a study of usage and discussion of the possible semantics, as well as some implementation issues. We have not formally proven that *feathering* is possible, but the reasoning above should be a starting point for such proofs.

We hope that the ideas will cause language designers to take *feathering* further. Perhaps the pull of the oar will take the ideas safely to port?

Acknowledgements

I am grateful to Professor Sverre Hendseth of NTNU, Trondheim, for his encouragement and enthusiasm about the theme. I also thank the peer review readers and the editor for thoroughly pointing out inconsistencies and more subtle matters.

References

[1] Tony Hoare, Concurrent programs wait faster. Microsoft Research, 2003. See http://research.microsoft.com/en-us/people/thoare (ref. notes on the first two pages)

[2] Øyvind Teig, XCHANs: Notes on a New Channel Type. In: P.H. Welch et al., *Communicating Process Architectures 2012 (CPA-2012), Proceedings of the 34th WoTUG Technical Meeting*, ISBN 978-0-9565409-5-9. Open Channel Publishing Ltd., 2012, pp. 155-170. See http://www.teigfam.net/oyvind/pub/pub_details.html#XCHAN

[3] 'Feathering', defined in OneLook, see http://www.onelook.com/?w=feathering&ls=a

[4] Google Go programming language by Robert Griesemer, Rob Pike, and Ken Thompson. See "The Go Programming Language Specification" at http://golang.org/ref/spec

[5] INMOS Limited, Occam 2 Programming Manual, Prentice-Hall International, ISBN 0-13-629312, 1988. See http://www.wotug.org/documents.shtml

[6] Peter H. Welch. An occam Model of XCHANs, https://www.cs.kent.ac.uk/research/groups/plas/wiki/An_occam_Model_of_XCHANs , 2013.

[7] Publish-subscribe pattern. See http://en.wikipedia.org/wiki/Publish–subscribe_pattern

Communicating Process Architectures 2013
P.H. Welch et al. (Eds.)
Open Channel Publishing Ltd., 2013

The Distributed Application Debugger

Michael Quinn JONES [1] and Jan Bækgaard PEDERSEN

Department of Computer Science, University of Nevada, Las Vegas, NV, USA

Abstract In this paper we introduce a tool for debugging parallel programs which utilize the popular MPI (message passing interface) C library. The tool is called The Distributed Application Debugger and introduces distinct components around the code being debugged in order to play, record, and replay sessions of the user's code remotely. The GUI component incorporates the popular GDB debugger for stepping through code line by line as it runs on the distributed cluster as well as an analysis tool for stepping through the executed portions of code after the session has completed. While the system is composed of multiple components, the logistics of coordinating them is managed without user interaction; requiring the user only to provide the location of the program to debug before starting.

Keywords. distributed, debugger, debugging, parallel, MPI, TCP

Introduction

Anyone who has developed software before will likely agree that bugs will be encountered along the way. No matter how careful you are, logic errors, memory mismanagement, race conditions, inaccurate execution path assumptions and a whole host of other issues will be encountered from time to time. These issues are acceptable, understandable and expected when designing software. As a result of the popularity of sequential programs, and the common understanding that debugging will always be part of their development, many exceptional debugging resources have been created to help the developer analyze and step through sequential code.

Developing parallel programs which run on distributed computer clusters introduces additional challenges to those present in traditional sequential programs. When debugging parallel programs, one needs to be able to inspect both the sequential code executing on each node and track the flow of messages being passed back and forth between them in order to infer where a problem actually lies. One such framework for distributed programming is called MPI [1]. It stands for Message Passing Interface (MPI) and is a C library which utilizes the power of distributing work across processors which communicate with each other by passing messages.

The tool introduced in this paper, and described in greater detail in [2], known as The Distributed Application Debugger, was developed to debug code which runs in parallel within all versions of MPI including MPI-1, MPI-2, and MPI-3. Among other things, the tool gives the user a centralized and organized space to examine the output of each node while being able to also inspect and match the messages being passed between them. It provides them with a way to not only replay a recorded session but also to halt a session in order to step through each node's execution while inspecting all the variables of a system. Finally, because clusters of computers are often housed a universities or super computer centers around the world,

[1]Corresponding Author: *Michael Quinn Jones, University of Nevada Las Vegas, 4505 Maryland Parkway, Las Vegas, NV, 89154, United States of America.* E-mail: `mjones112000@gmail.com`.

the tool is fundamentally structured to run remotely and behind any number of impeding computer servers that needed to be logged into first before being able to access the actual computer cluster.

1. Related Work and Motivation

The Distributed Application Debugger bases much of its foundation from two sets of related work. First was the results from two surveys given to two different sets of graduate students at the University of Nevada, Las Vegas over the course of two years. The purpose of the surveys was in response to research by Pancake [3] who suggests that tools for parallel programming and debugging are often found to be unhelpful for users because their developers do not spend enough time trying to understand what the root problems that their users really need addressed are. The surveys, reported in [4] and [5], attempted to classify bugs found by students learning to program with MPI into 5 distinct categories. The first four categories were based on the the *Partitioning* and *Communication* aspects of the parallel construction model known as PCAM [6].

Partitioning deals with the task of partitioning both the data and functionality of the algorithm being implemented. It can be further sub-categorized as *Data Decomposition*, which covers developing code structured to deal with managing memory, modeling data structures etc., and *Functional Decomposition* which deals with the organizational side of defining the responsibilities of each node and establishing roles for architectures such as pipelining and master/slave relationships. The second category, *Communication*, is the task of implementing interprocess communication. It can also be broken down into two sub-categories: *Messaging* which deals with correctly addressing individual send and receive calls between destination and source nodes, and *Protocol Specifications* which deals with the overall communication pattern.

The survey asked the students to report each time they encountered a bug while developing their projects for the semester. The bug report asked them to estimate the time they spent on it and to classify it as a *Data Decomposition*, *Functional Decomposition*, *Messaging*, *Protocol Specifications* or a fifth category, *Sequential*, type bug. In both surveys, the *Sequential* type, which deals with the traditional bugs found in sequential programs such as mistakes with conditionals, method calls, pointers, pre and post conditions, and algorithm modeling to name a few, made up the vast majority of bugs encountered while writing parallel programs as displayed in Figure 1.

The second motivational related work were previous tools presented by Tribou [7] and Basu [8]. These tools, built to debug PVM [9] and LAM-MPI [10] parallel programming libraries respectively, both focus on debugging from a multilevel approach as introduced in [11]. This approach focuses on allowing the user to first inspect the code from a *Sequential* level by organizing print statements together by issuing node, as shown in Figure 12, and allowing the user to attach GDB to their program as they would debug any sequential program. Secondly, users are able to inspect their code from a *Message* level which allows users to inspect data that was sent in a message and then compare it to the data that was received. Finally, the tool introduced a third level of debugging, the *Protocol* level which concerns itself with the messaging system as a whole by matching send and receive pairs and alerting the users when a message was not matched at all.

While the graduate student surveys and the tools introduced by Tribou and Basu influenced the layout and features included in the Distributed Application Debugger, the need for a new tool for debugging MPI programs became apparent when the results of the survey of graduate students showed that most of their debugging was done using print statements. Although commercial debuggers, described in Section 2, are available that can monitor dis-

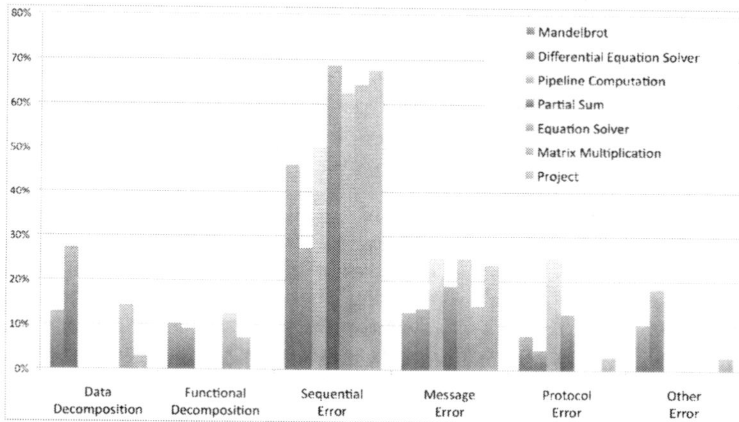

Figure 1. Error classifications from a survey [5] of graduate students taking a parallel programming class.

tributed processes at the petascale level, research by [12] found that 80 percent of developers used less then 4 processes when debugging their code. With this in mind we felt that a debugging tool focused on the common debugging needs of its target audience, rather than on extreme scalability, could still be very effective. The tool consists of debugging messages, detailed at great detail in [2], sent between the components detailed in Section 3. It includes features allowing the users to play, record, and replay their code running in a distributed system with a front end which allows them to analyze it in a sequential way.

2. Other Tools

The Open MPI Project's frequently asked questions website [13] characterizes how to debug applications running in parallel as a difficult question to answer. In their words

> *"Debugging in serial can be tricky: errors, uninitialized variables, stack smashing, ... etc. Debugging in parallel adds multiple different dimensions to this problem: a greater propensity for race conditions, asynchronous events, and the general difficulty of trying to understand N processes simultaneously executing – the problem becomes quite formidable."*

The project recommends two enterprise level debuggers, *DDT* (short for the Distributed Debugging Tool) by Allinea Software [14] and *TotalView* by Rogue Wave Software [15], to aid in the complicated task of debugging MPI programs. Because of this endorsement, we felt that it was important to touch on some of their features and how they may be a better tool of choice at times than the Distributed Application Debugger.

DDT and TotalView have many features in common. Both GUIs are very *debugger-centric* in the way that the view is always focused on source code and the node specific view is simply the line of the source code currently being executed for any specific node. Because of this, other features, such as a graphical views of the messages being sent, are done with pop up windows. The Distributed Application Debugger took great lengths to not have views presented in pop ups so that the user is not tasked with juggling them. The Distributed Application Debugger by contrast, is more of an analysis tool which integrates a debugging feature when needed.

Both *DDT* and TotalView offer great features that allow inspection of the code at both the process level and the thread level. They display their message stacks in graphical form, whereas the Distributed Application Debugger displays its messages in tabular form. Also, both not only are able to record and replay MPI sessions, but also allow the user to 'rewind' a session, whereas the Distributed Application Debugger always goes forward. Most impressively, they both can scale to over 100,000 processes which the Distributed Application Debugger would not be able to presently. This scalability comes with a large price, however, with DDT costing over $600 for an academic or government workstation license, and TotalView costing in the range of $1,000.00.

The Distributed Application Debugger's strengths lie in its simplicity. It can be configured to run remotely, and does not require any applications to be running on the remote machines prior to the session beginning. It copies, compiles and launches any component needed in the communication line outlined in Figure 6 as part of its start-up process. It also leverages the extremely popular and powerful GDB debugger with which most students are already familiar. This cuts down on its learning curve and further assists the students in focusing on their parallel programs as just a set of running sequential programs. The Distributed Application Debugger's layout presents a useful layout for viewing 2 or 3 nodes at a time because their data are displayed side by side without have to switch between them, while DDT and TotalView display the main code and require the user to request from which node they want to see data. The Distributed Application Debugger's weakness is in its scalability of processing messages across many processes. The processing time for a sample program which distributed 800 messages across 8 nodes each passing 100 messages, for instance, took only 3 seconds to process. The processing time for the same sample program distributed between 80 nodes each passing just 10 messages, however, took 37 seconds.

Figures 2 and 3 display the main front end GUI for DDT and TotalView respectively.

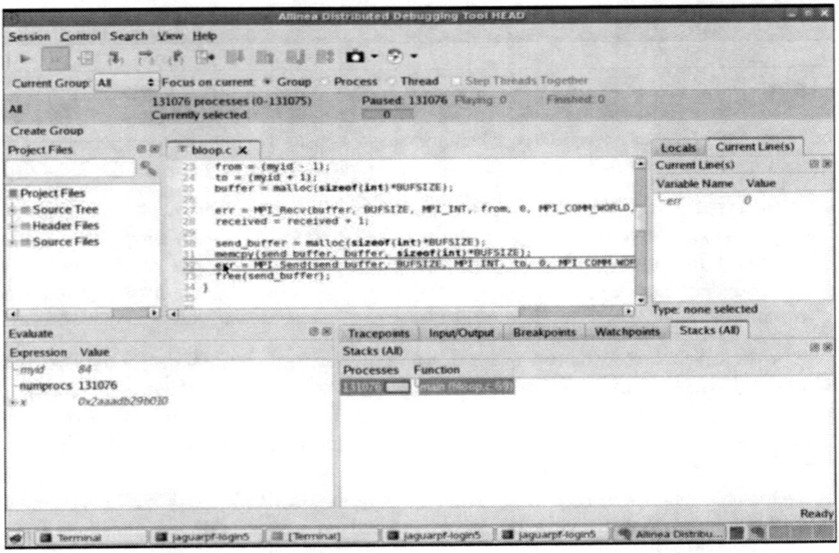

Figure 2. The GUI display of Allinea Software's DDT application.

3. Architecture

This section focuses on the architecture of the Distributed Application Debugger which is broken into 4 distinct components focused on giving the ability to instantiate and debug MPI

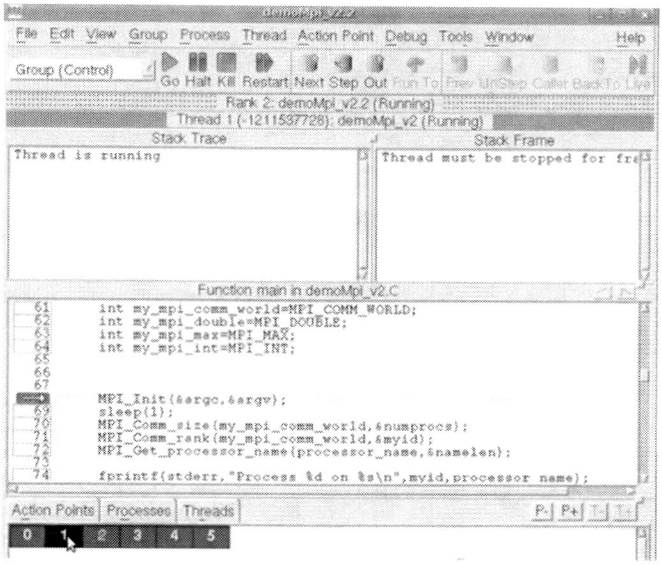

Figure 3. The GUI display of Rogue Wave Software's TotalView application.

programs remotely. The first of the components is a graphical user interface known as The Client. The Client, displayed in Figure 4, allows the users to input their remote credentials, instantiate a connection to a machine within a distributed cluster, and then initiate debugging sessions on their parallel programs until they choose to disconnect from the cluster. Each node of the distributed system conveys debugging data in its own panel which consists of a Console tab displaying output to that node's `standard out`, a Messages tab displaying all sent and received messages by that node, and an MPI tab which displays every MPI command run on that node along with its parameter values (as illustrated in Figure 5). More details regarding the debugging information presented in these panels are described in Section 5.

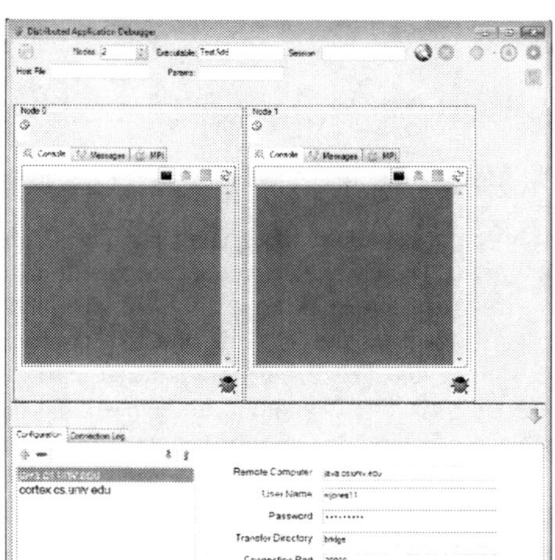

Figure 4. A debugging session configured for two nodes with both displayed.

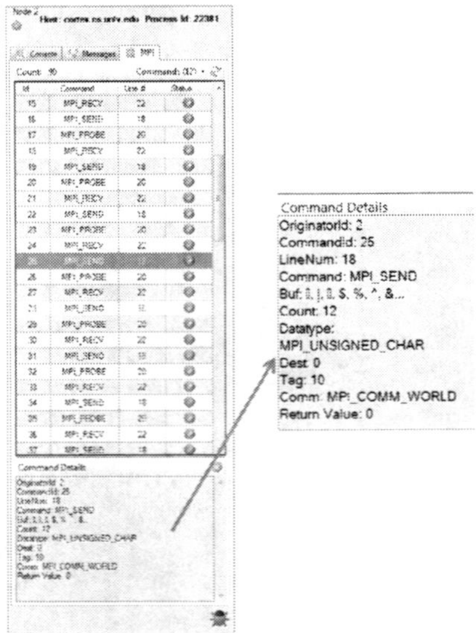

Figure 5. The extra information displayed in the Command Details panel of each node.

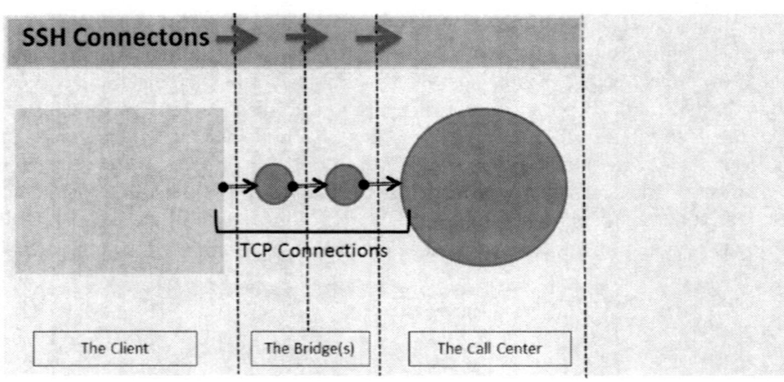

Figure 6. The system connected via TCP sockets.

The second and third components of the system are The Bridge and The Call Center which are depicted in Figure 6. The Bridge is a relay component meant to move data without inspecting it. The Bridge reads from one TCP port, the 'A' side, and writes to a second TCP port, the 'B' side. It also reads from the 'B' side and likewise writes what is read to the 'A' side. It is important to note that The Bridge is not necessary in the case that the parallel cluster is accessible directly from The Client but, in the case that the cluster is housed on a machine whose network is not accessible publicly, becomes crucial in connecting The Client to the cluster. The third component, The Call Center, however, is a crucial component in order to make debugging possible. The Call Center is a program which runs outside of the MPI system and must be run on a computer within the cluster. Its listens for commands from The Client in order to start a debugging session, and is responsible for starting and retrieving debugging sessions for The Client and relaying status information back to it. The connection from The Client to the remote Call Center is done in a two step process. First,

Figure 7. Compile time changes made from including The Runtime's `mpi.h` file.

after the user has supplied credentials of which machines to connect to (as shown in the configuration section of Figure 4), The Client uses the order in which these credentials are entered to SSH into each destination machine in sequence, copy component libraries to the directory specified as the 'Transfer Directory' in the configuration section, compile them and then launch the appropriate component (Another Bridge or the Call Center). The last destination gets the Call Center launched on it and all of the other ones get a Bridge. After each of these nodes has either a Bridge or The Call Center running, The Client makes a TCP connection to the first node, which makes a connection to the second node etc., until The Client and The Call Center are connected via TCP either directly or through a sequence of Bridges as illustrated in Figure 6.

The last component of the architecture is The Runtime. The Runtime is a C library that redirects the user's MPI code to a set of intermediary methods which connect back to the Call Center and relay debugging information on behalf of the code being debugged. In order to incorporate The Runtime, the user needs to include a local MPI header file that contains macros which change the file's MPI calls at compile time from the true MPI functions to The Runtime's methods (This is illustrated in Figure 7). When the user starts a new MPI debugging session from The Client, The Bridges relay data to The Call Center, which launches an MPI session and then relays debugging data from The Runtime as displayed in Figure 8. The system's Bridge and Call Center applications also clean themselves up upon detecting that the program that they are reading from or the program that they are writing to has closed their socket connections. This reliably helps the system close in series when the user disconnects The Client from the session or if one of the components should crash.

It should be noted that the concept of each component cleaning itself up is one reason why The Bridge component was developed to relay messages which could have essentially been implemented by network administrators configuring port forwarding on their routers. When The Client closes its TCP connection, each Bridge detects that its input connection has closed and then properly closes their outgoing TCP connections in series until The Call Center detects that its incoming TCP connection has closed and can then properly clean itself up as well. If port forwarding has been used, when The Client closes its outgoing TCP

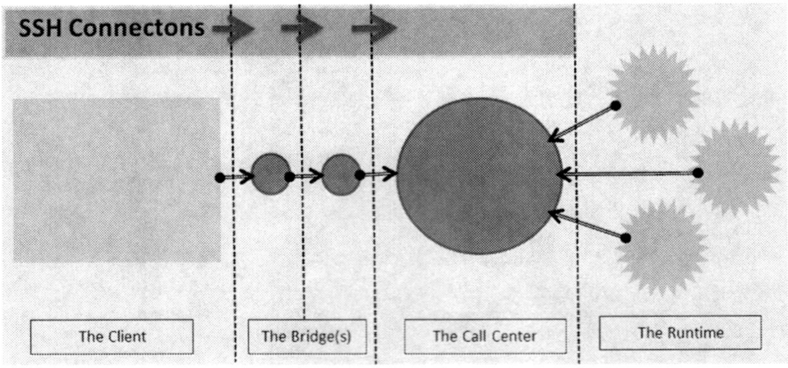

Figure 8. The system connected fully connected from The Client to the MPI nodes via The Call Center.

connection to an intermediary node, The Call Center would not get any more messages, but also would not know that it had been disconnected, and thus not know that the session is over and it should terminate itself.

4. Supported Commands

The Distributed Application Debugger does not support all of the 300 commands found within the MPI library [16], but does support 12 core commands making analyzing the initializing, message passing, synchronizing and finalizing of a typical MPI program possible. Table 1 lists the commands available for debugging within the Distributed Application Debugger along with their signatures.

Table 1. The MPI commands supported by the Distributed Application Debugger.

Command	Description
MPI_Init(int *argc, char ***argv)	Initializes the MPI environment
MPI_Comm_rank(MPI_Comm comm, int *rank)	Determines the rank of the process in the cluster
MPI_Comm_size(MPI_Comm comm, int *size)	Determines the size of the cluster
MPI_Send(void *buf, int count, MPI_Datatype datatype, int dest, int tag, MPI_Comm comm)	Performs a blocking send
MPI_Recv(void *buf, int count, MPI_Datatype datatype, int source, int tag, MPI_Comm comm, MPI_Status *status)	Blocking receive for a message
MPI_Isend(void *buf, int count, MPI_Datatype datatype, int dest, int tag, MPI_Comm comm, MPI_Request *request)	Begins a nonblocking send
MPI_Irecv(void *buf, int count, MPI_Datatype datatype, int source, int tag, MPI_Comm comm, MPI_Request *request)	Begins a nonblocking receive
MPI_Probe(int source, int tag, MPI_Comm comm, MPI_Status *status)	Blocking test for a message
MPI_Iprobe(int source, int tag, MPI_Comm comm, int *flag, MPI_Status *status)	Nonblocking test for a message
MPI_Wait(MPI_Request *request, MPI_Status *status)	Waits for an MPI request to complete
MPI_Barrier(MPI_Comm comm)	Blocks until all processes in the communicator have reached this routine.
int MPI_Finalize(void)	Terminates MPI execution environment.

5. Features

The Distributed Application Debugger introduces several features for inspecting data collected from The Runtime. First a (debugging) session is run in one of three modes: PLAY, RECORD, or REPLAY. PLAY is a standard MPI session in which The Runtime executes each line of the real MPI code and sends back the line number, though without the source file name, the parameters, and the result of each command run on each command.

Although one may think that PLAY mode would be the only mode necessary in order to run the debugger in order to inspect results, research by Leblanc and Mellor-Crummey [17] prompted us to add two other modes: RECORD and REPLAY. Leblanc and Mellor-Crummey illustrated that since parallel programs often include nodes passing messages asynchronously, that the execution behavior of a parallel program in response to a fixed input can be nondeterministic. Given this information we felt that it was crucial that the Distributed Application Debugger be able to record the execution of an MPI session and allow the user to both inspect it by hand and replay it by re-executing the code. During RECORD, the Call Center runs like it does in PLAY but also records an XML log file of all the data sent back to The Client during the session (illustrated in Figure 10). The data contains the name of the RECORD session, along with the number of nodes on which then program was executed, the host file used (which nodes on the cluster participated in the execution), the location of the executable, and the parameters passed to it. As seen in Figure 9, these recorded session are available to the user and can be recalled to run in the third and most useful debugging mode, REPLAY.

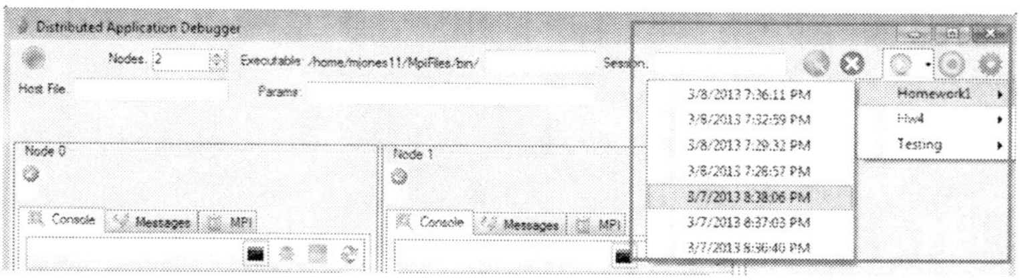

Figure 9. The names of the recorded sessions available for replay.

Nodes run in REPLAY mode do not execute MPI commands, but rather reload the results of each MPI command from the log files saved during a RECORD session. Running in REPLAY mode has several advantages. First, while in REPLAY mode, the user is able to playback the results of a session that has been recorded in exactly the same way as when it was executed (and recorded) before. This feature becomes especially helpful for users who are investigating exceptional cases which may be difficult to reproduce due to race conditions or other asynchronous factors. Secondly, the user has the ability to run a 'mixed mode' replay, in which designated nodes actually invoke the MPI framework (and send and receive actual messages) while the other nodes continue recalling their values from the log files. This allows the user to systematically narrow down which node or nodes may be causing the error being investigated. When a session is running in mixed mode replay, however, each node must consider if it will just relay back the results from the recorded log file, or if it must execute the actual MPI commands. As an example, the simple command MPI_Comm_size command does not affect any other nodes and thus does not have to ever be issued to the real MPI framework from a node running in REPLAY mode. MPI_Recv, however, must consider if the sender is replaying from a log file, or if it is a node that is invoking the real MPI framework. In the case that they are both replaying, the results from the MPI_Recv may just be read from the log file and sent back to The Call Center. If, however, the sender is actually issuing real MPI commands the receiving node must execute the MPI_Recv command so that the buffers of the MPI system do not fill up and cause the application to freeze. If a node running in REPLAY

```
<MPI_SIZE rank="0" commandId="3" dateTime="Mon Mar 04 08:55:15 2013 ">
  <parameters>
    <comm>1140850688</comm>
  </parameters>
  <returnvalue>0</returnvalue>
</MPI_SIZE>

<MPI_SEND rank="0" commandId="4" dateTime="Mon Mar 04 08:55:15 2013 ">
  <parameters>
    <buf>
      <value>0</value>
      <value>-2147483648</value>
      <value>2147483647</value>
      <value>356456</value>
      <value>765</value>
      <value>68378376</value>
      <value>67787</value>
      <value>17636</value>
      <value>585356</value>
      <value>253636</value>
    </buf>
    <count>10</count>
    <datatype>MPI_INT</datatype>
    <dest>1</dest>
    <tag>10</tag>
    <comm>1140850688</comm>
  </parameters>
  <returnvalue>0</returnvalue>
</MPI_SEND>

<MPI_RECV rank="0" commandId="5" dateTime="Mon Mar 04 08:55:15 2013 ">
  <parameters>
    <buf>
      <value>H</value>
      <value>e</value>
      <value>l</value>
      <value>l</value>
      <value>o</value>
      <value> </value>
      <value>W</value>
      <value>o</value>
      <value>r</value>
      <value>l</value>
      <value>d</value>
      <value></value>
    </buf>
    <count>12</count>
    <datatype>MPI_CHAR</datatype>
    <src>1</src>
    <tag>10</tag>
    <comm>1140850688</comm>
    <status>
      <MPI_SOURCE>1</MPI_SOURCE>
      <MPI_TAG>10</MPI_TAG>
      <MPI_ERROR>0</MPI_ERROR>
    </status>
  </parameters>
  <returnvalue>0</returnvalue>
</MPI_RECV>
```


Figure 10. A portion of the values recorded within a log file from a RECORD session.

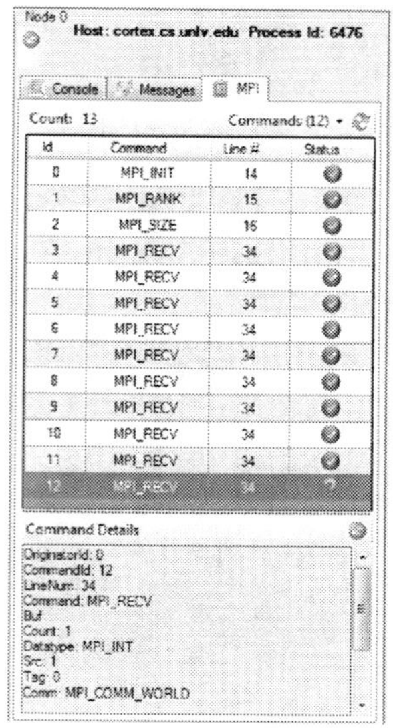

Figure 11. A message for which no POST has been received.

mode does issue a real MPI command, it will compare the results of the command to that of the values recorded in the log file: if a discrepancy is found, a warning will be relayed back to The Client, which will display a special symbol in the Status column of the MPI tab (see Figure 11).

Within each node's 'node panel', The Client organizes the debugging data sent back from The Runtime among three distinct tabs: the Console tab, the Messages tab, and the MPI tab. The Console tab displays a dedicated area for monitoring the `standard out` of each node. This allows the user to use print statements when needed. Such output is presented in the order which it was produced by the node. This is illustrated in Figure 12.

The Messages tab displays each message sent and received and matches sends and received as shown in Figure 13. When the user is in RECORD or REPLAY mode, the Messages tab also provides the option of recalling the buffer values which were sent or received in order to further investigate the message content during the debugging session.

The MPI tab, which displays every MPI command executed during the session, shows the order in which the MPI commands were executed while the code ran. An especially useful detail featured on this tab is that The Runtime reports each MPI command with the combination of a PRE message and a POST message. The PRE message indicates that an MPI command is about to be executed and what parameters were passed to the method. When the MPI command completes, The Runtime sends back a POST message which contains the results of the MPI command. The MPI tab displays an entry for each PRE message received. This informs the user about which MPI command is about to be executed, but the Status will be shown as 'Incomplete' until a corresponding POST message is received. This allows a user who is debugging a program which is not completing, to view the line number and parameters passed into the last command which started execution in order to investigate why it did not finish (no POST message was ever received, so the MPI command never finished).

The Distributed Application Debugger offers two different types of filters to cut down on the overwhelming amount of data that is present within each tab. The first type pertains to identifying mismatched messages as described above. The messages tab offers a filter button, , which removes all matched messages from the node's messages tab. (A matched message

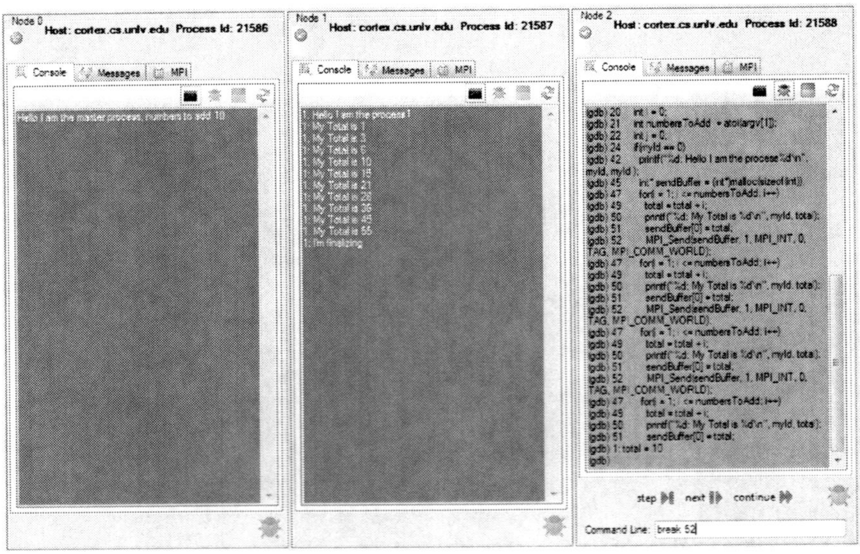

Figure 12. An MPI debugging session where GDB is attached to node 2.

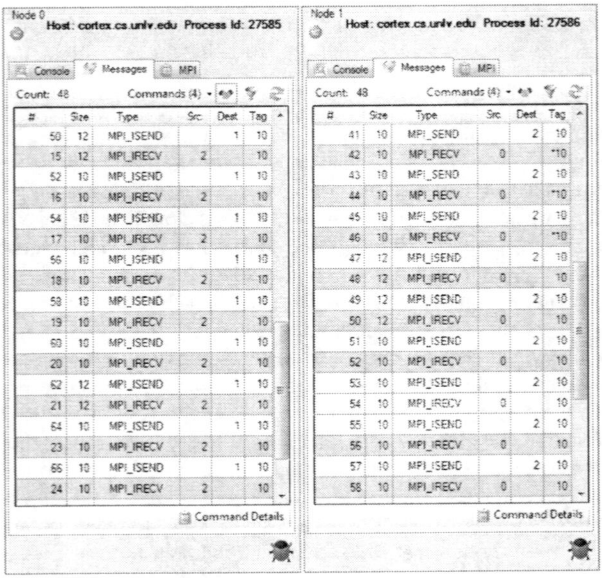

Figure 13. Two nodes displaying automated message matching.

is a message for which The Client has received POST messages from both the sender and receiver). Invoking this filter leaves the user with just the mismatched messages to investigate. The second type is aimed at helping the user cut down on the number of commands displayed within the *Messages* and *MPI* tabs. It is a drop-down which allows the users to check or uncheck the commands that they want displayed within the tab (see Figure 14).

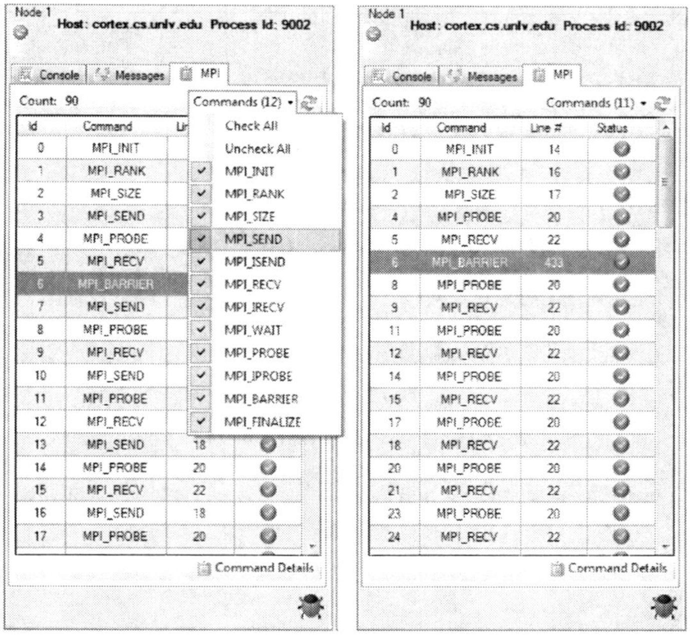

Figure 14. The MPI panel before and after a command filter was applied.

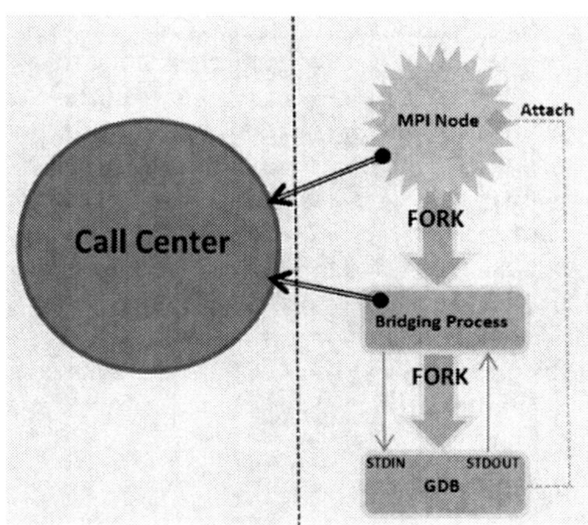

Figure 15. A Runtime node with an extra connection to the Call Center from GDB.

Finally, the Distributed Application Debugger incorporates the popular sequential command line debugging tool *GDB* [18]. The Client allows the user to attach GDB to any of the running nodes regardless of the mode in which the session is running. When the Call Center detects that a node is to be run with GDB, it instructs The Runtime to attach GDB to the selected nodes through a series of process forkings, and then waits for an additional connection back from the Bridging nodes controlling GDB. This is illustrated in Figure 15. The Client can then supply command line instructions to the Call Center which will route these commands to the Bridging Process controlling the GDB process. The Bridging Process then writes the commands to the GDB's stdin (standard in). The Call Center routes, in addition to the standard data described, additional messages back from The Runtime. This additional data is the stdout (standard out) from all GDB processes (which contains other debugging data such as line numbers, variables' values and breakpoint information) which The Client can present to the user as displayed in Figure 12.

6. Conclusion

The Distributed Application Debugger has many features for debugging distributed MPI applications. It works remotely from home, even when the parallel cluster is not accessible directly. It graphically displays the nodes of a cluster in a way that represents both the sequential nature of the code being executing within each process, as well as the parallel nature of the messages being passed between them. It reliably handles all of the messages which are passed back within any given session to give valuable debugging data to the user about the MPI commands that are starting and completing as well as all of the data being printed to `standard out`. It offers features for recording and replaying MPI sessions which can help programmers focus on a problem that may be hard to recreate. It offers buffer value inspection to aid in debugging common sequential bugs along with message matching to cut down on the time for message error debugging. Finally it seamlessly integrates GDB to encourage alternatives to inefficient print statements.

7. Future Work

Some work should be done to extend the number of commands supported by the Distributed Application Debugger in the future. While the 12 commands it currently supports, presented in Table 1, are a start, MPI programs which implement commands outside of this initial scope will not get the full debugging features of the application at this time.

The next logical progression for the Distributed Application Debugger would be to integrate a development environment within it to compliment its debugging features. Integrated Development Environments (IDE), like Eclipse [19] or Microsoft Visual Studio [20] have become very popular with developers as a place that they feel comfortable both developing and debugging their programs in. In the current system, the user is expected to develop their code as they would normally do so, and to attach the debugger when they encounter bugs and need more data to resolve them. The system should offer a command line feature to take advantage of the persistent SSH session maintained with the cluster so that the user could add, update and delete file and directory names. In the future an interface with popular source control software such as Subversion [21] would be useful as well to let users index the various versions of their project files.

References

[1] Jack Dongarra. MPI: A Message Passing Interface Standard. *The International Journal of Supercomputers and High Performance Computing*, 1994.

[2] Michael Q. Jones. The Distributed Application Debugger. *M.Sc. Thesis, University of Nevada, Las Vegas, United States*, May 2013.

[3] Cherri M. Pancake. Why Is There Such a Mis-Match between User Need and Parallel Tool Production? *Keynote address, 1993 Parallel Computing System: A Dialog between Users and Developers.*, April 1993.

[4] Jan B. Pedersen. Classification of Programming Errors in Parallel Message Passing Systems. *In* Proceedings of Communicating Process Architectures 2006 (CPA'06) *IOS Press*, September 2006.

[5] Jan B. Pedersen and Michael Q. Jones. Error Classifications for Parallel message Passing Programs: A Case Study. *Proceedings of Parallel and Distributed Processing techniques and Applications (PDPTA'12)*, July 2012.

[6] Ian Foster. Designing and Building Parallel Programs: Concepts and tools for parallel software engineering. *Addison Wesley*, 1995.
[7] Erik H. Tribou. Millipede: A Graphical Tool for Debugging Distributed Systems with a Multilevel Approach. *M.Sc. Thesis, University of Nevada, Las Vegas, United States*, August 2005.
[8] Hoimonti Basu. Interactive Message Debugger for Parallel Message Passing Programs Using LAM-MPI. *M.Sc. Thesis, University of Nevada, Las Vegas, United States*, December 2005.
[9] Parallel Virtual Machine, Oak Ridge National Laboratory. `http://www.csm.ornl.gov/pvm`, 2011.
[10] LAM/MPI Parallel Computing. `http://www.lam-mpi.org`, 2012.
[11] Jan B. Pedersen. Multilevel Debugging of Parallel Message Passing Systems. *PhD Thesis, University of British Columbia, Vancouver, British Columbia, Canada*, June 2003.
[12] Susanne M. Balle and Robert T. Hood. Global Grid Forum User Program Development Tools Survey. *Global Grid Forum*, 2004.
[13] The Open MPI Project. FAQ: Debugging applications in parallel. `http://www.open-mpi.org/faq/?category=debugging`, 2013.
[14] Allinea Software. DDT product page. `http://www.allinea.com/products/ddt`, 2013.
[15] Rogue Wave Software. TotalView product page. `http://www.roguewave.com/products/totalview.aspx`, 2013.
[16] Mathematics Argonne National Laboratory and Computer Science Division. Web pages for MPI Routines. `http://www.mcs.anl.gov/research/projects/mpi/www/www3`, 2013.
[17] Thomas J. Leblanc and John M. Mellor-Crummey. Debugging Parallel Programs with Instant Replay. *IEEE Transactions on Computers*, April 1987.
[18] GDB - The GNU Debugger. `http://www.gnu.org/directory/gdb.html`, 2013.
[19] The Eclipse Foundation. `http://www.eclipse.org/`, 2013.
[20] Microsoft Visual Studio. `http://msdn.microsoft.com/vstudio/`, 2013.
[21] Apache Subversion. `http://subversion.apache.org`, 2013.

Communicating Process Architectures 2013
P.H. Welch et al. (Eds.)
Open Channel Publishing Ltd., 2013

BPU Simulator

Martin REHR, Kenneth SKOVHEDE, and Brian VINTER
Niels Bohr Institute, University of Copenhagen, Denmark

Abstract. A number of scientific applications start their life as a Matlab prototype, which are later re-implemented in a low level programming language, typically C++ or Fortran for the sake of performance. Bohrium is a project that seeks to eliminate both the cost and the potential errors introduced in that process. The Bohrium runtime currently supports multiple execution platforms, such as CPU, clusters and GPGPU, and in this work, we introduce the Bohrium Processing Unit, BPU, which will be the FPGA backend for Bohrium. We model the BPU as a PyCSP application, and describe the clear advantages of using CSP for simulating this new soft-CPU. The current PyCSP simulator is able to simulate 2^{20} Monte Carlo Pi simulations in less than 35 seconds in the smallest BPU simulation.

Keywords. scientific byte code, FPGA, PyCSP

Introduction

From a quick glance on the TOP500 list [1], it becomes clear that every high performance system in the world efficiently utilizes multi-level parallelism. The parallelism is used to divide the computation on multiple machines, multiple computation units inside the machines, multiple cores inside the computation units, and on the instructions in each core. This multi-level parallelism is a prerequisite for utilizing the machines efficiently, but also hard to correctly implement. To further complicate things, different generations of the same type of processor may have widely different constructions, which require low-level changes to the code when the hardware is upgraded or the vendor is changed. One example of this problem can be seen in [2] where an attempt to run an OpenCL [3] kernel tuned for Nvidia on AMD hardware shows a 60 times speed difference.

Each of the different levels of parallelism has their respective set of tools, such as MPI [4], pthreads [5], SSE [6], and OpenCL [3]. To achieve acceptable performance from such a system, the application developer must master all of the tools individually as well as combined. For any non-trivial application, this requires substantial effort, and often involves multiple specialists spending time tuning the parts.

This stands in sharp contrast to the application development methods used by scientists and non-computer professionals, who describe the algorithms mathematically, in terms of matrix and vector operations. The solution is then commonly prototyped in a programming language that naturally maps these constructs, such as Matlab [7] or NumPy [8]. Once the prototype works correctly, there is a multitude of different ways to convert this to efficient code. However, all the current methods require rewriting the program into something that works efficiently with each of the tools for the different levels of parallelism. This naturally increases both the cost of the final executable, as well as introduces another source of errors. In many cases, the persons that developed the prototype are not the same that implement the efficient version, which further increases the costs and prohibits the prototype implementer from updating the program.

1. Bohrium

In the Bohrium runtime system [9,10], shown in Figure 1, the prototype is also the final implementation, which significantly reduces the costs associated with high-performance program development. The core idea in Bohrium is the use of *Vector Bytecode*, which enables separation between the implementation language and the execution.

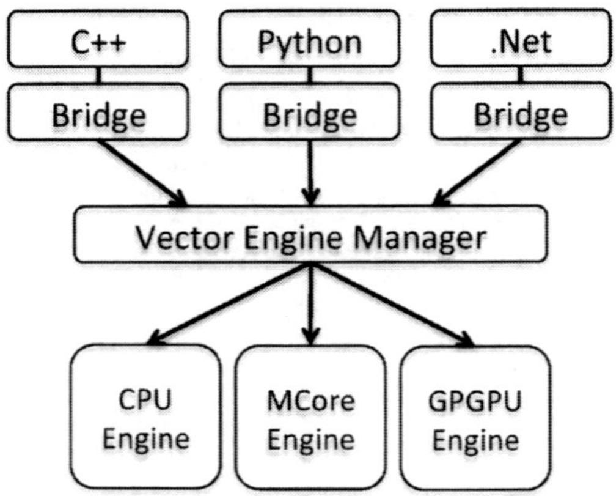

Figure 1. Overall Bohrium Architecture.

To provide a high productivity environment for the end user, the Bohrium system provides a number of language bridges. The language bridges attempt to integrate naturally into the host language. As the name implies, they provide a bridge from the programming language to the Bohrium runtime. The overall programming method is known as vector programming, and similar to the programming model found in Matlab [7] and NumPy [8].

In the current implementation of Bohrium, we provide language bridges for C++, Python, and the Common Intermediate Language, also known as .Net. While these are the only three bridges implemented, they show that the vector programming approach can be applied to different classes of programming languages, including: static and dynamic typed, compiled, interpreted, procedural, and functional. All implementations are provided as libraries, so that they may be used without customized tool-chains, compilers, or special runtime environments. Both the Python and CIL implementations include fallback code, that allows execution, even in the case where Bohrium is not present on the system. This makes it possible to develop and verify applications in an environment the user is comfortable with, and then later add the option of efficient execution.

The Python implementation [9] mimics the NumPy libraries and uses NumPy for fallback, so any existing NumPy program can be executed with Bohrium by changing the import statement. The C++ bridge uses templates and operator overloads to provide an elegant interface to vectors and matrices. The CIL implementation uses the NumCIL [11] library as a frontend and provides simple access to vector and matrix operations from any of the CIL languages, such as C#, F#, IronPython and Visual Basic.

Figure 2 shows how a 5-point stencil operation is performed with multiple operations on the same underlying 2-dimensional data structure. The idea is to define views that select only a subset of the elements in the full matrix. As all the views are of equal size, it is possible to apply a series of element-wise additions followed by a multiplication to obtain the result. As an example of code the user can write, we have included a simple 5-point stencil operation, performed on a variable named `grid`, written in C++, Python and C#.

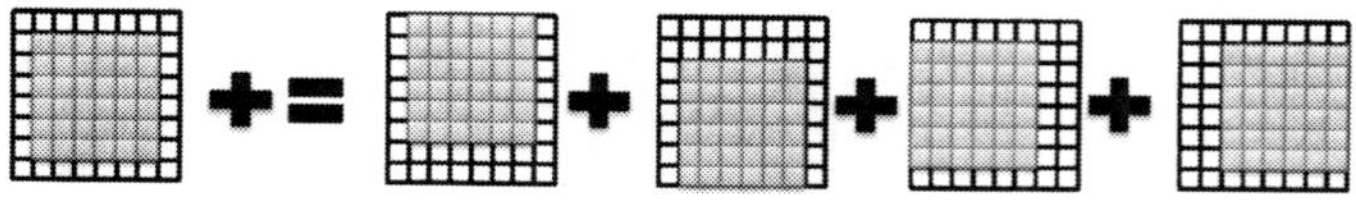

Figure 2. 5-point stencil operation.

C++:

```
multi_array<double> center, north, south, east, west;
auto w = grid.shape(0), h = grid.shape(1);
center = grid[_(1,-1,1)][_(1,-1,1)];
north =  grid[_(0,-2,1)][_(1,-1,1)];
south =  grid[_(2, h,1)][_(1,-1,1)];
west =   grid[_(1,-1,1)][_(0,-2,1)];
east =   grid[_(1,-1,1)][_(2, w,1)];
center(0.2( center+north+east+west+south ));
```

Python:

```
center = grid[ 1:-1, 1:-1]
north =  grid[  :-2, 1:-1]
south =  grid[ 2:   , 1:-1]
west =   grid[ 1:-1, 2:  ]
east =   grid[ 1:-1,  :-2]
center[:] = 0.2 * (center+north+east+west+south)
```

C#:

```
var center = grid[R.Slice(1,-1), R.Slice(1,-1)];
var north =  grid[R.Slice(0,-2), R.Slice(1,-1)];
var south =  grid[R.Slice(2, 0), R.Slice(1,-1)];
var west =   grid[R.Slice(1,-1), R.Slice(0,-2)];
var east =   grid[R.Slice(1,-1), R.Slice(2, 0)];
center[R.All] = 0.2 * (center+north+east+west+south);
```

Even though the matrices in the examples are of the same size, the bridges allow some flexibility in this regard. The rules of NumPy broadcasting are supported, meaning that operands of different sizes may be applied to each other, if one of the dimensions can be repeated, so the number of elements in each dimension matches. If a scalar is applied to a vector, the scalar value is converted to a vector of the same length, with the value repeated. If a vector is applied to a 2-dimensional matrix, similarly, an extra dimension is added, which repeats the vector. This broadcasting mechanism is similar to what is found in NumPy, and greatly increases productivity in the bridged languages. This broadcasting is performed by manipulating the views, so that a broadcast does not involve copying data. The data structure used to describe views is shown in Figure 3.

To reduce complexity in the Bohrium system, the broadcasting must be performed before the bytecodes are formed, so that each following component can assume an equal number of elements for each operand in the instructions.

To limit the overhead of invoking the Bohrium system, and increase the possibility for optimizing the execution, each of the bridges perform lazy evaluation of the operations performed on the data structures. When a side effect is noticeable, such as when accessing a scalar value, a batch of instructions are coded as *Vector Bytecode* and forwarded to the underlying *Vector Engine Manager*.

The *Vector Engine Manager*, VEM, is responsible for splitting a batch to multiple underlying components. In the simplest case, the VEM simply forwards the instruction batch to and underlying *Vector Engine*, VE, which executes the batch. The VEM and VE components share the same API, which makes it possible to change the *Vector Engine* used to execute the

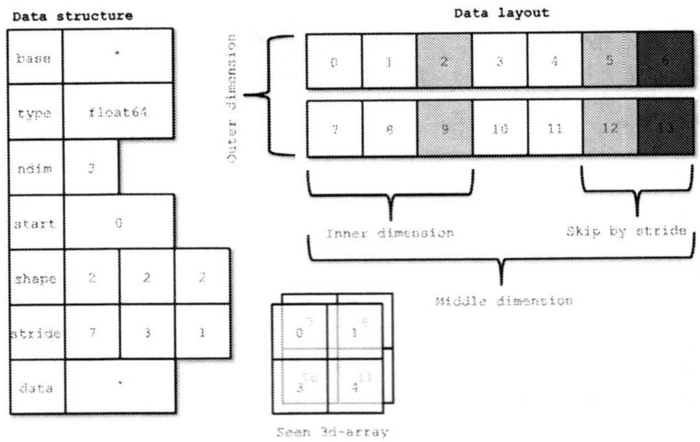

Figure 3. Data structure for describing a multidimensional array.

batch, but also to nest multiple VEM components. This interchangeability ensures that the user can have a very simple development setup with only a single core CPU VE, and then change the execution setup to use a multi-node cluster with GPGPUs without changing the source code. This works because the bridge can consider the underlying components as a *Virtual Machine*, which executes *Vector Bytecode* and is thus not tied to a particular setup.

In the current implementation, Bohrium supports a simple *pass-through* VEM that can handle a single attached VE, as well as an MPI based VEM that supports any number of VEMs underneath. Work is in progress on a VEM that can handle multiple attached VEs and perform runtime scheduling for setups with multiple GPGPUs, possibly mixed with CPUs.

The VEs that target the CPU are currently working in an interpreted manner, where they execute a single instruction at a time. For the multi-core CPU VE, the target elements for each instruction are divided among the participating cores. Even with this inefficient approach, Bohrium is able to outperform NumPy with up to 2 times speedup on a single core [10].

The VE that targets GPGPUs is based on OpenCL [3] and performs a form of Just-In-Time compilation to produce execution kernels. The kernels drastically reduce the overhead associated with invoking a method on the GPGPU, as well as reduces the amount of data that needs to be transferred. This approach has shown more than 90 times speedup, for unmodified NumPy code, when moving from the CPU to the GPGPU [12].

In this work, we introduce the Bohrium Processing Unit, BPU, a planned soft-CPU for Field Programmable Gate Arrays (FPGAs), which implements a subset of the Bohrium instruction set. The goal of the work is thus to add a FPGA VE to Bohrium. Since FPGA development is slow and does not motivate the developer to test new ideas quickly, we wish to work in a simulation environment that allows us to quickly test ideas in a manner that can easily be migrated to a FPGA environment.

We expect the FPGA VE to perform on par with GPGPUs, but with a different set of drawbacks and benefits. One such difference is that FPGAs are equipped with a network port, which allows us to process network data without invoking the CPU or transferring data over the PCI bus. Another difference is that the FPGAs usually contain an ARM based hard processor, which can be used to orchestrate and pre-process data before it reaches the BPU.

2. Related Work

The vector programming approach is widely used, in projects as different as Chapel [13] and CoArray Fortran [14]. It is widely used because it is a natural way to express many algorithms, and at the same time expresses parallelism. The use of vectors removes the traditional loops that would otherwise traverse the vectors and process each element. Without the loops present, the execution system can implement the loops in a more suitable fashion, such as using tiled memory access or instruction-fused execution.

Some systems are built on vector programming and thus often contain special constructs in the language that enable operations on vectors in an intuitive way, such as those found in Chapel [13] and Cilk [15]. There is also a multitude of libraries built for C++ that all seem to favor using C++ templates to produce a domain specific language that enables compile time optimizations. In the Thrust library [16], the class system is used to facilitate data copy between devices through *host* and *device* classes.

The project that appears to be most similar to Bohrium is Intel's Array Building Blocks [17], which defines an intermediate code and a virtual machine for execution [18]. The Array Building Blocks system performs some compilation together with the C++ program compilation, and some compilation in the virtual machine at runtime.

What makes the Bohrium runtime stand out from these existing projects is that it targets the entire chain from language to execution, and that implementations exist for multiple languages. One notable supported language is Python, with an interface that mimics the NumPy libraries, such that any program written for NumPy can execute with Bohrium simply by changing the import statement. Where other approaches tie themselves to a particular language, they place the burden of learning a new language onto the programmer. Because multiple languages are supported, programmers can reuse their existing skills and support libraries when using Bohrium.

Another notable difference is the use of shaped views, which not only reduces the need for data copies, but also supports matrix views with higher ranks than three. The views and reshaping mechanisms used in Bohrium are similar to the slicing and broadcast concepts in NumPy.

Previous work using CSP to model FPGAs have focused on a direct translation between the CSP model and an existing FPGA programming language such as Handel-C [19] which increases our confidence that the approach is viable. However, the idea of using CSP to simulate and develop all aspects of an FPGA-based CPU is novel.

3. Vector Byte Code

The vector bytecode is designed to be the glue between the vector engines and the implementation language. In its current form, it represents a completely register based execution model with no stack. Each shaped multidimensional array is considered a register, and there is no imposed limit to either size of a register, nor the number of registers used. Similar to traditional register based architecture, the vector bytecodes are encoded with an opcode, an output register and one or two input registers. The element-wise add operation can thus be expressed as:

```
add $c, $a, $b
```

After executing the above operation, the register `$c` will contain the results of the element-wise addition of `$a` and `$b`. In the Bohrium binary representation, a constant value can be used instead of one of the input operands. The operands are defined as registers, but implemented as simple memory locations in the local memory space. For devices such as

GPGPUs and FPGAs, the data must be moved or remapped into another address space, and the instructions must then also be re-encoded prior to execution on the other devices.

The major part of the Bohrium bytecode instructions is element-wise operations that operate on two inputs and produce a single output. The instruction set also includes a few non-element-wise instructions, such as the operations that perform reductions, together with some management instructions, such as the SYNC instruction which moves data back into the local memory space.

The BPU code loader will rewrite each instruction to better fit the micro-code architecture and issue appropriate `load` and `store` operations that move data from the memory storage into the registers in the BPU core. To support operands that are larger than the actual ALU widths, the registers will be indexable into lines that are the same width as the ALU. This approach also requires the code loader to rewrite single instructions into multiple instructions that work on register lines.

This rewrite is fairly simple, and is mostly limited by the speed with which data can be written into the register files, and is mainly introduced to keep the ALU as simple and efficient as possible.

4. The Bohrium Processing Unit (BPU)

The BPU is designed as a RISC processor using the Harvard Architecture [20] where the data and code segments are separated. A schematic overview of the BPU is shown in Figure 4. The BPU consists of a number of code segments named `Code A-D`. These segments and

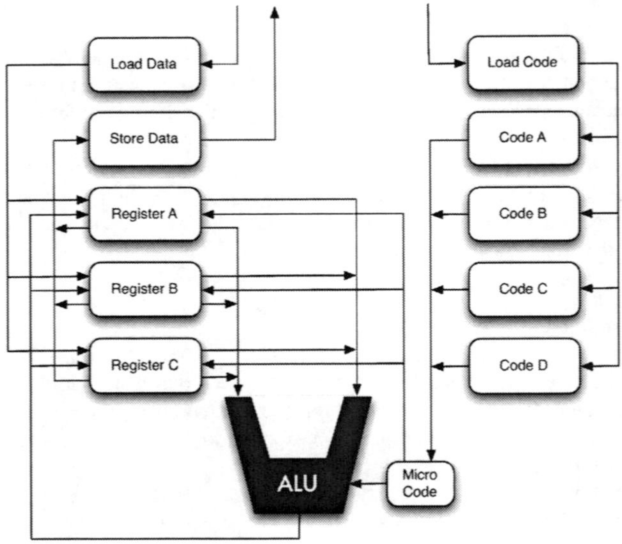

Figure 4. Schematic overview of the BPU.

the instructions within them are executed sequentially, however, code segments are fetched from main memory asynchronously with execution to ensure that at least one code segment is available for execution at any given time. To overlap communication and data transfer triple buffering is used by having one register file [21] for computation, one for fetching new data from main memory and one for storing the computed results in main memory. These are named `Register A-C` in Figures 4 and 5. Memory transfers between registers and main memory is also overlapped with execution such that previously calculated data is stored in main memory and data for future instructions are loaded from main memory while the ALU execute instructions using the data present in the execution register file. Each instruction is executed sequentially and the number of instructions in each kernel may exceed the number of lines in the register files. When the ALU has processed all the data in the `execution` register file then the register files are rotated, that is, the `execution` register file changes state to a `store` register file and the calculated results are pushed to main memory. Likewise the `load` register file is changed to an `execution` register file and the `store` register file is changed to a `load` register file for retrieving data for the next computation. This is sketched in Figure 5.

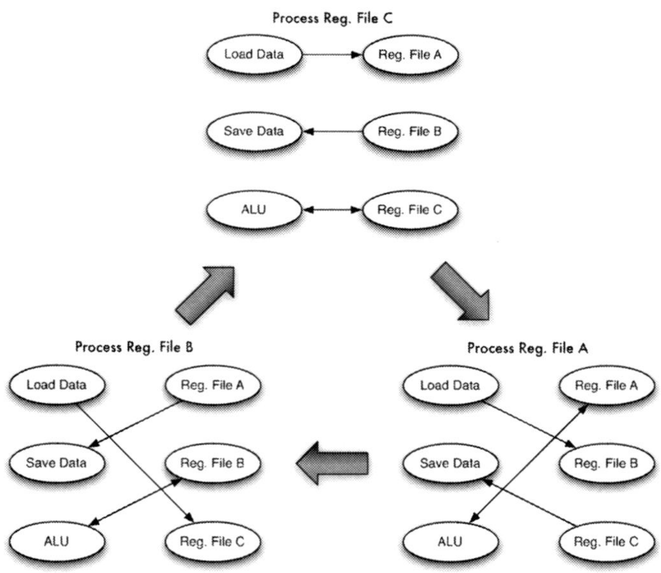

Figure 5. Schematic overview of the BPU register rotation.

4.1. BPU Hardware

Building the BPU in actual hardware will be done using FPGAs. Modern FPGAs have an integrated general purpose ARM processor which can be used for orchestrating data between the FPGA memory elements and main memory. This comes in handy because the ARM processor can be used as an input and output DMA handler for the BPU. The BPU is a vector processor meaning that all computation units are performing the same operation in each clock-cycle on each data entry in the register file, which is commonly known as Single Instruction Multiple Data (SIMD). When creating the BPU for a specific FPGA device we are going to use all the arithmetic units for the same instruction in each clock cycle which defines the width of our ALU shown in Figure 4. The width of the ALU then defines the width of the register files such that a full register line is processed in each clock-cycle. The total number of memory elements available on a specific FPGA is divided between code segments and register data for highest throughput.

Programming FPGAs can be done in several ways. The classical sequential programming methods such as VHDL [22] and Verilog [23] are the most used languages, but a less common language called Handel-C [19], supports CSP. By using Handel-C we can keep a close relation between the BPU simulator and the actual hardware implementation [24], but we still need to look further into the relevant FPGA devices before deciding on a development tool.

The obvious way to apply the Bohrium Vector Byte Code to FPGAs would be simply creating a VHDL, Verilog or Hande-C code generator back-end and then burn each kernel directly to the FPGA. However, creating hardware kernels on a FPGA may take hours where our Bohrium Vector Bytecode is generated on the fly at runtime in a JIT manner, so burning kernels directly to the FPGA device is only feasible if the same kernel is used over and over or if you have dedicated FPGAs for all the kernels one would like to execute. Even with one FPGA card for each kernel, it would still require a very close relationship between the Bohrium runtime environment and the hardware, which is not feasible. By using the BPU approach we make a clean interface between the Bohrium runtime environment and the actual hardware that executes the vector bytecode. Keeping the BPU simple by implementing only SIMD arithmetic and handling data with the built-in general purpose ARM processor, we expect good performance from the bytecode vector processor approach.

5. The BPU Simulator

The BPU is still in its design phase, and for testing design choices PyCSP [25,26] is used to implement a functional simulation. Each of the logical components in Figure 4 are implemented as processes in PyCSP. The register file and code components consist of a number of processes to handle the three register files and four code modules, the number of register files is fixed to three while the number of code modules is read from a configuration file to allow easy experimentation with the design. An example of a PyCSP network of the BPU simulation during execution can be seen in Figure 8 where the processes, with their PyCSP names, can be seen to match the design in Section 4. The figure is acquired with the PyCSP tracing module [27].

The `LoadCode` component is simple and it merely load static kernels into the `Code` modules. The `Code` modules either become active and pass the program sequence to the `micro-code` component or simply wait for termination. We do not support multiple kernel-loads at the present time, however, the internal structures are prepared such that a small extension of the `CodeInput` module is sufficient. The `micro-code` component continuously receives an instruction and splits it into an `ALU` operand and a register file selection signal. Each register file is in turn connected to the `InputDMA`, `OutputDMA` or `micro-code` and `ALU` in line with Figure 5. The 'active' register file refers to the one connected to the `micro-code` and `ALU` processes. The `ALU` is straightforward and executes the operation that is passed from the `micro-code` process on the values that are passed from the register file. The CSP network is shown in Figure 6.

5.1. DMA

Since the DMA components are highly dependent on the environment that the BPU will run in, and this environment is still quite ill-defined, the DMA components currently do not implement any functionality but simply works as a sink for all incoming messages.

5.2. Kernel Memory

The kernel memory will eventually receive its input from the DMA module, but for now the application is hardcoded into the process. We plan an intermediate version where the kernel

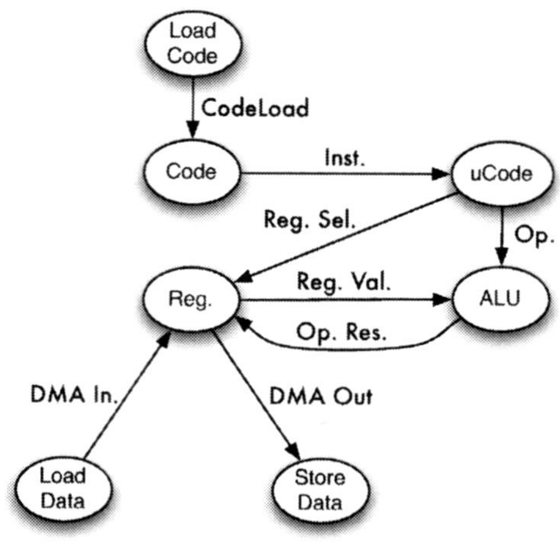

Figure 6. CSP overview of the BPU simulation.

memory is read from files, which will be trivial but will allow us to test more kernels. The entire implementation is shown below. Note that while all four kernels will be identical only one of them will in fact be sending instructions to the uCode unit.

```
@process
def kMemBlock(cout, cin):
    target = 2**20

    pgm = [('rnd', 0, 0, 0),
           ('mul', 0, 0, 0),
           ('rnd', 1, 1, 1),
           ('mul', 1, 1, 1),
           ('add', 2, 0, 1),
           ('ones', 3, 0, 0),
           ('lt', 0, 2, 3),
           ('sum', 4, 4, 0),
           ]

    if cout:
        _ = cin()
        for _ in xrange(target/constants.register_size):
            for op in pgm:
                cout(op)
        cout(('halt', 0, 0, 0))
    else:
        while True:
            _ = cin()

@process
def kMem(cout, codein):
    Parallel(kMemBlock(cout, codein.reader()),
             kMemBlock(None, codein.reader()),
             kMemBlock(None, codein.reader()),
             kMemBlock(None, codein.reader())
             )
```

5.3. Micro Code

The uCode unit is in the design to enable BPU implementation of multi-cycle instructions, but for now it simply translates an instruction into the respective control messages for the register file and the ALU. In the below code the uCode unit receives a message from the kernel-memory in the format: operation, destination register, source register one, and source register two. The register choices are then sent to the register file and the operation to the ALU.

```
@process
def uCode(cin, cout, src1, src2, dest):
    while True:
        op, d, s1, s2 = cin()
        dest(d)
        src1(s1)
        src2(s2)
        cout(op)
```

5.4. Register File

The register file consists of three identical register windows as described in Section 4. Since the DMA components are inactive only one register window is actually active, but since all three are identical they are all instantiated, although the input and output windows are connected to the DMA component those channels are actually inactive.

```
@process
def RegisterFileElement(
    out1, out2, in1, selectout1, selectout2, selectin1, cin, cout, mode):

    registers = numpy.zeros((constants.number_of_registers,
                             constants.register_size),
                            dtype=numpy.float64)

    if mode == 'Execution':
        while True:
            output1sel, output2sel, input1sel = ParallelRead([selectout1,
                                                               selectout2,
                                                               selectin1])

            out1(registers[output1sel])
            out2(registers[output2sel])
            registers[input1sel] = in1()

    elif mode == 'Output':
        while True:
            cout(registers)
    elif mode == 'Input':
        while True:
            registers = cin()

@process
def RegisterFile(out1, out2, in1, selectout1, selectout2, selectin1, cin, cout):

    Parallel(RegisterFileElement(out1.writer(),
                                 out2.writer(),
                                 in1.reader(),
                                 selectout1.reader(),
                                 selectout2.reader(),
```

```
                         selectin1.reader(),
                         None,
                         None,
                         'Execution'),
    RegisterFileElement(None,
                         None,
                         None,
                         None,
                         None,
                         None,
                         cin.reader(),
                         None,
                         'Input'),
    RegisterFileElement(None,
                         None,
                         None,
                         None,
                         None,
                         None,
                         None,
                         cout.writer(),
                         'Output')
    )
```

5.5. ALU

The ALU receives the instruction, operation, as a mnemonic string and for each clock cycle simply, reads two inputs, calls the actual function on those data and sends the result back to the register file through the destination channel. A special *halt* will force the ALU out of its main-loop, after which it prints out statistics on the simulation and finally poisons the channels that it is connected to, these poison operations will propagate throughout the system and terminate the simulation in full.

```
def add(a, b):
    return a + b

def sub(a, b):
    return a - b

def mul(a, b):
    return a * b

def div (a, b):
    return a / b

def rnd(a, b):
    return numpy.random.random(len(a))

def lt(a, b):
    return a < b

def gt(a, b):
    return a > b

def eq(a, b):
    return a > b

def sum(a, b):
```

```
    return numpy.ones(len(a)) * a.sum()

def ones(a, b):
    return numpy.ones(len(a))

def halt(a, b):
    return a

func = {'add': add, 'sub': sub, 'mul': mul, 'div': div, 'rnd': rnd, 'lt': lt, \
        'gt': gt, 'eq': eq, 'sum': sum, 'ones': ones, 'halt': halt}

cost = {'add': 1, 'sub': 1, 'mul': 2, 'div': 2, 'rnd': 4, 'lt': 1, 'gt': 1, 'eq': 1, \
        'sum': numpy.log2(constants.register_size), 'ones': 1, 'halt': 0}

@process
def ALU(INST, SRCA, SRCB, DEST):
    cycles = 0
    inst = None
    while inst != 'halt':
        a, b, inst = ParallelRead([SRCA, SRCB, INST])

        result = func[inst](a, b)
        DEST(result)
        cycles += cost[inst]

    print 'Simulation halted after',cycles,'cycles'
    print 'BPU config:'
    print 'Register length:', constants.register_size
    print 'Number of registers per file:', constants.number_of_registers
    poison(INST)
    poison(SRCA)
    poison(SRCB)
    poison(DEST)
```

6. Proof of Concept

The simulation is evaluated with a Monte Carlo Pi simulation. The following example represents a somewhat naive way of estimating π, based on a Monte Carlo simulation. The model is simple; if we take a square with two units side length and inscribe that by a circle with unit radius then we can throw a dart at the drawing multiple times, each time the dart falls within the circle we will increase a counter. We know that the area of the square is four unit-squared and the area of the circle is Pi, thus the ratio of darts that fall within the circle is the ratio of four unit-squared that Pi represents. For simplicity we reduce the problem to the upper right quadrant of the circle, e.g. one fourth of the original problem, as shown in Figure 7.

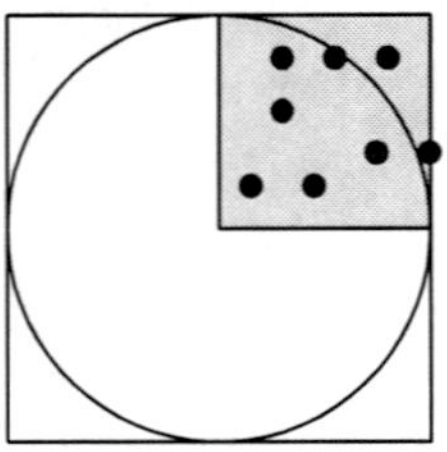

Figure 7. Diagram showing the Monte Carlo Pi method.

Obviously this is an extremely inefficient way of determining the value of π, however, Monte Carlo methods and the closely related Las Vegas methods are quite valuable tools in various fields of science and the simple model that we present here may be replaced by a set of real problems within physics, chemistry and biology.

The executed BPU Monte Carlo Pi code is the following:

```
rnd  0        #Fill reg-0 with random numbers
mul  0, 0, 0  #Square the values in reg-0
rnd  1        #Fill reg-1 with random numbers
mul  1, 1, 1  #Square the values in reg-1
add  2, 0, 1  #Add reg-0 and reg-1 into reg-2
ones 3        #Fill reg-3 with 1s
lt   0, 2, 3  #Less-than test reg-2 against reg-3 and store result in reg-0
sum  4, 4, 0    #Add reg-0 to reg-4
```

The code is repeated a number of times, depending on the width of the ALU so that 2^{20} samples are processed. Figure 8 shows a snapshot of the PyCSP execution. The simulation execution time scales almost inversely proportionally with ALU width, i.e. linearly with the number of instructions that are simulated. Only when the width becomes very large are the actual operations large enough to impact performance. The simulated time and the time to execute the simulation can be seen in Figure 9.

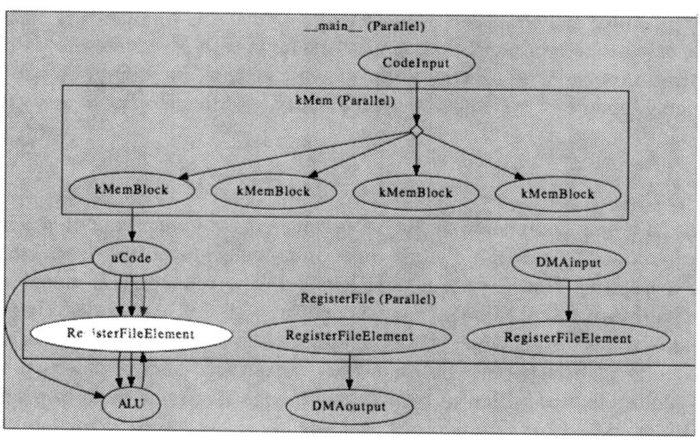

Figure 8. Snapshot of the PyCSP simulation.

6.1. CSP Observations

It is not surprising that CSP turns out to be ideal for CPU simulation, since a CPU consists of logically concurrent components, and several hardware definition languages apply CSP-like features such as Handel-C. One component that we observe will be needed in the next refinements of the BPU simulator however is a convenient distribution of the clock-signal, which should have broadcast semantics, i.e. one clock-process sends a message that must then be read by all attached processes. Since PyCSP does not have one-to-all channel semantics, i.e. a channel that when receiving a message ensures that all connected readers have received the message before re-opening the channel for writing[1] this must be implemented through a process.

[1] Such a channel type is not trivial to implement

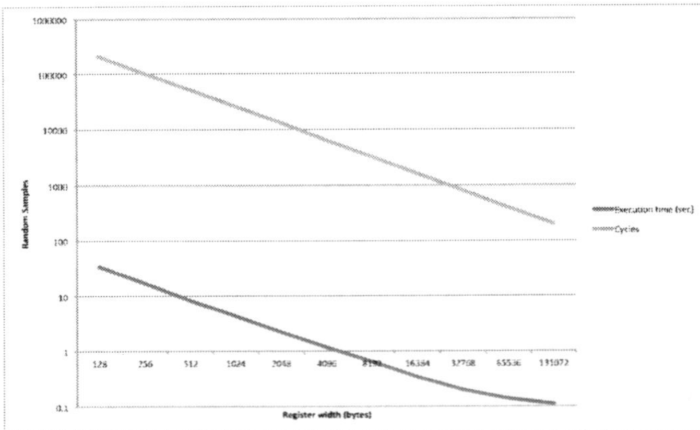

Figure 9. Simulation and execution time of 2^{20} Monte Carlo Pi samples.

7. Conclusion

In this paper, we have introduced the Bohrium Processing Unit. We have shown how it is very easily modelled and simulated using PyCSP and that this simulation is efficient enough to facilitate extensive experimentation. A small example with 2^{20} Monte Carlo Pi simulations took less than 35 seconds to execute in the slowest version. In addition, we found that the graphic tracing module in PyCSP has helped identify coordination errors very efficiently.

8. Future Work

The current BPU implementation in PyCSP is quite coarse-grained and in the future it will be iteratively refined from a pure Python function implementation to a model that translates directly to a Register Transfer Level (RTL) model. The refined RTL model will then be ported to a hardware description language, which supports FPGAs, and finally implemented in actual hardware. We expect the PyCSP implementation to go from the current model where tens of processes are used to one where hundreds or even thousands of processes are involved and are expecting that transition to be painless, though that remains to be seen. We know that PyCSP can indeed run thousands of processes [25]. As the BPU model is specified in greater detail it will be reimplemented using more and even smaller processes, a method that may be considered as recursive refinement, given the obtained speed of the simulation we expect that even a fine grained simulation model will run at a speed that does not prohibit experimentation.

References

[1] Jack J. Dongarra, Piotr Luszczek, and Antoine Petitet. The LINPACK Benchmark: Past, Present and Future. *Concurrency and Computation: practice and experience*, 15(9):803–820, 2003.

[2] Peng Du, Rick Weber, Piotr Luszczek, Stanimire Tomov, Gregory Peterson, and Jack Dongarra. From CUDA to OpenCL: Towards a Performance-portable Solution for Multi-platform GPU Programming. *Parallel Computing*, 38(8):391–407, 2012.

[3] OpenCL. The OpenCL Specification, Khronos Group. `http://www.khronos.org/opencl`. [Online: accessed 26 July 2013].

[4] Jack J. Dongarra, Rolf Hempel, Anthony J.G. Hey, and David W. Walker. A Proposal for a User-level, Message Passing Interface in a Distributed Memory Environment. Technical report, Oak Ridge National Lab., TN (United States), 1993.

[5] IEEE. Threads Extension for Portable Operating Systems (Draft 6), February 1992, P1003.4a/D6.

[6] Alex Klimovitski. Using SSE and SSE2: Misconceptions and Reality. *Intel developer update magazine*, pages 1–8, 2001.

[7] Matlab Programming Environment. `http://www.mathworks.com`. [Online; accessed 26 July 2013].

[8] Travis E. Oliphant. Python for Scientific Computing. *Computing in Science and Engg.*, 9(3):10–20, May 2007.

[9] Mads R. B. Kristensen, Simon A. F. Lund, Troels Blum, and Brian Vinter. cphVB: A System for Automated Runtime Optimization and Parallelization of Vectorized Applications. *CoRR*, abs/1210.7774, 2012.

[10] Kenneth Skovhede. Combining High Productivity with High Performance on Commodity Hardware, chapters 7 and 8. `http://www.nbi.ku.dk/english/research/phd_theses/phd_theses_2013/kenneth_skovhede/Kenneth_Skovhede.pdf`, 2013. [Online; accessed 26 July 2013].

[11] Kenneth Skovhede and Brian Vinter. NumCIL: Numeric Operations in the Common Intermediate Language. *Journal of Next Generation Information Technology*, 4(1), 2013.

[12] Troels Blum. Generalizing Execution of Vectorizable Computations by Generating Vector Oriented Byte Code. `https://cphvb.googlecode.com/files/cphvb-speciale.pdf`, 2011. [Online; accessed 21 February 2013].

[13] Cray Inc. Chapel Language Specification. `http://chapel.cray.com/spec/spec-0.91.pdf`, 2012. [Online; accessed 9 August 2013].

[14] Robert .W. Numrich and John Reid. Co-array Fortran for Parallel Programming. In *ACM Sigplan Fortran Forum*, volume 17, pages 1–31. ACM, 1998.

[15] Robert D. Blumofe, Christopher F. Joerg, Bradley C. Kuszmaul, Charles E. Leiserson, Keith H. Randall, and Yuli Zhou. *Cilk: an Efficient Multithreaded Runtime System*, volume 30. ACM, 1995.

[16] Christian DeLozier. GPU acceleration for the C++ Standard Template Library. [Online; accessed 12 March 2013].

[17] Chris J. Newburn, Byoungro So, Zhenying Liu, Michael McCool, Anwar Ghuloum, Stefanus Du Toit, Zhi Gang Wang, Zhao Hui Du, Yongjian Chen, Gansha Wu, et al. Intel's Array Building Blocks: a retargetable, dynamic compiler and embedded language. In *Code Generation and Optimization (CGO), 2011 9th Annual IEEE/ACM International Symposium on*, pages 224–235. IEEE, 2011.

[18] Intel Array Building Blocks Virtual Machine. `http://software.intel.com/sites/default/files/m/b/6/a/arbb_vm.pdf`. [Online; accessed 12 March 2013].

[19] Handel-C Synthesis Methodology. `http://www.mentor.com/products/fpga/handel-c/`. [Online; accessed 26 July 2013].

[20] Myke Predko. *Programming and Customizing the 8051 Microcontroller*. McGraw-Hill, Inc., New York, NY, USA, 1999.

[21] Mazen A. R. Saghir and Rawan Naous. A Configurable Multi-ported Register File Architecture for Soft Processor Cores. In *Reconfigurable Computing: Architectures, Tools and Applications*, pages 14–25. Springer, 2007.

[22] Peter J. Ashenden. *The Designer's Guide to VHDL*. Morgan Kaufmann Publishers Inc., San Francisco, CA, USA, 2nd edition, 2001.

[23] Samir Palnitkar. *Verilog HDL: a Guide to Digital Design and Synthesis (second edition)*. Prentice Hall Press, Upper Saddle River, NJ, USA, second edition, 2003.

[24] Jonathan D. Phillips and Gardiner S. Stiles. An Automatic Translation of CSP to Handel-C. In *in Communicating Process Architectures 2004*. IOS Press, 2004.

[25] Rune Møllegaard Friborg, John Markus Bjørndalen, and Brian Vinter. Three Unique Implementations of Processes for PyCSP. In Peter H. Welch, Herman Roebbers, Jan F. Broenink, Frederick R. M. Barnes, Carl G. Ritson, Adam T. Sampson, Gardiner S. Stiles, and Brian Vinter, editors, *Communicating Process Architectures 2009*, pages 277–292, nov 2009.

[26] Brian Vinter, John Markus Bjørndalen, and Rune Møllegaard Friborg. PyCSP Revisited. In Peter H. Welch, Herman Roebbers, Jan F. Broenink, Frederick R. M. Barnes, Carl G. Ritson, Adam T. Sampson, Gardiner S. Stiles, and Brian Vinter, editors, *Communicating Process Architectures 2009*, pages 263–276, nov 2009.

[27] PyCSP Tracing Module. `https://code.google.com/p/pycsp/wiki/Getting_Started_With_Tracing`. [Online; accessed 26 July 2013].

Communicating Process Architectures 2013
P.H. Welch et al. (Eds.)
Open Channel Publishing Ltd., 2013

Exploring GPGPU Acceleration of Process-Oriented Simulations

Frederick R.M. BARNES [1], Thomas PRESSNELL and Brendan LE FOLL
School of Computing, University of Kent, UK

Abstract. This paper reports on our experiences of using commodity GPUs to speed-up the execution of fine-grained concurrent simulations. Starting with an existing process-oriented 'boids' simulation, we explore a variety of techniques aimed at improving performance, gradually refactoring the original code. Successive improvements lead to a 10-fold improvement in performance, which we believe can still be improved upon, allowing us to explore simulations with larger numbers of agents (30,000 rather than 2,000) interactively and without significant performance degradation.

Keywords. GPGPU, process-oriented simulations

Introduction

This paper considers the issue of using GPUs, SIMD-style hardware architectures, to accelerate fine-grained concurrent systems, that employ and encourage MIMD-style parallelism and that use a range of mechanisms for inter-process communication and synchronisation. As a non-trivial example we consider a simulation based on Reynolds' boids [1], whose structure and operation is described in section 1, together with a brief overview of the occam-π concurrency model used. The boids simulation used as an example is not trivially parallelised, and in general can be considered an *n-body* problem.

Our background is in the area of fine-grained concurrency, developing languages (occam-π [2]) and run-time kernels (CCSP [3]) supporting the construction of massively concurrent software systems[2], and their efficient execution on multicore and multiprocessor platforms. A wide range of software systems have been built using these techniques, including embedded control systems, agent-based simulations and operating systems. Concurrency is used as a fundamental design principle, with language (compiler) guarantees of freedom from race-hazard and aliasing error, permitting the building of predictably scalable systems [4], based on the semantics of CSP [5] and the π-calculus [6]. Furthermore, such systems can lend themselves to speedup on parallel hardware, where there is enough 'parallel slack' to ensure execution cores are kept busy.

The recent emergence of GPGPU (general-purpose graphics processor) programming, and its increasing availability, leads us to ask how we can take advantage of this in order to speed up computationally heavy process-oriented systems. The principle programming abstractions for GPUs are data-parallel and C based, with CUDA [7] and OpenCL [8] foremost. The issue of parallelising existing code and algorithms for GPU execution is largely

[1]Corresponding Author: *Fred Barnes, School of Computing, University of Kent, Canterbury, Kent, CT2 7NF, UK*. E-mail: `f.r.m.barnes@kent.ac.uk`

[2]By *massively concurrent* we mean large numbers (10^4–10^6) of communicating concurrent processes on a single processing node.

well-studied [9,10], though optimal implementations can be difficult and likely tailored to the specific problem and hardware.

Section 2 explores three different approaches that treat the GPU as an abstract computational resource within a process-oriented program, with successive modifications aimed at improving performance. For each, the design and performance of the approach is discussed. Section 3 draws some concluding remarks on the work, with some discussion of limitations, applicability to other simulations and directions for future work.

1. A Process-Oriented Boids Simulation

Figure 1 shows the process-network for part of the "boids" simulation [4], developed as part of the CoSMoS project [11]. The boxes represent processes — independent threads of control with their own (private) state. The lines connecting them represent channels or bundles of channels (those with arrowheads), and barriers (no arrowheads). Channel communication is synchronous and unbuffered — readers must wait for writers and vice-versa. Furthermore, there is no *shared data* or uncontrolled *aliasing* in this system (enforced by the occam-π [2] compiler), with data copied or references *moved* between processes. Several of the processes in Figure 1 utilise *many-to-one* channel structures — the many writers compete for access using a fair-queueing semaphore with correct use enforced by the compiler.

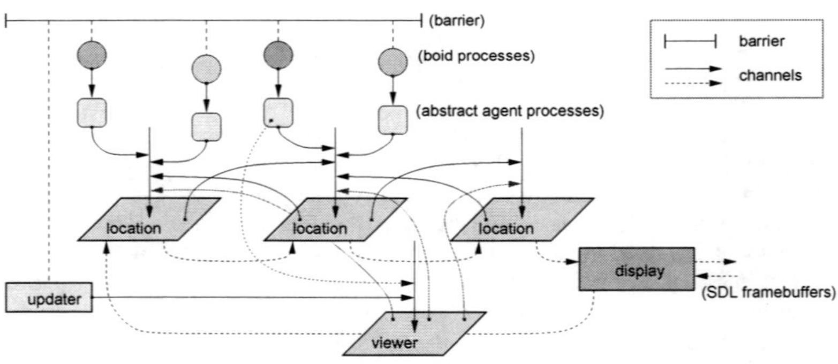

Figure 1. Process network for a dynamic agent-based simulation.

The topology of the world in which the boids exist is constructed from a toroidal 2-dimensional grid of 'location' processes (only 1 dimension is shown in Figure 1). Associated with each location are 'viewer' and 'updater' processes, the former keeping details of agents in the associated and neighbouring locations, the latter controlling updates to this view. Boid and updater processes are *enrolled* on a barrier that is used to control progression in the simulation, ensuring that all boids (and other processes) proceed at the same rate, and further ensuring that the view returned to all boids in any particular cycle is consistent.

The concurrent design of the simulation confers several advantages. Firstly, it is structured in a way that enables a consistent understanding of the system, its components and their interactions — we transition from design (Figure 1) to a concurrent implementation seamlessly. Secondly, modifying or extending the system, to change the behaviour of some or all boids, allow for the creation and destruction of boids at run-time, or to explore interesting non-Euclidean topologies, is largely straightforward.

It should be noted that this is the system we are starting with: a MIMD-style process-oriented CPU based multiprocessor implementation. If we were starting anew with the intention of GPU utilisation, the system would likely be designed and constructed in a very

```
1: procedure BOID(space link, barrier t)
2:     state me = initial_state ()
3:     while True do
4:         state[] all, vis, obs
5:         sync t                                        ▷ enter observation phase
6:         all = GET_VIEWABLE(link)
7:         vis, obs = PRUNE_VISIBLE(all, me)
8:         me = CENTRE_OF_MASS(vis, me)
9:         me = REPULSION(vis, me)
10:        me = MEAN_VELOCITY(vis, me)
11:        me = OBSTACLES(obs, me)
12:        UPDATE(link, me)
13:        sync t                                        ▷ enter update phase
14:    end while
15: end procedure
```

Algorithm 1. Boid agent algorithm.

different way: a data-parallel design implemented in the preferred GPU language (C/C++ with OpenCL or C-style with CUDA). For a fairly small system such as the one described, a complete re-implementation on the GPU would ultimately yield better performance, and this has been investigated by others previously [12]. This, however, is not the primary objective of this work, which is to explore ways to improve performance of process-oriented systems using a GPGPU.

1.1. Simulation Operation and Performance

The behaviour of the individual boid processes is shown in Algorithm 1. In each cycle, the boids first synchronise on the barrier (step 5) entering the *observation* phase. Here, the 'viewer' processes associated with each boid's current location are queried (step 6) to retrieve information about viewable agents in the local and neighbouring locations. The information obtained here is essentially an array of data structures containing agent information such as relative position (calculated by the 'viewer' process) and velocity.

The retrieved information is split (step 7) into visible boids and obstacles, based on relative position and velocity. A series of rules are then applied (steps 8 – 11) to compute an acceleration value for the boid, in turn producing a new velocity. This is communicated to the current 'location' process (step 12) and used to update the boid's position. If crossing a boundary, the current location instructs the abstract agent to move to a new location. The boids synchronise on the barrier again (step 13) entering the (empty) *update* phase, before looping and synchronising once again.

To put this code into context, Figure 2 shows the logical state (position and velocity) of boids in a 2 dimensional world. Each of the grid locations shown is implemented by a 'location' and 'viewer' process pair, with agents (both boids and obstacles) connected via their abstract agents to the particular 'location' they inhabit. The positions, velocities and other agent state are represented as floating-point numbers in the program — individual grid locations (the unit space of the simulation) can contain any number of agents at different positions within, up to a predefined maximum. For the boid shown near the centre, whose field-of-vision is indicated, the array generated in step 6 (*all*) will contain information for all the agents shown, including itself. At step 7, this is pruned down into two arrays containing information about the three visible boids (*vis*) and the single visible obstacle (*obs*). For typical flocking behaviours, the ratio of viewable agents to visible ones is generally high, so much of the observed state is discarded at step 7. In practice this is not too significant, since the size of the communicated agent state is relatively small (32 bytes). For denser simulations or with different parameter sets, however, this contributes significantly to computational overhead.

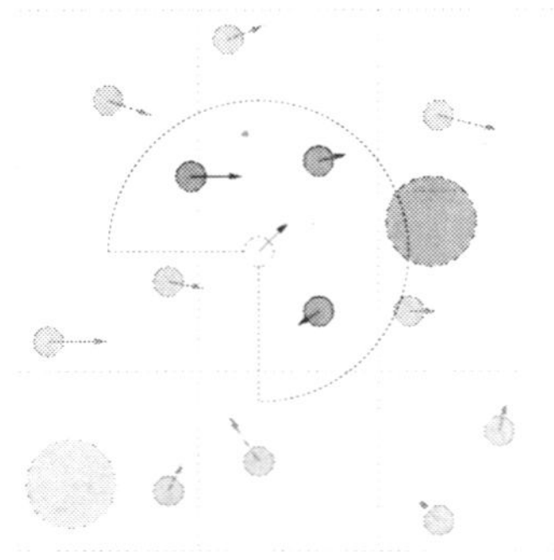

Figure 2. Boids and their world, from the perspective of the white boid near the centre.

In parallel with the boids and synchronising on the barrier are the 'updater' processes associated with each location. In the update phase these instruct the corresponding viewer to collect state from the connected location processes, providing a consistent view for the next cycle of execution.

Figure 3 shows three screen-shots of a system containing 2048 boids in an 8×6 grid at time steps 1, 100 and 500. As can be seen, from the initial random placement, flocks soon emerge. Figure 3(d) shows the speedup of this simulation on a typical desktop computer: an Intel Quad Core i7 (2600K) running at a fixed 3.4 GHz, with 4 physical cores and 4 hyper-threads. Speedup is only achieved when dealing with large numbers of boids and does not exceed the number of physical cores (4 in this instance). Figure 3(e) shows the simulation throughput, in thousands of boid cycles per second, for 2048 boids over a period of time. Increasing the number of cores has a measurable advantage, but the simulation does slow down once large flocks start to form — roughly from $t = 500$ onwards. Enabling the visualisation reduces performance proportionally with the requested display frame rate and size. To remove these (and other CPU-loading effects) the benchmarks are run with the visualisation disabled.

The performance figures reported are for a single run of the simulation only, with the simulation reporting the cycle time every 10 iterations (the average of the last 10 iterations). The difference reported by separate runs of the simulation, given the same parameters, is small — typically less than 1%. Only showing the results for a single run (instead of averaged over a number of runs) gives a better indication of the variability of performance over time. the 'noise' experienced by 4 and 8 threads in Figure 3(e) (and elsewhere) is due to two factors: the way processes are scheduled by the run-time system [3], that becomes significantly less stable when the CPU is forcibly shared between the simulation and other OS goings on; and the changing distribution of work as boids move between locations, that can create significant workload differences from one cycle to the next.

1.2. Comments on Performance

Little attempt has been made to optimise the code at this point — the simulation was developed in a domain-specific context, using individual concurrent processes to represent different physical artefacts, with the interactions between them suitably abstracted. As a result, the boid behaviour as implemented in Algorithm 1 is straightforward. The boids *see* the relative position and velocities of other agents, deriving their new state from that information plus

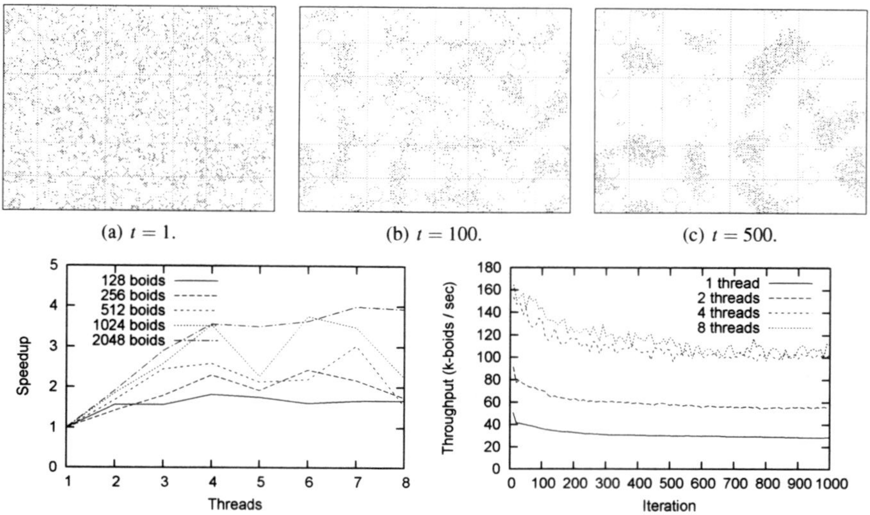

(a) $t = 1$. (b) $t = 100$. (c) $t = 500$.

(d) Speedup at $t = 500$. (e) Throughput for 2048 boids.

Figure 3. Boid simulation at successive time steps, showing flocking behaviour (a,b,c); multicore CPU speedup (d) and throughput (e).

their current state. Furthermore, the code is largely obvious and easily understood or adapted by domain experts, not necessarily computer scientists.

For the applications and investigations this simulation was originally created for, performance is sufficient. With the visualisation enabled, 2048 boids in an 8×6 grid can be interacted with in real-time at 25 frames per second, roughly half the speed of the system with visualisation disabled. Given the way it is implemented, the decrease in performance as flocks start to form is not unexpected. Each flock is essentially a localised n-body problem and each boid's own view of the flock is unique (as the implementation deals with relative positions and velocities).

2. Exploring GPGPU Acceleration

To remain compatible with a wide range of GPU hardware, we limit our assumptions about what might be available. Specifically, we consider the GPU as a SIMD processor that supports integer and floating-point operations, without requirements for inter-thread communication, atomic actions or complex data sharing. In practice, a GPU is often much more capable than this, supporting inter-thread synchronisation, atomic manipulation of shared data and possessing significant internal memory bandwidth, with a capability for asynchronous data transfer to/from the host's memory. Taking advantage of such facilities, however, greatly increases the complexity of the resulting system. For the time being, our objective is to retain some simplicity in the resulting code — certainly to the extent that it is still within the grasp of domain experts. Despite limiting the potential performance where GPUs are concerned, not assuming too much makes the approach potentially applicable to other SIMD devices, such as FPGAs and DSPs, though these are outside the scope of this paper.

For our initial experiments as reported here, we have chosen to use CUDA [7] (Compute Unified Device Architecture) to implement code for the GPU. Invoking external C calls in order to interface with the CUDA run-time library is straightforward from occam-π, as are considerations for data-type compatibility. To keep the implementation simple, the occam-π program handles all GPU interactions transparently via external function calls. Whilst there are other approaches that could perhaps be taken (see Section 3) this is certainly the simplest for the general programmer. The CUDA toolkit from NVIDIA provides a C-like langage for writing the actual GPU *kernel* (code that is executed in parallel on the available stream processors) and a run-time library for accessing the GPU. Although this limits our investigations to NVIDIA's devices, attempts to use other technologies including OpenCL [8] (that targets a wider range of SIMD-style processing hardware) have proven to be less successful (further investigation required).

The following sections explore some different strategies aimed at improving performance of the simulation. The first (2.1) treats the GPU as an "on-the-side" computational resource, with all GPU interactions handled through a server process to which other processes can connect and make requests. The second (2.2) considers a different approach: refactoring the simulation to use *shared-memory* to hold the simulation state, in increasing amounts, whose safe access is controlled by the existing barrier. With some guided optimisations to restrict the amount of data copied to or from the GPU, this starts to demonstrate a significant improvement in performance. Section 2.3 considers the combination of these two approaches, with further improved performance as a result. The third approach considered here (2.4) refactors the system in a way that turns the parallelism 'inside-out' — moving from many long-lived high-level concurrent processes (the boids) to a few high-level concurrent processes, with short-lived concurrent processes internally. In effect, collecting all the individual boid processes into one "super-boid" process. The resulting system, containing many short-lived repeating concurrent processes, is more suited to the GPU execution style than the other approaches we consider, and demonstrates considerably better performance as a result.

2.1. A Process-Oriented GPU Server

The first approach introduces a "GPU server" process, to which all similar boids are connected, that collects and executes computation requests — shown in Figure 4(a). The operation is straightforward: the server process collects requests (comprising the boid's state, the set of other visible boids, and a channel on which to return results), copies this to the GPU, executes the boid algorithms on the GPU, copies the results back to host memory, and finally communicates these results back to the individual boid processes. The specific parts of Algorithm 1 targeted for GPU execution initially are steps 8 – 10. The code that implements this is a direct by-hand translation from the occam-π, with each boid's algorithm executing in its own GPU thread with its own data.

Figure 4(b) shows the throughput for the basic CPU implementation ("no server") alongside a CPU implementation of the GPU server process, to measure the overhead of the infrastructure. The different lines (256, 1024 and 2048) are for different request-buffer sizes in the GPU server[3]. Since there is only one GPU server process, all the computation is necessarily serialised. Figure 4(c) shows the cycle times for the same basic "no server" implementation alongside those where the same algorithm is executed on the GPU — in this case, one half of an NVIDIA GTX 590 (which is in effect two GTX 580s side-by-side). This shows a general drop in performance, although the throughput is more consistent. Both sets of results suggest

[3]The server process buffers requests from boids (each 32 bytes) together with information about what other boids are visible (each 32 bytes). As such, the amount of data copied to the GPU is $((n*32)+(v*32))$ bytes, where n is the number of requests (512, 1024, 2048) and v is the number of viewable agents, which is typically $\leq (50 \times n)$, varying considerably. For 1024 requests, the total amount of data copied to the GPU is typically less than 2 megabytes. When the request buffer is full, or has timed out, these are sent to the GPU as a single job (to be distributed amongst the available stream processors).

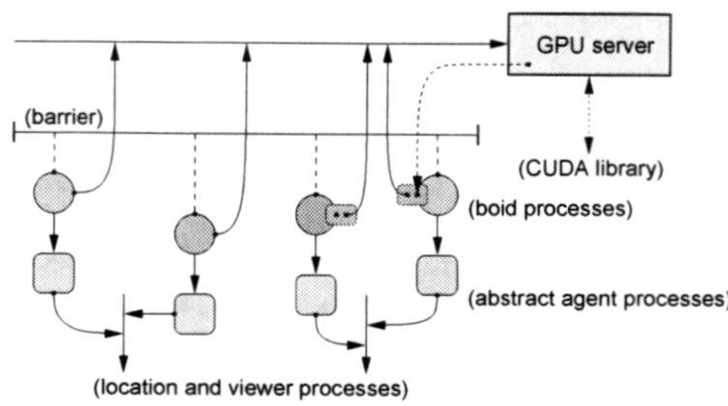

(a) Abstracting the GPU resource using a parallel process.

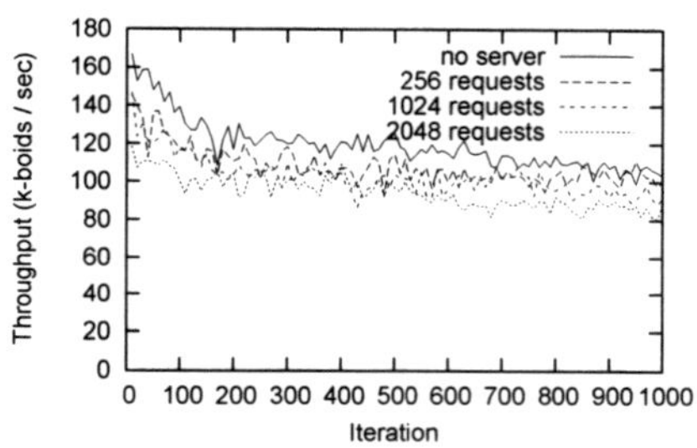

(b) Serial infrastructure only (no GPU).

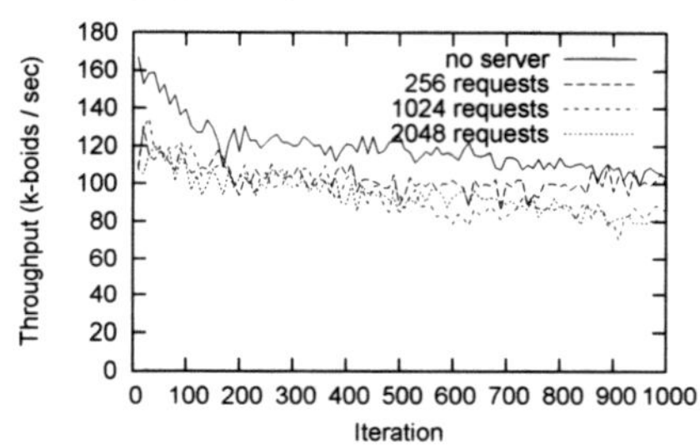

(c) Using half of a GTX 590 GPU.

Figure 4. Network structure and performance using a GPU server process (2048 boids).

that the targeted parts of Algorithm 1 (steps 8 – 10) account for only a small percentage of execution time.

With some basic profiling, it quickly becomes clear that the majority of time is spent in step 7, involving trigonometry based on the viewing angle and direction of the boids. Given this, the next logical might sensibly be to move more of the algorithm onto the GPU. In practice, however, this causes a significant drop in performance — down to between 20,000 and 50,000 boid cycles per second. The reason is a significant increase in the quantity of boid state that needs to be copied to the GPU. In the previous version (Figure 4) only the state of visible boids was copied (visibility having been determined on the CPU beforehand); here the state of all boids, provided by the viewer process, is copied. When multiplied by the number of boids this quickly becomes significant. In tests, only around 3 – 5 % of viewed boids are actually visible (as determined by typical simulation parameters). For a system of 2048 boids, this data is around 13 megabytes in size.

It is well known that copying data between the host and GPU memories is expensive, so the lack of performance here is not unexpected. Unfortunately, however, for this particular simulation the resulting performance is worse than the CPU-only version for 2 cores. Other similarly structured but operationally different simulations may perform better, however, particularly in cases where the amount of data copied between the host and GPU is small, and the computation is expensive.

2.2. *Rethinking the Problem: Shared Data*

The results shown so far demonstrate a reduction in performance — the quantity of data involved and nature of the problem means that naïvely transplanting chunks of code onto the GPU, surrounded by data copies, does not always work well (though this is well known already) and definitely does not work well in this particular case. Whilst there is plenty of opportunity for optimising what we have, such as parallelising the GPU server's operation for multicore CPUs and multiple GPUs, the limiting factor (and most significant overhead) will still be data copying.

The amount of data (and copies of) involved in the original simulation arises as a result of the way the simulation was implemented — communicating boid state between processes. Once updated, an individual boid's state will be copied for each other boid that can see it, modified to provide a relative position (a simple vector subtraction) as part of the observation. This eludes to a decision in the original implementation, that was to perform this computation for each viewable boid at the point of observation, rather than in each boid.

In an attempt to reduce the amount of data required by the algorithm when running on the GPU, we re-engineered the simulation to utilise shared data for the boid state. Whilst this required fairly extensive modifications to individual processes (reading and writing boid state, including positions that are now absolute) the overall structure of the system (shown in Figure 1) did not change. Furthermore, taking this approach (that we could have done initially) does not much weaken the flexibility of the simulation nor damage any of the existing concurrency abstractions present: global shared data can easily be modelled (or implemented) as a process containing that data, with which other processes communicate. Dynamically creating and destroying boids is no issue, nor is physically distributing the simulation. Instead of communicating boid state, however, the 'location' and 'viewer' processes now only need to be aware of which boids are present (an integer identifier). Race-hazards on the shared state are avoided by the distinct *phases* of operation, marshalled by the barrier.

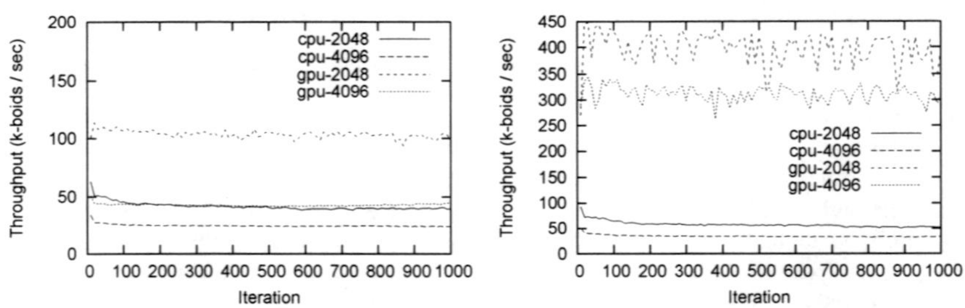

(a) Non-optimised shared agents only. (b) Shared agent and viewable data.

Figure 5. Benchmark results of initial shared-data implementations.

For the CPU only implementation, sharing the boid state improves performance by around 50%. Re-introducing the GPU server process infrastructure, but using shared boid state, produces varied results. As shown in Figure 5(a), performance overall for the GPU version is lacking — the CPU results are for a single server process only, thus all the computation is serialised on a single CPU core. With the GPU version, there is still a large amount of data copied to the device — many times the original state-space of the simulation — primarily consisting of the array of viewable boids (and obstacles) returned by the viewer processes. To reduce this, we share the array of viewable agents at each location. This necessitates a change in the boid algorithm implementation, that previously sorted the viewable and relative-position arrays into visible 'boids' and 'obstacles', to instead perform just one pass over the viewable array. The results of this are shown in Figure 5(b), demonstrating an improvement in performance when using the GPU (and for larger numbers of boids).

At this stage, the individual viewer processes extract lists of agents (boids and obstacles) from their accompanying and surrounding location processes, collate these and return to individual boids (via a pointer and size) when requested. As a result, the list of agents in each location is duplicated 8 more times in transit to the GPU — clearly inefficient.

Removing this overhead improves the best-case performance slightly — up to 450,000 boid cycles per second for 4096 boids, where all processing is done in a single GPU kernel run, with each GPU thread performing the computation cycle for a single boid. Although there is no duplicated data, there is some amount of redundant data in the global 'viewable' array that is copied to the GPU. Removing this redundant data (using a shared server process that allocates this space dynamically) had a negligible impact on performance.

An effect of making the viewable state shared is that the 'viewer' processes become redundant, since the contents of other locations can be observed directly. The 'updater' processes remain however, and are used to signal moves between locations, avoiding any races between observing and changing the contents of individual locations.

2.3. Reintroducing High-Level Parallelism

The graphs showing CPU performance in Figure 5 effectively measure the algorithm's execution on a single processor core, as there is only one GPU server process. A reasonable next step is to re-introduce parallelism at the application level, to support multiple CPU cores, multiple GPU devices or a mixture of both. The results of this are shown in Figure 6, with the average throughput against the number of GPU server processes (using the CPU or GPU), and specific throughput for the GPU (using both devices on the GTX-590). The results do not indicate a significant performance improvement — the infrastructure required to implement this (in a way that is compatible with the existing boid code) creates significant overheads, due to the need to synchronise the activities of the processes involved.

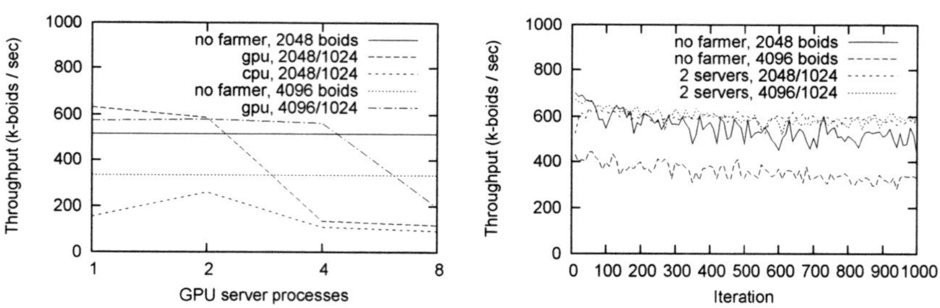

Figure 6. Parallelising the GPU server, 2048 and 4096 boids in varying batch sizes.

Despite the overall lack of performance (compared with a version "no farmer" where the boids execute their algorithms locally in parallel, scheduled over 8 CPU threads) some improvements are observed. Doubling the number of boids to 4096 produces significant improvements on two GPU devices in batches of 2048 (not shown on the graph) — as is partly expected, given where the overhead lies. An advantage of this particular approach is that the boids themselves still see the processing as an abstract resource, to which they send their current state and viewable neighbourhood (via global shared data) and receive back an updated state.

2.4. Removing Concurrency

So far, we have focussed on improving performance in a way that keeps the individual 'boid' processes intact — they are still in control of their own behaviours, even though computation is delegated. As a further optimisation we replace the GPU server process network with a single process that synchronises on the same barrier as the boids, but which performs all computation directly — either on the CPU or the GPU, in parallel as appropriate. To accommodate this we introduce an additional phase to the barrier, in which the newly computed velocities of individual boids are written back to the shared data. Removing the explicit channel communication between boids and a centralised server process has significant performance advantages, as shown in Figure 7, where a throughput of around 1.6×10^6 boids-per-second is attained when using two GPU devices at a density of 38 boids per location with 16,384 boids (in a 24×18 grid). This is a significant increase in the number of boids: previously we had been unable to explore such large systems interactively in real-time, except with fairly specialised hardware platforms such as the *display wall* used in [4]. Furthermore, we are now able to explore aspects such as how the density of a simulation affects throughput (and ultimately the resulting emergent behaviour) — interesting *flocking* behaviours occur between 20 and 60 boids per location; more than this and the simulation becomes over crowded, until it is completely "jammed" at around 120 boids per location.

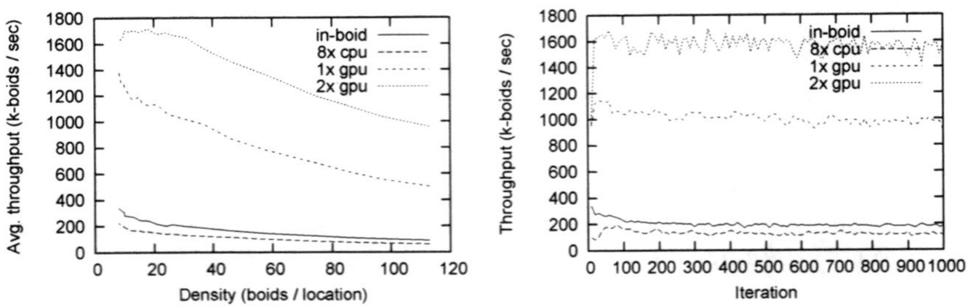

(a) Varying simulation density. (b) Throughput at 38 boids/location.

Figure 7. Further optimisations removing channel communication, 16384 boids.

The only remaining activity in the boid processes is initiating movement between locations. As a further optimisation we centralise this in a similar way. The boid processes now contain no activity other synchronising on the barrier — as such, we remove them entirely. The result is a further improvement in performance, at the (arguable) expense of some design clarity (the boids now solely exist as shared global data). Other agents in the system, such as the 'interactive' one, are still easily implemented as concurrent processes, however — as could any complex agents (without interfering with the execution of others on the GPU).

As a further optimisation, we remove the copy-back phase (where the boid state is updated) and instead double-buffer the entire agent state array (including obstacles). Figure 8 shows the high-level design of this version together with benchmark results for up to 30,000 boids. As the results show, performance is significantly improved, managing a throughput of around 2 million boids-per-second for (roughly) 10,000 boids and higher using both GPU devices available on the GTX-590.

Although we have removed the individual boid and obstacle processes, extending the simulation (as done with the interactive agent and GUI rendering processes) is still a largely straightforward task — simply adding parallel processes with access to the shared state and synchronising on the barrier. The grid of location processes remains for convenience, used only during initialisation (populating the virtual world with boids and obstacles) and for moving the interactive agent that behaves like an obstacle.

3. Conclusions and Future Work

In this paper we have explored some different approaches for speeding-up the execution of fine grained concurrent simulations, using a commodity GPU resource. Starting from a

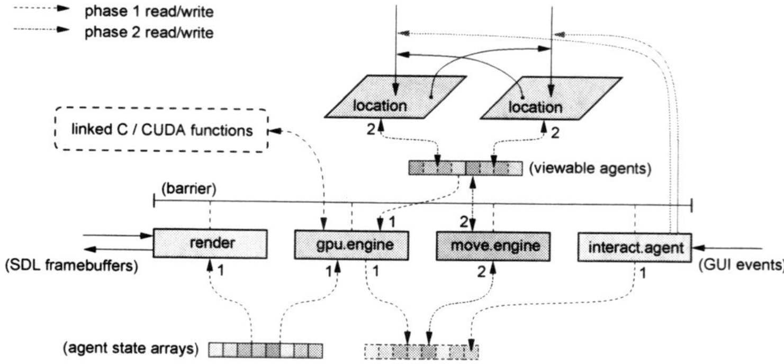

(a) Simulation structure showing two-phase reads and writes.

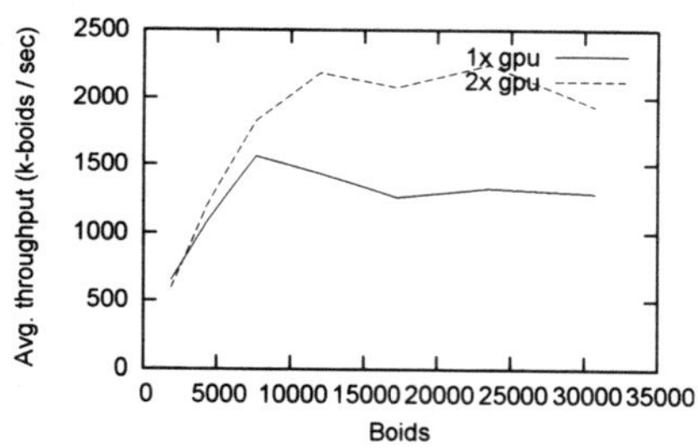

(b) Average throughput for varying grid sizes, 40 boids per location.

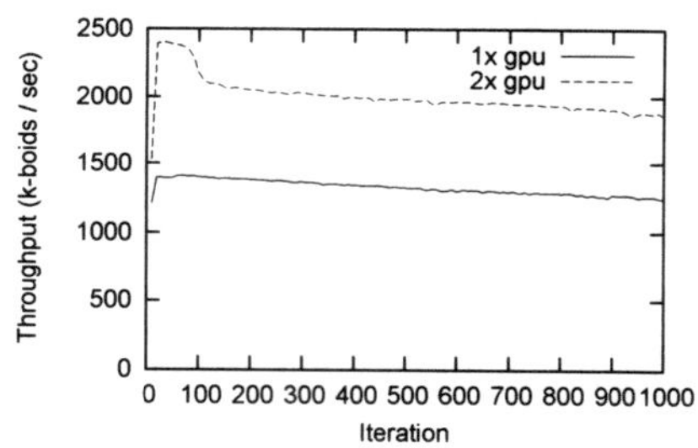

(c) Throughput for $30,720$ boids in a 32×24 grid.

Figure 8. Version of the simulation utilising reads and writes on shared data instead of channel I/O, using a barrier to coordinate access.

CPU-only highly concurrent process-oriented simulation — that scales acceptably well on multicore platforms — we applied successive and generally small modifications with varying results.

Initial attempts at abstracting the GPU resource using a parallel process produced unimpressive results (Figure 4), due primarily to the amount of data copying involved. Further changes to the system included the introduction of shared data, for the agent state (Figure 5(a)) and viewable data (Figure 5(b)), with only minor performance improvements. Reducing the amount of data copied and parallelising the GPU server's execution (Figure 6) gave a 3-4$\times$ speedup over the original, but only slightly improved on the best CPU results. Removing the agent-initiated computation (Figure 7) and then the agent processes entirely (Figure 8) produced the best results, with up to 10$\times$ speedup over the original.

Although the design of the system has changed somewhat, it is an evolution of an existing system (not a clean-room implementation). The concurrent design of the system, and support for that concurrency in the implementation language used (occam-π) admits a certain flexibility, e.g. changing from a message-passing to a shared-data implementation. The design presented in Figure 8(a) has a good balance between concurrent design, performance and flexibility, but is not perhaps an obvious design. But neither is it the most optimal in terms of execution performance. Despite this, the fact that we can explore and evolve systems in this way is of clear benefit, here and elsewhere.

Being able to run larger (and more complex) simulations was one of our original objectives, as a way of exploring interesting and more complex behaviours in real-time that only emerge when dealing with larger numbers of agent interactions. Following the same rules from Algorithm 1, Figure 9 shows results for different parameter sets, resulting in different emergent behaviours.

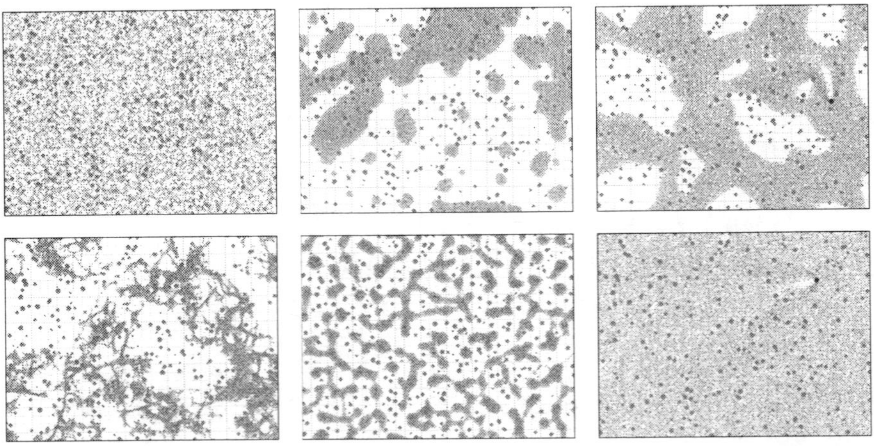

Figure 9. Screen-shots of 32,000 boids and 307 obstacles in a 32×24 grid, with varying parameters changing the behaviour; right-most images show the interactive agent.

The system in Figure 8 handles boid movement in the 'move.engine' process, iterating over the boids (sequentially) and manipulating the viewable arrays directly. This is computationally light and accounts for only a fraction of the overall cycle time ($< 0.05\%$). Running this on the GPU would *probably* be of some performance benefit, but doing so is not straightforward (due to the shared viewable state) — and would restrict the implementation to particular GPU platforms (because of the atomic increments or inter-thread synchronisations required to do this efficiently).

3.1. Lessons Learned

Here we reflect on some of the lessons learned in undertaking this work, and consider how the approaches can (or might) be applied in other contexts.

Perhaps the most obvious observation is that we chose a difficult problem: the way the system was originally implemented did not lend itself to a straightforward GPU implementation (abstracting and offloading the computation by way of a "GPU server" process). The principal overhead was one of memory copies to the GPU — that subsequent optimisations (shared memory) sought to overcome. In simulations where the amount of data involved (though not necessarily the number of agents) is significantly smaller or where the amount of computation performed is greater, we would expect the initial approach to perform substantially better.

In terms of applying communicating process ideas to GPU programming, we believe this has merit. In typical GPGPU simulations, the activities of the host and the GPU are interlocked, often with only one OS thread on the host side. The occam-π concurrency model provides a level of parallel flexibility that does not exist in ordinary (C-style) procedural programming — utilising all available CPU and GPU processing capabilities from within a single application becomes an easy prospect, saving the programmer from potentially complex thread synchronisation and race-hazard issues. Furthermore, adding extra functionality "on the side" is similarly straightforward, even if that extra functionality is largely independent of the existing system (e.g. a process that periodically samples the simulation for correctness).

3.2. Future Work

Whilst we have been successful in our efforts to improve performance, there remain several avenues of future work to be explored, some of which we would expect to yield a further improvement in performance.

The work presented here has focussed on one particular and quite simple approach to accessing SIMD-style GPU resources from MIMD-style process-oriented programs. Specifically, hiding all GPU interaction behind single external function calls, passing data to and from the GPU at that point. Whilst this is straightforward, and potentially lends itself to other SIMD architectures, it does not necessarily allow for optimal performance — every cycle of the simulation requires the entire simulation state to be copied to the GPU, and the changed state back again, before the next cycle can start.

In light of this, an alternative approach (and as a further modification to the simulation) is to maintain the simulation state on the GPU foremost, with all the computation there, transferring the resulting state to the CPU when necessary. Although this would likely result in better performance for this particular simulation, it is harder to guarantee in the general case, particularly where a range of different agent types is involved (not just boids and data-only obstacles).

No significant attempts have yet been made at optimising our hand-written CUDA. Some work with the CUDA profiling tools would reveal any deficiencies in our CUDA kernels, pointing the way to further optimisation. Similarly, we do not attempt to make efficient use of constant memory on the GPU, nor change the way CUDA threads are arranged, both of which can have a significant (and positive) impact on performance [13]. To better understand the performance ceiling, it would be worthwhile implementing a GPU-only version (with no occam-π concurrency at all). There has been past work looking specifically at GPU and other accelerated implementations of Reynolds' boids [14,12], that perform well with a variety of optimisation techniques, but little in recent years that considers the same fundamental algorithm.

Given the facilities present in modern GPUs, such as inter-thread synchronisation, atomic operations and direct access to host memory, one possibility would be to implement channel-style communication algorithms on the GPU itself, perhaps in each GPU thread, presenting the GPU resource as an array of parallel processes that can be communicated with directly. How this would work in practice is perhaps unclear, as different GPU devices provide different capabilities in differing amounts (e.g. CUDA GPU threads can only be synchronised within *thread blocks*, not across all threads). The general (developer community) view is that this sort of approach does not work well, given that the GPU infrastructure is foremost intended for, and largely assumes, highly-parallel short-lived tasks — i.e. pixel shading.

The work presented here has required the existing occam-π code to be re-implemented in CUDA, although the structure and operations on data are a direct hand translation. Making this automatic, by having the compiler generate CUDA kernels and appropriate interfaces for identified portions of the source program, would make the process of accessing GPU and other similar SIMD resources significantly simpler. Whether such a mechanism would prove to be useful (in terms of improved performance with minimal effort) remains to be seen, and would require some consideration of automatic refactorisation to be of benefit to simulations such as our original 'boids'.

Whilst not directly relevant to the benchmarks reported in this paper, there is scope for additional optimisation when visualisation is enabled. This is currently handled from the CPU, via the 'display' process (Figure 1). At each cycle of the display (which we allow to run slower than the simulation if desired) the painted framebuffer is communicated to the GPU, that then displays it on the screen (via the host's windowing system). Where the GPU being used for computation is the same as that being used for display, there is a clear benefit in generating the framebuffer (and keeping it) on the GPU. Although not entirely straightforward, this is well documented and can be done through the use of OpenGL textures.

3.3. Source Code

For the CUDA-enabled occam-π user, this simulation is now part of the standard KRoC distribution (since version 1.6.0), together with a CUDA wrapper library "ocuda" and other small example programs. KRoC is now hosted on GitHub: `https://github.com/concurrency/kroc/`.

Acknowledgements

The authors would like to thank the anonymous reviewers for valuable feedback on an earlier version of this work, and acknowledge the support of the Faculty of Sciences at the University of Kent (who provided funding for the GPU hardware used).

References

[1] Craig W. Reynolds. Flocks, herds and schools: A distributed behavioral model. *SIGGRAPH Comput. Graph.*, 21(4):25–34, August 1987.

[2] P.H. Welch and F.R.M. Barnes. Communicating mobile processes: introducing occam-pi. In A.E. Abdallah, C.B. Jones, and J.W. Sanders, editors, *25 Years of CSP*, volume 3525 of *Lecture Notes in Computer Science*, pages 175–210. Springer Verlag, April 2005.

[3] C.G. Ritson, A.T. Sampson, and F.R.M. Barnes. Multicore scheduling for lightweight communicating processes. *Science of Computer Programming*, 77(6):727–740, June 2012.

[4] Adam T. Sampson, John Markus Bjørndalen, and Paul S. Andrews. Birds on the wall: Distributing a process-oriented simulation. In *2009 IEEE Congress on Evolutionary Computation (CEC 2009)*, pages 225–231. IEEE Press, 2009.

[5] C.A.R. Hoare. *Communicating Sequential Processes*. Prentice-Hall, London, 1985. ISBN: 0-13-153271-5.

[6] R. Milner. *Communicating and Mobile Systems: the Pi-Calculus*. Cambridge University Press, 1999. ISBN: 0-52165-869-1.

[7] NVIDIA. CUDA C programming guide 4.2, April 2012. `http://www.nvidia.com/content/cuda/cuda-developer-resources.html`.

[8] Khronos OpenCL Working Group. The OpenCL specification 1.2, November 2011. `http://www.khronos.org/registry/cl/`.

[9] Shuai Che, Michael Boyer, Jiayuan Meng, David Tarjan, Jeremy W. Sheaffer, and Kevin Skadron. A performance study of general-purpose applications on graphics processors using cuda. *Journal of Parallel and Distributed Computing*, 68(10):1370 – 1380, 2008.

[10] Richard Vuduc, Aparna Chandramowlishwaran, Jee Whan Choi, Murat Efe Guney, and Aashay Shringarpure. On the limits of GPU acceleration. In *Proc. USENIX Wkshp. Hot Topics in Parallelism (HotPar)*, Berkeley, CA, USA, June 2010.

[11] S. Stepney, P.H. Welch, J. Timmis, C. Alexander, F.R.M. Barnes, M. Bates, F.A.C. Polack, and A. Tyrrell. CoSMoS: Complex systems modelling and simulation infrastructure, April 2007. EPSRC grants EP/E053505/1 and EP/E049419/1. URL: `http://www.cosmos-research.org/`.

[12] Alessandro Ribeiro da Silva, Wallace Santos Lages, and Luiz Chaimowicz. Boids that see: Using self-occlusion for simulating large groups on GPUs. *Comput. Entertain.*, 7(4):51:1–51:20, January 2010.

[13] Jason Sanders and Edward Kandrot. *CUDA by Example: An Introduction to General-Purpose GPU Programming*. Pearson Education, 2011.

[14] Craig Reynolds. Big fast crowds on PS3. In *Proceedings of the 2006 ACM SIGGRAPH symposium on Videogames*, Sandbox '06, pages 113–121, New York, NY, USA, 2006. ACM.

Communicating Process Architectures 2013
P.H. Welch et al. (Eds.)
Open Channel Publishing Ltd., 2013

A Personal Perspective on the State of HPC in 2013

Christopher C.R. JONES
BAE Systems, Preston, England.
chris.c.jones@baesystems.com

Abstract. This paper is fundamentally a personal perspective on the sad state of High Performance Computing (HPC, or what was known once as Supercomputing). It arises from the author's current experience in trying to find computing technology that will allow codes to run faster: codes that have been painstakingly adapted to efficient performance on parallel computing technologies since around 1990, and have allowed effective 10-fold increases in computing performance at 5 year HPC up-grade intervals, but for which the latest high-count multi-core processor options fail to deliver improvement. The presently available processors may as well not have the majority of their cores as to use them actually slows the code – hard-won budget must be squandered on cores that will not contribute. The present situation is not satisfactory: there are very many reasons why we need computational response, not merely throughput. There are a host of cases where we need a large, complex simulation to run in a very short time. A simplistic calculation based on the nominal performance of some of the big machines with vast numbers of cores would lead one to believe that such rapid computation would be possible. The nature of the machines and the programming paradigms, however, remove this possibility. Some of the ways in which the computer science community could mitigate the hardware shortfalls are discussed, with a few more off the wall ideas about where greater compute performance might be found.

Keywords. HPC, electromagnetic simulations, non-scalability, programmability, crisis

Introduction

The point of parallel computing from an HPC (High Performance Computing) perspective is to be able to solve bigger problems faster than can be achieved with a single processor – this is about memory size and computing performance. It would be nice to be able to treat the combination of parallel computer and parallel code as the ultimate in object orientation – to add a new feature to the solver capability would be to add a new object. Sadly, this is not how it works in practice. Despite superficial resemblances, networks of objects are not a good match for networks of computers. Objects can interact in ways that are far more subtle and complex than message passing over wires, even with our best (software engineering) efforts to ensure low coupling. The result is almost the opposite of what is needed - to add a new feature requires new, additional code to be integrated into the existing object code under the current paradigm. Code maintenance and development, therefore, is seriously non-trivial - its cost depends on the size of the whole system, not the size of the new feature. Furthermore, obtaining good parallel computing performance and scalability to large numbers of cores is very hard to achieve in the first place, and yet harder to maintain through cycles of code development.

Hard as code performance is to get on large parallel machines, the hardware is becoming finer grained and intrinsically less homogeneous as the HPC industry slavishly tries to keep up with what it thinks Moore's law was, claiming to supply us with ever more processing power via more and more cores on single silicon substrates. This is creating new problems:

1. Memory addressing for cores on a processor is different from that for separate processors. In practice, this is even more complicated as will be described briefly later.
2. Often a different parallel approach and library is used for core to core communications to that for processor to processor – e.g. OpenMP for the first and MPI for the second
3. Bandwidth and latency for communications is similarly different for the various layers.

This adds to the existing problem for the user of large parallel computers, that more processors and cores does not necessarily make a given problem run faster, and in general speed of solutions does not scale with number of cores – we cannot run a problem on a bigger computer and expect it to run as much faster (if at all) as we might naively expect. So a solution that might take 250 hours on 100 processors in the old currency will not run in 25 hours on 1000 cores or 2.5 hours on 10,000 cores, etc. In fact, we may find that without completely rewriting the solver, and possibly rethinking the numerical approach, we cannot speed up an existing problem at all. We might be able to run bigger problems, but not speed up the existing range of problems. While being able to run bigger problems is important, we also need to speed up the solution time. Yet, because of the interaction between processing time and the amount of data to be communicated between parallel processes, and the fact the individual cores are not contributing significantly to the continuing (apparent) increase in processor performance, each problem size for a particular solver has a *sweet spot* in core count where the performance is optimal, and above and below this core count the performance drops off. We may be criticised for following a data-parallel route to parallelising our solvers, but remember the earlier point – one of the driving reasons for using parallel computing is the access to more memory – this is not merely nice to have, it is absolutely essential.

It is therefore, the combination of the microprocessor designers' failing to comprehend the real issues of multi-core, i.e. parallel processing, and the failure of the computer scientists to offer a better programming paradigm that we can use that has led us to this position.

The present parallel computer hardware approach is wrong, but there are few options, so we just have to make it work. There are three major hardware issues to overcome:

1. the inhomogeneity of the communications which makes it harder to get real performance and requires special programming,
2. the contention between many cores trying to gain access to a single block of memory with limited memory access bandwidth, and
3. the lack of reliable, high bandwidth, low latency, homogeneous interconnect.

Hand-in-hand with the failure of HPC platforms to deliver realisable computing performance, knowledge in computer science (CSP [1,2,3,4,5], the π-calculus [6], occam [7, 8,9,10,11], etc.) has overtly not been adopted to attempt to mitigate the deficiencies in the hardware – or failed equally spectacularly to be disseminated to users.

A real source of frustration is that we did have technology, both hardware and software in the Transputer and occam, that with development over the intervening years since Inmos' demise would/could have avoided these issues. Development of programming potential has continued without a *"good"* hardware platform for implementation, and ignored by the HPC industry at large. Good science has continued in a few special places, but funding sources have not recognised the dangers lurking ahead and have failed to put sufficient priority and resources in those special places. Now, the writing is on the wall for scalable general high performance computing unless and until these issues are recognised and serious investment can be focused here.

1. Our Need for High Performance Computing (HPC)

HPC is increasingly being used in the Aerospace industry. Initially this has been in the risk reduction of the design function to remove ineffective solutions and optimise towards a best compromise. Aerodynamics and structural analysis were probably the areas where earliest HPC effort was directed. In the late 1980s and early 90s, the HPC platform of choice was the CRAY XMP followed by the YMP. These were multi-purpose, multi-user systems, set up for very many, relatively small jobs, while electromagnetic (EM) simulations were large jobs requiring the whole machine for periods of ten days and more at a time. This disjoint in job pattern resulted in only one partial EM simulation ever being accomplished at that time. From the early 1990s, it became feasible to do realistic electromagnetic simulations of lightning strikes to aircraft and their effect on coupling into aircraft systems cables. This became feasible at that time because of the subject-dedicated affordable parallel computing offered by Transputer technology followed by their emulation using consumer microprocessors.

One hundred and ninety two T800 processors provided equivalent performance to that achievable from the Cray YMP, but as a machine dedicated to electromagnetics, and with a single user provided the required access to long run times. With one more foray into shared supercomputing resources which failed, we have maintained EM resources since those early 1990s with increases in compute performance of about a factor of 10 at each upgrade at 5 yearly intervals. It is a sad but true fact that these increments in performance of the computing platform have been largely absorbed by more refinement in the representation of our aircraft, in improvements in the modelling of features and materials, and improvements in the code accuracy through better boundary conditions and switching from single to double precision. Hence, while our present computer is about 10,000 times faster than the first, the same simulation now takes 3 days instead of 10, but with significantly improved accuracy. The new technology and ideas growing from CSP and Transputer had opened up new opportunities on which we were able to capitalise. But the HPC Industry were quite happy to let people think everything was carrying on along the upward trend of ever-increasing computing power, even though the achievable performance was rising more slowly or was static. We have now got to a point where the technologies available actually cannot solve some large problems because of lack of scaling.

Developments in-hand at the present will introduce increased requirements as we move from Cartesian *"sugar cube"* discretisation of the problem and the space around it, to unstructured, body-conforming cell models. Another development is to model the generation of voltage arcs and thermal sparks when the very high lightning currents flow through joints in carbon fibre composite (CFC) structures. This modelling work would increase the compute time up to thousands of days without some acceleration techniques[1], but even so may take from 10 to 20 days per run on the next 10-fold upgrade in HPC power.

Installed antenna performance (IAP) modelling and electrostatic discharge (ESD) have become, and high intensity radiated fields (HIRF) simulations are also becoming, part of the virtual armoury with comparative problem sizes and run times, but which add to the overall amount of computing required. See Table 1.

Table 1. Typical problem sizes in EM on current HPC.

Simulation	Cells	Time-Steps	Run Time
Lightning & HIRF	150M to 600M	1M – 4M	1 – 4 days
ESD	4.6B	160k	4 – 5 days
IAP	<12B	<60k	1 day

[1]An adaptation of Holland's method [12] for reducing the value of the speed of light to reduce the number of time steps required to cover an elapsed electromagnetic interaction time.

It should be noted that HPC procurements for this EM simulation purpose are always done against a compute performance requirement, not a number of cores.

2. Opportunities that cannot be tapped

At present, the turn-around time for just the computing element of the simulations we do (for instance) for lightning, when a significant airframe or system design change is proposed (e.g. one that might change the performance of the lightning protection), is between 1 and 3 months[2]: that is, there is a solid one to three months of computing on an entire 512 core computer. The latest processors improve on this marginally. Ideally, we would like to reduce this by a factor of 10, not only to reduce delivery times on the results, but also to give us some latitude for re-runs.

It is inevitable that we frequently choose sub-optimal solutions because we cannot evaluate multi-disciplinary trade-offs between the requirements for various design solutions. In many cases, we *"know"* that a decision is wrong, but cannot offer the solid evidence. Sadly, electromagnetic hazards are what we describe as *"unwanted performance"* issues and is considered more as a constraint on design than a driver. On the other hand, the low observability and antenna installed performance issues are more *"wanted performance"* and are perceived more positively. The ability to do *good* analysis very fast would enable us to provide solid evidence quickly, and influence decisions more and more often and thus reach nearer optimal solutions. Cost benefits would arise from this, especially if we could prevent eventual re-design and re-work, which is exceptionally expensive.

However, there are many other opportunities, such as the use of virtual or augmented reality, that would substantially reduce design, manufacturing and modification times and costs, that we cannot reap advantage from as these would require near real-time simulations. More processors/cores, beyond a limit, not only does not help, it starts to hinder. This cannot be described as scalable computing.

3. EM Simulations on HPC Platforms

We have recently tested one of our major codes on a new parallel cluster based on AMD 16-core processors. Each node was equipped with a quantity of memory, and four processor sockets. Each processor was made up of two pieces of silicon, each with 8 cores. Even on a piece of silicon, the communications bandwidth was not uniform between all cores, and that between cores on different silicon and between processors on the node were different again. That between nodes was governed by the system interconnect, in this case, Infiniband at 40GByte/s. This architecture is shown in Figure 1.

Even more recently, the code has been tested on different, single substrate, 8-core devices. Figure 2 shows the effect on the number of iterations of the EM solver per hour by using more cores rather than processors to solve a given problem. In each case 24 cores were used, but using 2, 4 and 8 cores per processor on 12, 6 or 3 processors. Everything else remained the same. Thus, more cores does not lead to greater performance. In fact, much of the current HPC offerings may not be capable of delivering needed computing performance, though many-core processors are an attractive option to those who pay for the space, cooling and electricity consumption. The truth is, however, that if the computing power is needed, the bills just have to be paid.

[2]This is likely to be about the limit of time available. Testing in these cases is not an option as it would require an aircraft to be taken from the assembly line for a period of weeks for testing, and then requiring some degree of refurbishment with consequences for increased cost and delays in delivery.

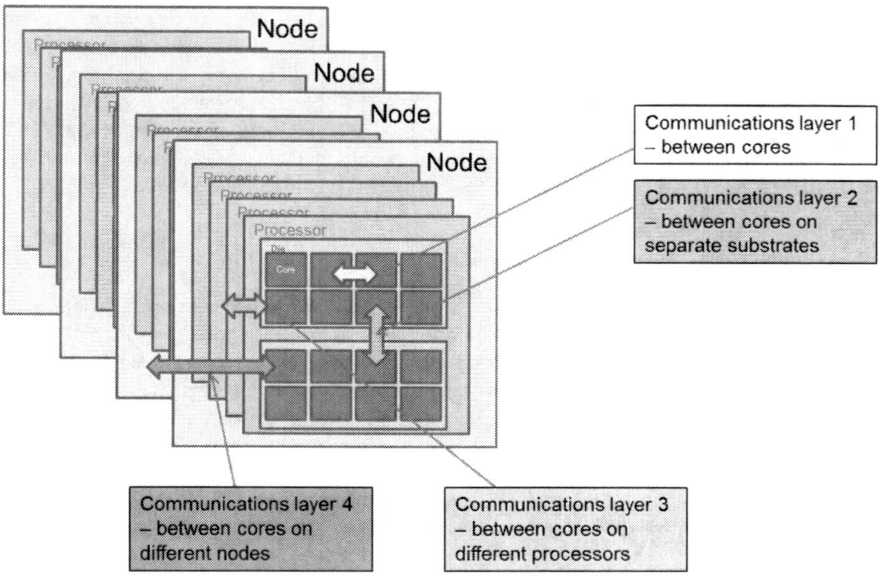

Figure 1. Typical communication hierarchy on modern HPC. Each layer has different bandwidth and latency. The concurrency programming model for layers 1, 2 and 3 is OpenMPI. For layer 4, it is MPI.

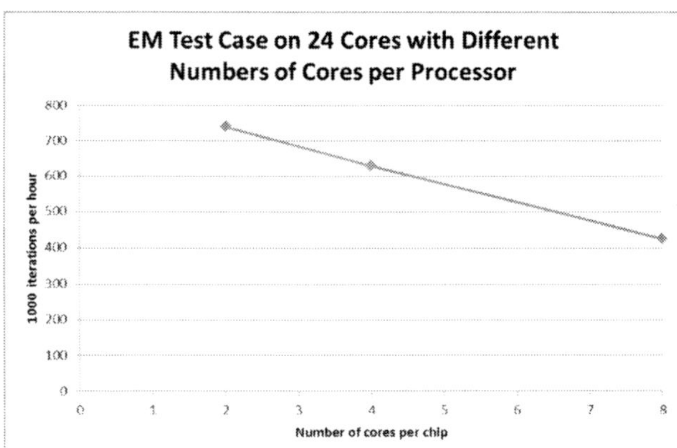

Figure 2. 24-core processing with various cores per processor.

This inhomogeneity introduces considerable complexity to understanding performance issues and load-balancing for optimisation. It also, usually, imposes at least two programming methods and libraries for message passing – OpenMP for inter-core/intra-node and MPI for inter-node communications – which itself adds complexity to understanding the parallel potential of processes and algorithms, as well as to implementation and maintenance, validation and documentation.

When we started out to use what was then known as *Massively Parallel Computing*, to run simulations of electromagnetic interactions, the introduction of Transputers was changing the world and offering affordable computer power of previously unavailable levels. This was transformative in that what had not been economically viable suddenly became eminently realisable. At this time, the Transputer was thought of as the gateway to affordable High Performance Computing (HPC). The Transputer's role as a very good embedded device was a later and secondary consideration. occam was similarly considered. The demise of the Transputer had the unwanted side-effect of side-lining occam as a mainstream HPC programming language, which the developers of occam and its derivatives have been unable to counter. This is part of the reason we are in the HPC pickle that we are in.

4. What is needed for usable, scalable HPC

First, we are not asking for parallel computing nirvana: we are not asking that our poorly written serial codes should be able to run optimally on arbitrary parallel architectures and HPC implementations, nor even that our well sculpted parallel codes should always run optimally without some intervention. However, we should not have to implement multiple parallel programming paradigms within a single code and revisit the code for every new compute machine offered by the purveyors of hardware. We should certainly not have to abstract our physics to alternative numerical algorithms to try and demonstrate that we can benefit from HPC platform claims that they continue to adhere to Moore's law.

It is worth noting here that our primary code was, 10-15 years ago, considered to be embarrassingly parallel. It was originally written in Fortran, rewritten from the mathematical algorithms in occam and analysed using a CSP formal description. After the demise of Transputer, the occam code was translated into Fortran with some parts written in C to manage run-time array sizing. Since then there have been some minor modifications to optimise its performance for Power PC, Compaq Alpha, AMD and Intel processors in array sizes from 92 to 512 cores and increasing its performance in increments of approximately 10 times every 5 years.

However, for the next HPC upgrade, while it is easy to find solutions that offer nominal performance gains, it is somewhat harder to maintain the improvement in actual, delivered performance running our codes on our data. In short, while a typical simulation currently takes 3 days on 512 cores of five year old technology, it is hard to find a solution that will run the same problem in any less time, let alone say 12 hours; and impossible to find one that will solve the problem in 10 minutes or even 1 hour, no matter how much money might be available to be spent on the computer. Ten minutes is perhaps too severe a requirement, but surely, 1 hour is not. This is not scalable computing.

What I need is:

- occam or occam-*like* communications constructs[3] embodied in Fortran or C/C++. (Personally, I would prefer occam, but it is very hard to convince others to switch from a programming language they have come to love – it is an emotional thing.)
- A smart means of launching parallel processes[4], whether many of the same, some of a number of varieties, or different from each other.
- Programming, profiling, performance monitoring and debugging tools.

Some will tell me straight away that I can have all of these things now, but that isn't really quite accurate. I am sure that I could have much of what I ask for in tools that I must assemble and/or compile myself, and with no kind of support other than a personal telephone

[3]However, the constructs are not enough: an understanding of the underlying CSP process *model* (which does not mean an understanding of CSP itself) together with an appreciation of the occam parallel security rules concerning shared references and aliasing (automatically enforced by the occam compiler, but not by those for the other languages) is also needed.

[4]Looking further to the future, an ability for the *dynamic* construction of process networks – i.e. not just at start-up – may be useful.

call to someone I happen to know. Actually, I work in industry where the software must pass some formal stages of review to ensure it works, does what it is supposed to do, is supported to assist those using it now and indefinitely into the future, and will transfer to or be available on other hardware in case of disaster or replacement, etc...

Ideally, I would like hardware, or equivalent software implementation, that incorporates programmable interconnect logic that allows a number of defined point-to-point links to be established for a run-time, to remove the delays involved in arbitrary routing, plus the tools to devise the routing plan and to program it.

Why are the computer science departments around the country not researching in these areas? Why indeed do the most recent *strategy* reports [13,14] have nothing to say about these aspects of the problem of high performance computing?

The above is by no means a complete wish list: other aspects of HPC that should be seriously researched include:

1. Analogue computer co-processors and the language extensions to program them. Analogue computers are a faster and a nearer exact means of modelling many physical phenomena. The means to create them should be more accessible now, with lower noise and more stable operational amplifiers, than was the case in days of the greatest interest in them (1950s and early 1960s). Imagine what could be programmed with a large parallel array of conventional processors each with a digitally controlled analogue computer.
2. FPGA co-processors and the language extensions to program them. This is not such a pipe dream. Two such companies, AlphaData and Nallatech, have already contributed to a significant computing capability of this type at Edinburgh's EPCC. However, when we wanted to consider the technology for our HPC role, our code could not be benchmarked because of the difficulties in translating it to the different technology. Either computing hardware is usable or fated to disappear. The potential benefits of this approach are so significant, however, that more effort should be given to a coherent processor/FPGA co-processor programming language extension.

5. Conclusion

HPC (or supercomputing) has become a poor relation in the computing world. It is apparently not a large enough endeavour globally to command specific technology developments. Instead, it hangs on the coat tails of the consumer, even the games market. HPC platforms declare nominal performance from which few, if any, problems can possibly gain advantage and even fewer will ever be able to invest in the coding to do it.

The really disappointing thing is that UK led the world into the parallel computing age, both conceptually and practically. Short-sighted government and the computer industry let that lead disappear, leaving the world to the HPC cul-de-sac in which we now find ourselves. An extraordinary note is that we were aware of this at least as far back as 1995 and warned ourselves and any who would listen, at the *Crisis in HPC* workshop [15,16,17]. That crisis is upon us now.

Even so, the occam community still has some of the ingredients to mitigate the present problems, but has mostly lain down and merely purred since the demise of the Transputer, rather than boldly addressing the current issues. By all means carry on research into the intricacies of pure occam-ite programming, but also have a good look at how this elegant paradigm can be raised into useful functioning parallel computing tools despite the serious shortcomings of the hardware world.

References

[1] Charles A. R. Hoare. Communicating Sequential Processes. *CACM*, 21(8):666–677, August 1978.

[2] Charles A. R. Hoare. *Communicating Sequential Processes*. Prentice-Hall, 1985.

[3] Andrew W. Roscoe. *The Theory and Practice of Concurrency*. Prentice Hall, 1997.

[4] Andrew W. Roscoe. *Understanding Concurrent Systems*. Springer, 2010.

[5] Steve Schneider. *Concurrent and Real-time Systems — The CSP Approach*. Wiley and Sons Ltd., UK, Baffins Lane, Chichester, UK, 1999. ISBN: 0-471-62373-3.

[6] Robin Milner. *Communicating and Mobile Systems: the π-Calculus*. Cambridge University Press, 1999. ISBN-10: 0521658691, ISBN-13: 9780521658690.

[7] Dick Pountain and David May. *A Tutorial Introduction to occam Programming*. McGraw-Hill, Inc., New York, NY, USA, 1987.

[8] M. D. May. CSP, occam and Transputers. In Ali E. Abdallah, Cliff B. Jones, and Jeff W. Sanders, editors, *25 Years of CSP*, volume 3525 of *Lecture Notes in Computer Science*, pages 75–84. Springer Verlag, April 2005.

[9] P. H. Welch and F. R. M. Barnes. Communicating Mobile Processes: introducing occam-π. In Ali E. Abdallah, Cliff B. Jones, and Jeff W. Sanders, editors, *25 Years of CSP*, volume 3525 of *Lecture Notes in Computer Science*, pages 175–210. Springer Verlag, April 2005.

[10] Peter H. Welch and Jan B. Pedersen. Santa claus: Formal analysis of a process-oriented solution. *ACM Trans. Program. Lang. Syst.*, 32(4):14:1–14:37, April 2010. `http://doi.acm.org/10.1145/1734206.1734211`.

[11] Peter H. Welch, Kurt Wallnau, Adam T. Sampson, and Mark Klein. To Boldly Go: an occam-π Mission to Engineer Emergence. *Natural Computing*, 11(3):449–474, September 2012. `http://dx.doi.org/10.1007/s11047-012-9304-2`.

[12] Richard Holland. FDTF analysis of nonlinear magnetic diffusion by reduced c. *IEEE Transactions on Antennas and Propagation*, 43(7), July 1995.

[13] Peter Coveney. Strategy for the UK research computing ecosystem. `http://www.clms.ucl.ac.uk/sites/default/files/ResearchComputing-glossy.pdf`, October 2011.

[14] Dominic Tildesley. A strategic vision for UK e-infrastructure. `https://www.gov.uk/government/publications/e-infrastructure-strategy-roadmap-for-development-of-advanced-computing-data-and-networks`, November 2011. Accessed: 2013-08-24.

[15] Preston Briggs. Crisis in HPC (post-workshop discussion), October 1995. `http://wotug.org/parallel/groups/selhpc/crisis/discussion/12229.html`.

[16] Chris P. Wadsworth and Roger G. Evans. Crisis in HPC (personal article in Flagship Bulletin), September 1995. `http://wotug.org/parallel/groups/selhpc/crisis/articles/flagship.html`.

[17] Crisis in HPC (summary of conclusions), September 1995. `http://wotug.org/parallel/groups/selhpc/crisis/conclusions.html`.

Communicating Process Architectures 2013
P.H. Welch et al. (Eds.)
Open Channel Publishing Ltd., 2013

An Evaluation of Intel's Restricted Transactional Memory for CPAs

Carl G. RITSON and Frederick R.M. BARNES
School of Computing, University of Kent, UK
{c.g.ritson, f.r.m.barnes}@kent.ac.uk

Abstract. With the release of their latest processor microarchitecture, codenamed Haswell, Intel added new Transactional Synchronization Extensions (TSX) to their processors' instruction set. These extensions include support for Restricted Transactional Memory (RTM), a programming model in which arbitrary sized units of memory can be read and written in an atomic manner. This paper describes the low-level RTM programming model, benchmarks the performance of its instructions and speculates on how it may be used to implement and enhance Communicating Process Architectures.

Keywords. Transactional Synchronization Extensions, TSX, Restricted Transactional Memory, RTM, transactional memory, performance

Introduction

With the release of their latest processor microarchitecture, codenamed Haswell, Intel added new *Transactional Synchronization Extensions* (TSX) to their processors' instruction set [1,2]. These extensions provide *Hardware Lock Elision* (HLE) and *Restricted Transactional Memory* (RTM). RTM allows the atomic manipulation of arbitrary size units of memory. Without RTM the largest unit of memory that can be atomically manipulated on x86 architectures is a single memory word, 32 or 64 bits, using for example a *compare-and-swap* (CAS) operation [1]. This restriction has necessitated the development of entire classes of non-blocking, lock-free and wait-free algorithms [3,4] along with complex architectural memory models [5,6]. Programmers using a *Communicating Process Architecture* (CPA) such as that provided by occam-π [7], Communicating Scala Objects [8] or ProcessJ [9], are freed from considering the details of atomic synchronisation on shared memory systems. However, the developers of high-performance CPAs for modern multi-core and many-core *must* concern themselves with these details.

Transactions are commonly found in database systems. These systems are, at least conceptually, providing a large shared mutable memory [10]. In a database system transactions provide: atomicity, consistency, isolation and durability [11]. In practice this means that all operations in the same transaction should be enacted together or at-least seen to be done so (atomicity), preserve system invariants (consistency), not interfere with other transactions in non-deterministic ways (isolation) and be persistent once committed (durability). Most relevant to the developers of operating systems and language run-time systems are atomicity and isolation, whereas consistency and durability are more higher level concerns. Transactions have the potential to vastly simplify the implementation of lock-free and non-blocking algorithms which required for efficient communicating process architectures [12,13].

[1]Compare-and-swap updates a memory location M with a new value Y iff its current value matches another value X, reporting the success or failure of the update.

The concept of using transactions for system memory was first described by Herlihy and Moss in 1993 [12]. Although hardware support for transactional memory has been implemented in specialised processors such as IBM's BlueGene/Q [14] and the cancelled Sun Rock processor [15], it has taken 20 years for it to become available in a commodity processor. In the intervening timespan much work has been done on *Software Transactional Memory* (STM) [16], where transactional memory is provided by a language run-time system rather than dedicated hardware. The literature has focused on how memory transactions (using STM) can be exposed to programmers; however, as CPAs do not typically expose shared memory this work is not relevant. Coupled with this, the lack of hardware support and the poor performance of STM [17], transactional memory has not seen significant uptake in the implementation of CPAs. One notable exception is Brown's *Communicating Haskell Processes* which uses STM to implement multiway choice and synchronisation [18].

This paper focusses on what Intel's hardware RTM offers the designer or implementor of a *run-time system* (RTS) or *virtual machines* (VM) for a *communicating process architecture* (CPA). Content is broadly divided into three sections: section 1 describes the instruction set and associated *programming model* for RTM; section 2 evaluates the *performance* of RTM on a recently released Core i7-4770 processor; and section 3 looks at the potential *applications* of RTM in the implementation of CPAs. Conclusions and future work are reviewed in section 4.

1. Programming Model

This section briefly describes the instruction set and programming model associated with Intel's RTM facilities.

At an instruction level RTM is simple to use. A transaction is initiated with an `XBEGIN` instruction. Computation proceeds normally; memory can be read and written with common instructions and other activities such as branching and arithmetic may also be used. At the end of the transaction an `XEND` instruction commits any changes to memory.

During the transaction read and write sets are constructed. These sets have cache line granularity and are based on the memory addresses read or written by the transaction's body. If these sets conflict with memory being read or written by other hardware threads then the transaction is aborted. If a transaction is aborted, all changes to memory and registers are discarded and execution jumps to a *fallback handler* supplied to the initial `XBEGIN` instruction. Before invoking the fallback handler, status flags are set in the processor's `EAX` register. These flags allow the fallback handler to determine what caused the transaction to abort.

It is worth clarifying that read and write sets are specific to a given hardware thread. We use the term hardware thread to describe the smallest unit of parallel execution in a computer system. In Intel's Haswell architecture a computer may have multiple processors, each of which has multiple cores, each of which has multiple hardware threads. Hardware threads on the same core share components such as arithmetic units and cache, but have their own registers and instruction pointer.

1.1. Instructions

There are only four machine instructions used to access RTM functionality.

- `XBEGIN` initiates a transaction. An instruction pointer (relative to the current instruction pointer) to a fallback handler is passed. `XBEGIN` is also valid within a transaction, permitting *nesting* (see 1.3).
- `XEND` completes a transaction, flushing changes to memory, assuming this is not a nested transaction (see 1.3). Following this instruction it can be assumed that changes are visible to other hardware threads.

- XABORT aborts the current transaction. It takes an 8-bit failure code which is accessible to the fallback handler.
- XTEST updates the processors comparator flags such that a subsequent branching instruction can determine if the processor is executing within a transaction or not.

1.2. Restrictions

The term *restricted* in restricted transactional memory refers to the fact that not all processor instructions and functions may be used within a transaction. If a restricted operation is attempted then the transaction is aborted and the fallback handler invoked. Essentially any instruction that causes changes to the processor or system state that cannot (trivially) be reverted induces a transaction to abort.

There are several restrictions worth noting:

- X87 floating-point and *multi-media extensions* (MMX) are not supported; however, this is a not a significant concern as *streaming SIMD extensions* (SSE) and *advanced vector extensions* (AVX) are supported permitting floating-point and vectorised computation.
- Instructions that halt the processors execution such as PAUSE and MWAIT are not supported.
- Debugging mechanisms are not supported, while not verified in this work essentially breakpoints cannot be placed within a transaction.
- An interrupt within a transaction will cause the transaction to abort before invoking the interrupt handler.
- Changes in privilege level will cause a transaction to abort and disregard the source of the invocation. This will prevent calls into the operating system kernel.
- Any exception will cause a transaction to abort and disregard the source of the exception. This includes memory addressing exceptions and page faults, meaning all memory accessed during a transaction must be mapped.

1.3. Nesting

Nesting of transactions is supported. Each invocation of the XBEGIN instruction increases a *nesting count* and XEND decrements the same count. The transaction is only completed and committed to memory when the nesting count is reduced to zero. However, any conflict causing the transaction to abort will invoke the outermost fallback handler. Essentially nested transactions are simply folded into the outermost transaction.

1.4. Failures

When a transaction is aborted and fails all changes to the processor state made within the transaction are discarded. This includes any exceptional conditions raised, such as a page faults. Thus if an exception is triggered within a transaction, the transaction will abort and the exception handler *will not* be invoked. A special case is interrupts which cause the transaction to abort, but still invoke the interrupt handler once the transaction state has been discarded.

The cause of a transaction's failure is reported to the fallback handler by setting flags in the EAX register. At the time of writing six flags are defined:

0 *XABORT*: An XABORT instruction aborted the transaction, in which the EAX register also carries the 8-bit code pass to the XABORT instruction.
1 *Retry*: The transaction might succeed if retried.
2 *Conflict*: Interference from another processor, core or hardware thread caused the transaction to abort.

3 *Overflow*: Overflow of buffers caused the transaction to abort.
4 *Debug*: A debug break point was encountered causing an abort.
5 *Nested*: The transaction was aborted within a nested transaction.

2. Performance

This section evaluates the performance of RTM on generally available hardware. The goal is to provide data which can inform the design of language run-time systems and CPA implementations.

When implementing a language run-time it is important to minimise run-time cost in terms of space and time complexity. However, without meaningful data it is not possible to make reasoned choices. For example, when implementing for performance it would not make sense to use a memory transaction if the same operation could be implemented with three CAS instructions if the overhead of a transaction is ten times that of a CAS.

With respect to Intel's RTM the critical questions to answer are:

- What is the cost of setting up a transaction? (see 2.3)
- What is the cost of committing a transaction? (see 2.3)
- Is there an overhead on operations within a transaction? (see 2.3 and 2.5)
- How big can a transaction reasonably be? (see 2.4)
- What is the cost of failed transactions and conflicts between transactions? (see 2.3 and 2.6)

2.1. Test Setup

All tests and benchmarks were performed using a system with a single Core i7-4770 processor running at 3.4GHz, with 16GiB of RAM running at 1600MHz and CAS timing of 9-9-9-24. Stock Ubuntu Linux 12.04.2 LTS was used with kernel 3.5.0-23-generic. No specific compiler support for RTM was employed, instead assembly macros were used in the standard GCC 4.6.3 installation.

The processor *Turbo Boost* was disabled in the system's BIOS to allow for consistent results regardless of hardware temperature. Additionally, CPU frequency scaling was disabled by modifying the `scaling_min_freq` parameter in the `sysfs` entry to match the `scaling_max_freq`. While executing POSIX threads were bound to specific hardware threads (or processor cores) using the `sched_setaffinity` API. The scheduling quantum was not adjusted.

2.2. Test Strategy

We define four different test operations:

- *read*: load words from increasing memory locations.
- *write*: store words to increasing memory locations.
- *cas*: compare-and-swap words at increasing memory locations.
- *abort*: store words to increasing memory locations, but abort the transaction after n words.

Each operation (with the exception of *abort*) can operate in either untransactional (u) or transactional (x) *mode*. Additionally the *word size* for these operations can be either 32-bits or 64-bits. For each combination of operation, mode and word size we test increasing numbers of words per operation, or *operation size*.

At each operation size we recorded the amount of time (and clock cycles) required to complete a number of operations (between 16384 and 131072), in order to derive the mean speed in bytes per nanosecond. Within a test run each operation uses a distinct memory area (aligned to a cache line boundary) to minimise the effect of L2 and L3 processor caches. The total size of the memory area is 512MiB, hence 16384 operations are performed when the total operation size is 32KiB.

We also recorded the success or failure of operations allowing calculation of failure rates. We do not retry failed transactions, but simply move on to the next operation. Failed transactions are disregarded when computing performance.

Untransactional reads and writes will always succeed; however, untransactional compare-and-swap (using `LOCK` prefix) can fail if multiple threads are using the same memory. Transactional aborts will always fail, although the failure reason may not be the abort if a transaction was interrupted before the `XABORT` instruction.

For abort operations we define two variants:

- *abortn* which sets up to do an operation of a given size, but aborts before performing any memory access,
- *abortm* which performs all memory access before aborting the transaction.

Both have the same compiled structure and use a counter to trigger the transaction abort. This is intended to minimise pipelining effects in the processor.

The source code for our benchmark is available on GitHub: `https://github.com/perlfu/rtm-bench`.

2.3. Base Transaction Cost

To determine the base cost of transactions we examine the time taken to complete untransactional and transactional operations with a size of zero. The overhead of our instrumentation and test framework will be the same between untransactional and transactional versions allowing us to determine the cost of invoking transactional mode on the processor.

Table 1. Cost of zero size test operations.

Operation	Time 32-bit	Time 64-bit	Cycles 32-bit	Cycles 64-bit
u_read	2.0ns	2.4ns	7.0	8.0
u_write	2.1ns	2.1ns	7.0	7.0
u_cas	2.0ns	2.0ns	7.0	7.0
x_read	15ns	15ns	50	50
x_write	15ns	15ns	50	50
x_cas	15ns	15ns	49	49
x_abortn	47ns	47ns	160	161
x_abortm	47ns	47ns	161	161

With respect to table 1 the base cost of invoking transactional mode appears to be 13ns or 43 clock cycles. We also observe that triggering a transactional abort is relatively expensive. If we assume this is because a pipeline and cache flush occurs then we can infer that successful transactions are effectively pipelined.

Before plotting trends it is desirable to closely review the cost of small numbers of operations. Table 2 and table 3 show these costs for 32-bit and 64-bit operations respectively. The critical observation is that once a transaction reaches two words in size it has equivalent cost to compare-and-swap operations on the same number of words. Beyond this (very low) tipping point a transaction is more efficient than compare-and-swap with the benefit of overall atomicity.

Table 2. Clock cycles for operations on small numbers of words for 32-bit mode.

Operation	1 word	2 words	3 words	4 words	5 words	6 words	7 words	8 words
u_read	23	14	13	14	15	16	19	20
u_write	29	26	25	22	22	22	22	23
u_cas	39	59	74	93	113	132	151	170
x_read	60	61	62	62	65	66	67	69
x_write	62	59	59	59	59	59	60	60
x_cas	60	64	66	71	73	78	77	80
x_abortn	66	161	160	161	161	161	162	162
x_abortm	170	175	173	175	179	182	182	181

Table 3. Clock cycles for operations on small numbers of words for 64-bit mode.

Operation	1 word	2 words	3 words	4 words	5 words	6 words	7 words	8 words
u_read	18	11	11	12	14	16	16	17
u_write	37	27	24	23	22	22	22	22
u_cas	37	58	72	92	110	129	148	167
x_read	60	62	67	67	67	67	67	67
x_write	63	62	61	61	60	60	60	61
x_cas	60	64	64	71	72	78	75	80
x_abortn	66	162	162	162	162	163	163	163
x_abortm	170	175	173	175	179	182	182	181

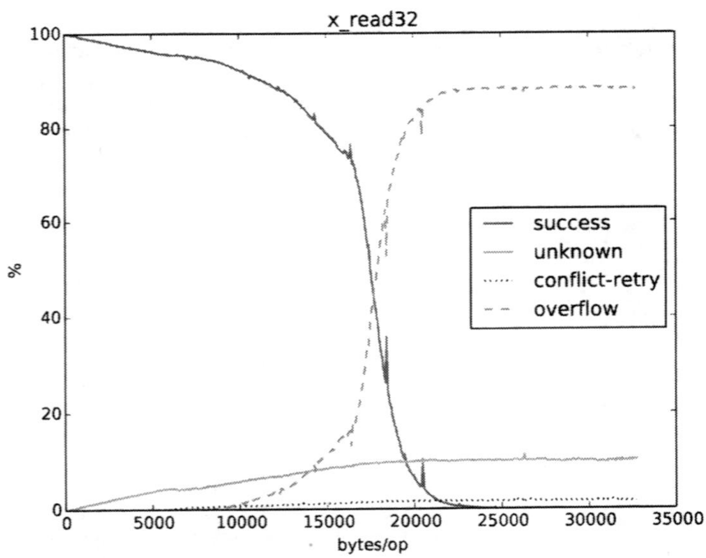

Figure 1. Success and failure rates for *read* transactions of varying sizes.

2.4. Transaction Size and Failures

We examine transaction completion rates at different sizes to determine both optimal and maximal transaction size. Figure 1 shows transaction failure and success rates for increasing operation (and transaction) sizes for 32-bit *read* operations. Figures 2 and 3 show the same information for *write* and *cas* operations respectively.

Up to 10KiB transactions performance degrades from success to unknown failure in an

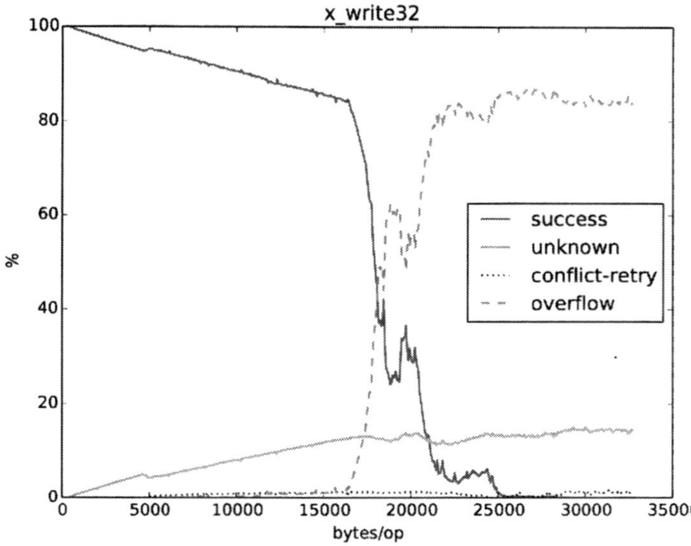

Figure 2. Success and failure rates for *write* transactions of varying sizes.

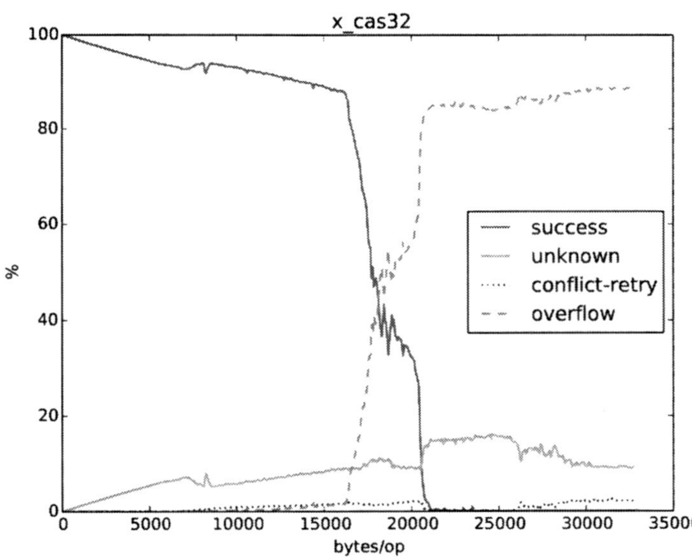

Figure 3. Success and failure rates for *cas* transactions of varying sizes.

almost linear manner. These unknown failures indicate that no transactional abort cause was indicated by the processor, which we believe relates to operating system context switches. This behaviour would be consistent with increased transaction length resulting in increased probability of an operating system interrupt. We could have adjusted the operating system scheduling quantum, but chose not to do so in order provide results representative of normal operating conditions.

There is a pronounced phase shift at 16KiB where transaction success rapidly gives way to overflow related transaction aborts. It is also worth noting that there is a small, but

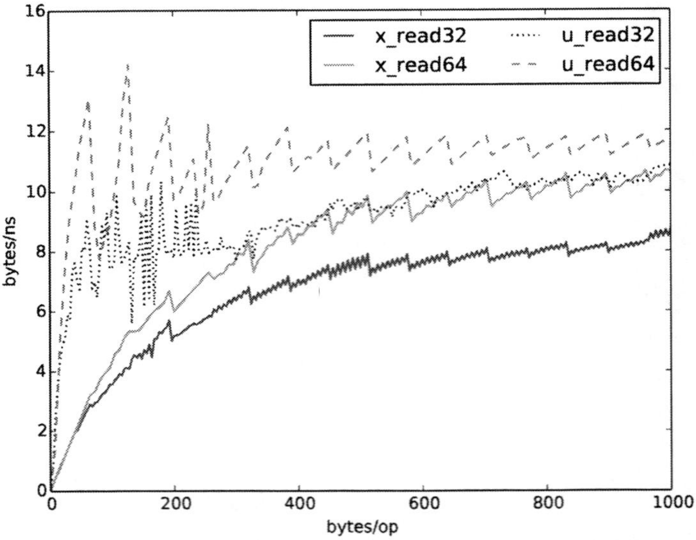

Figure 4. Performance of transactional reads compared with untransactional reads.

observable, set of failures related to memory conflicts despite the fact that no shared memory has been used or accessed. We assume this is either incorrect reporting by the processor or operating system operations (such as page table manipulations) occuring on another core.

It is our observation that the largest stable transaction size is 16KiB, where there is a 85% chance of success (on our test setup and without contention). We speculate that the transaction buffer is the processors L1 data cache; however, parts of this are also used to shadow any modified registers reducing the available buffer size. We also observe that no transactions complete once the transaction size exceeds 26KiB.

2.5. Transaction Performance

Figure 4 shows the performance (in bytes per nanosecond) of memory reads occuring with increasing transaction sizes. At small sizes we observe transactional reads to have significantly lower performance than untransactional reads; however, at larger transaction sizes transactional reads reach 80% of the performance of untransactional reads. A sawtooth pattern occurs in these results (and others) at a frequency of 64 bytes, which corresponds to the size of a cache line. With respect to the data, optimal performance is achieved when the operation is a multiple of the cache line size. We infer that this occurs because the processor loads a cache line of data at once, thus an operation loading 68 bytes of data will have the same cache fill overhead as a 128-byte operation.

Figure 5 shows the performance of memory writes, transactional versus untransactional. Unlike memory reads we observe rapid convergence of transactional and untransactional write performance. At operation sizes 256 bytes and larger the overheads of transaction setup and teardown have been completely amortized. We propose that this suggests that for reads there is a constant overhead per-read when operating within a transaction, but there is not for writes.

Figure 6 compares the performance of transactional compare-and-swap operations with untransactional compare-and-swap operations synchronised using the `LOCK` prefix instruction. This locks the memory bus, making changes to memory by a following (single) instruction atomic. In practice modern processors do not lock the memory bus, but use cache coherence

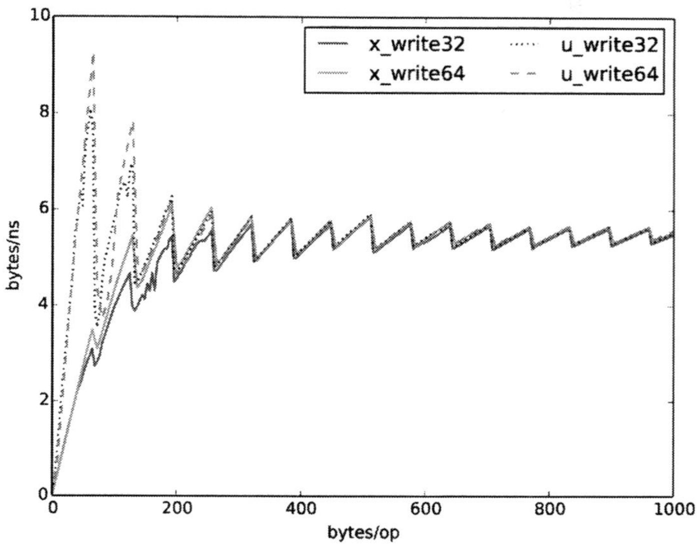

Figure 5. Performance of transactional writes compared with untransactional writes.

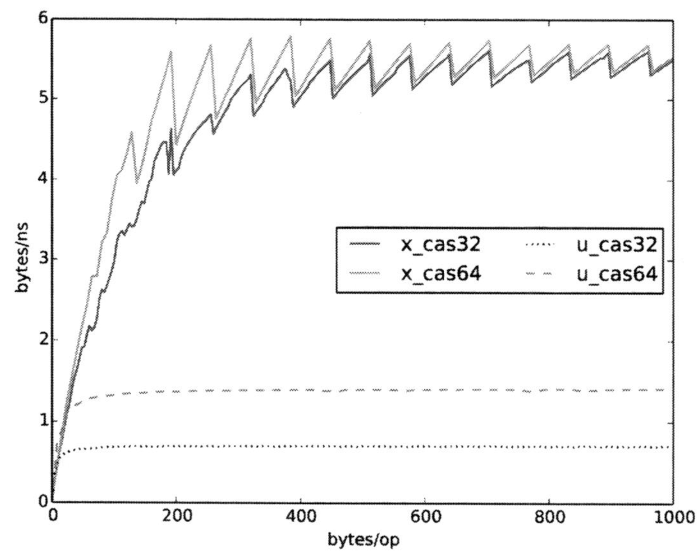

Figure 6. Performance of transactional *cas* compared with untransactional *cas* (synchronised with LOCK prefix instructions).

mechanisms to achieve the same effect. We observe that transactional compare-and-swap quickly surpasses the (constant) performance of its untransactional equivalent, achieving approximately five times the performance.

Figure 7 compares the performance of different kinds of transactional operations. What can be observed is that the cost of transactional write and compare-and-swap operations converge once they reach 256 bytes in size.

Finally, Figure 8 illustrates the cost of explicitly aborting transactions. The *abortn* op-

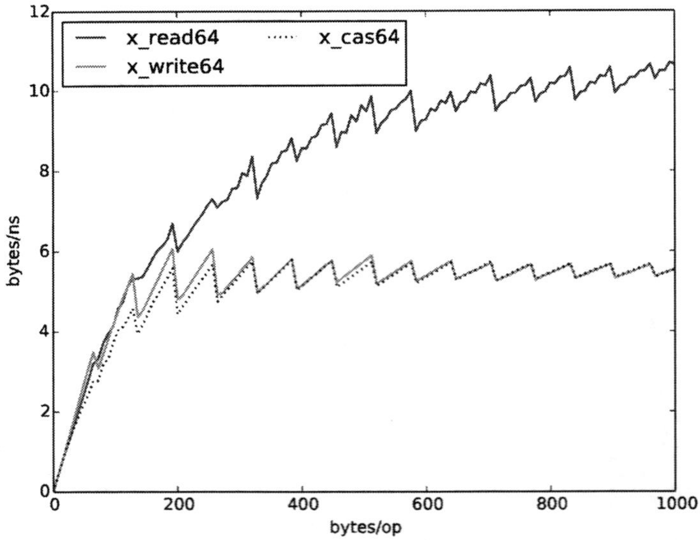

Figure 7. Performance of three different types of transactional operations: 64-bit read, 64-bit write and 64-bit *cas*.

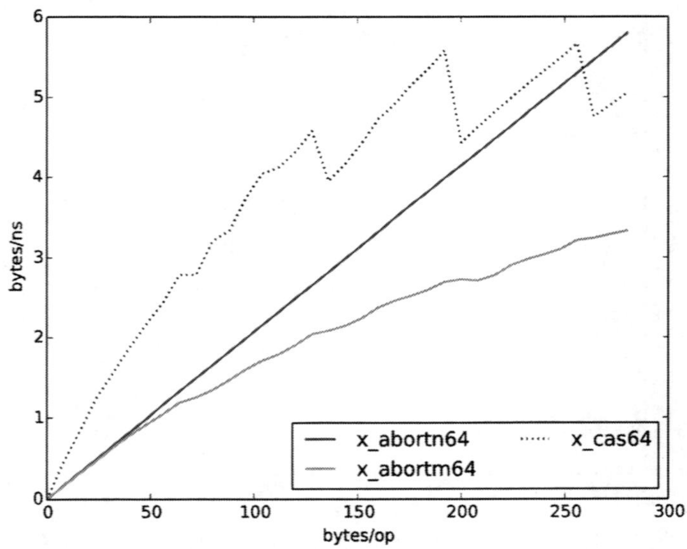

Figure 8. Performance/cost of explicitly aborting a transaction, 0 to 300 bytes in size.

eration aborts the transaction without explicitly modifying any data, thus shows a linear ($y = mx$) relationship as we would expect. Performance of the *abortm* operation which writes all memory words of the operation size before aborting the transaction has significantly worse performance than compare-and-swap operations which equally write all memory words before completing. That said, there are signs of convergence, and looking at larger operation sizes (Figure 9), *abortm* performance surpasses compare-and-swap performance around 600 bytes. From these results we can conclude that for small transactions successful completion

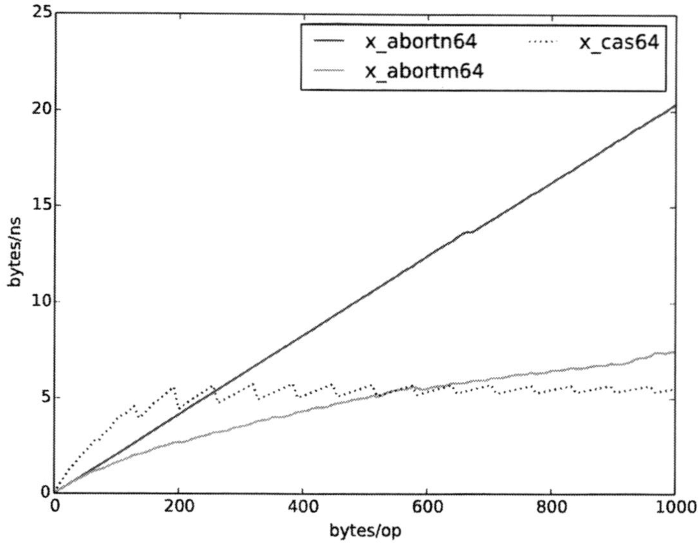

Figure 9. Performance/cost of explicitly aborting a transaction, 0 to 1000 bytes in size.

is significantly cheaper than unsuccessful completion. We infer that this cost arises from the restoration of register state within the processor, a cost not incurred by successful transactions which can simply discard old state data.

2.6. *Transaction Contention*

We performed a very preliminary evaluation of the interactions between concurrent transactions by different cores on a shared area of memory. Results from these evaluations may be less accurate than those presented so far as threads may progress at different rates artificially reducing contention. That said, for larger operation sizes these results should still be indicative of real-world behaviour.

Initially we tested the effect of two threads both reading from the shared memory. We observed no change in the completion rate of either thread. This is what we would expect and shows that shared read sets do not cause transactions to abort.

Figure 10 shows the results of two threads (0 and 1) performing competing write transactions on the same memory buffer. For smaller operation sizes conflicts do not arise, but beyond 192 bytes (four cache lines) the success rate rapidly degrades in both threads. There appears to be a slight bias toward thread 1 succeeding, but this is probably an artifact of synchronisation within the test harness. The stepped changes in success rates reinforce the notion that read and write sets of transactions are of cache line granularity [1].

Figure 11 shows the results of two threads (0 and 1) performing competing compare-and-swap transactions on the same memory buffer. Here *failure* indicates that the transaction completed successfully, but that the memory had already been compare-and-swapped by the other thread. The interesting observation in this test is that almost 10% of transactions fail with transactional buffer overflows despite the relatively small size of transactions. This suggests that either the processor is misreporting failures as overflows or there is another factor aside from L1 cache size which is involved when multiple cores are performing transactions.

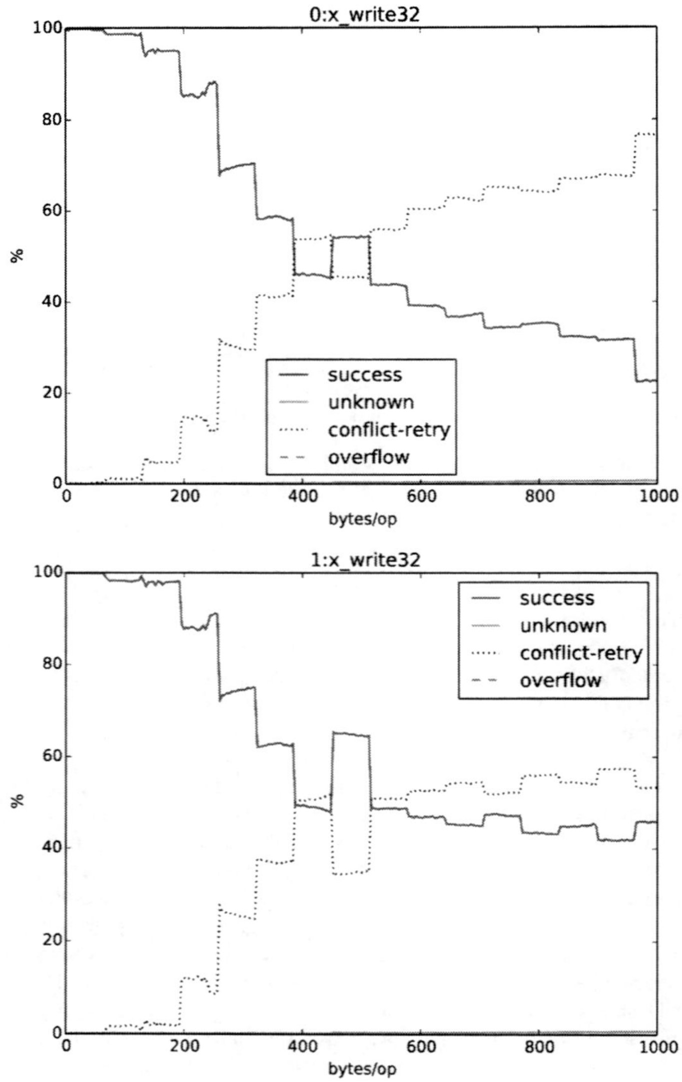

Figure 10. Success and failure rates for competing *write* transactions of varying sizes.

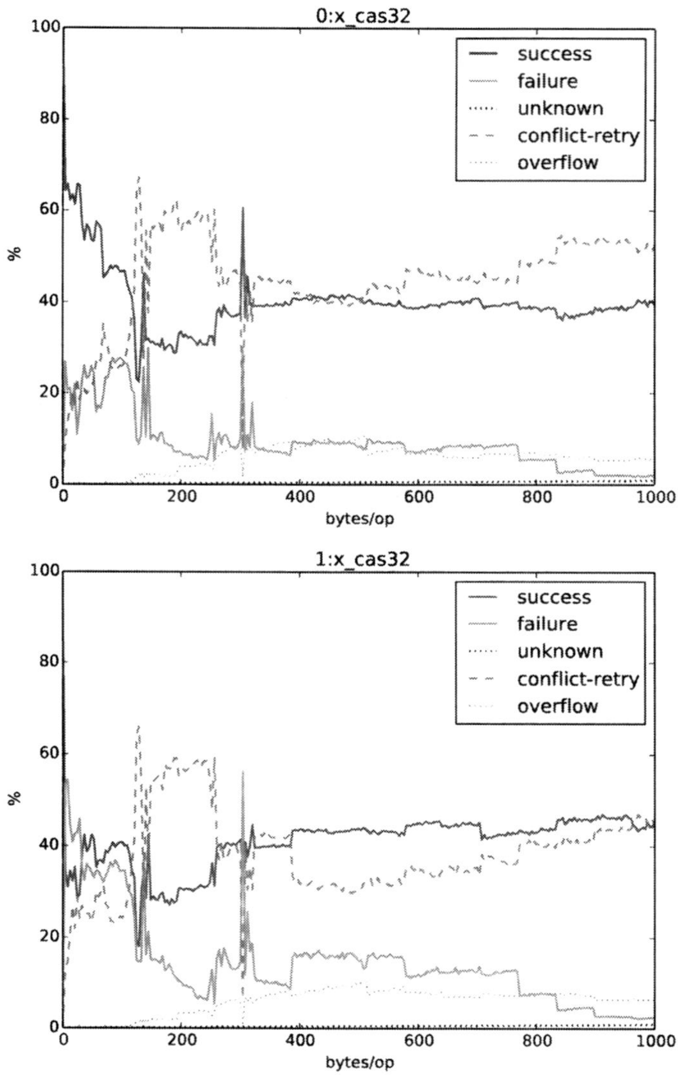

Figure 11. Success and failure rates for competing *cas* transactions of varying sizes.

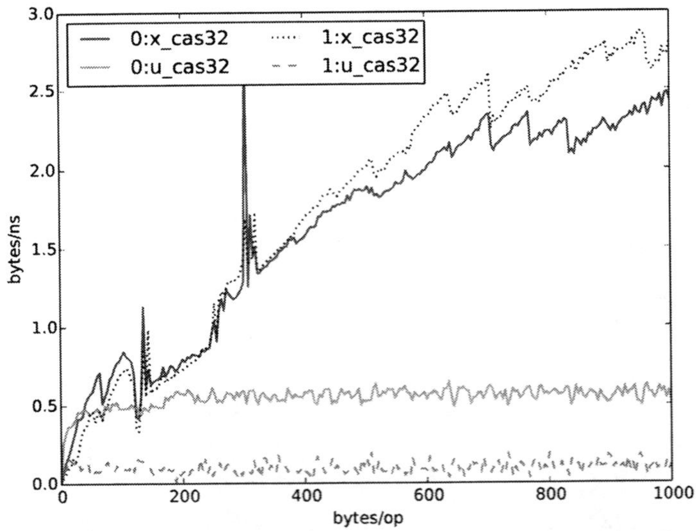

Figure 12. Performance of competing transactional and untransactional *cas* operations.

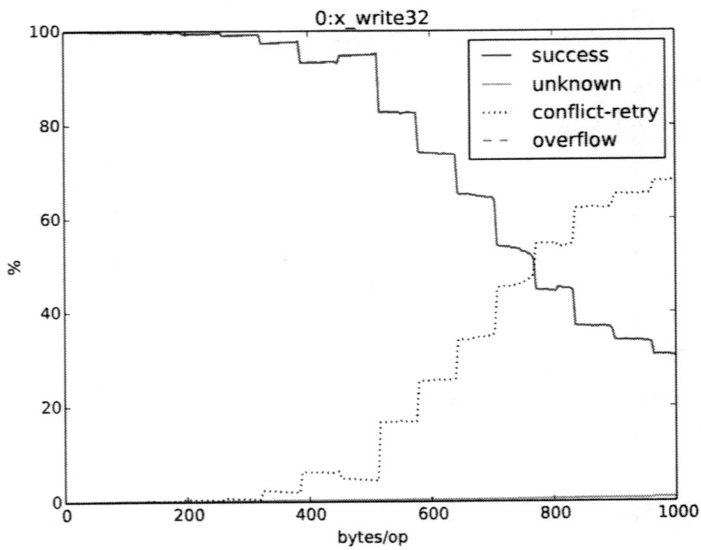

Figure 13. Success and failure rates for transactional *write* operations competing against untransactional *write* operations.

Figure 12 illustrates the performance of these competing threads and compares them to equivalent competing untransactional threads. The performance of untransactional compare-and-swap operations is essentially constant, whereas transactional compare-and-swap operations have better performance with larger operation sizes.

Figure 13 shows the competition of transactional memory writes with untransactional memory writes. As untransactional writes always succeed, success and failure is only shown for the transactional thread. The results show a lower level of interference than competing

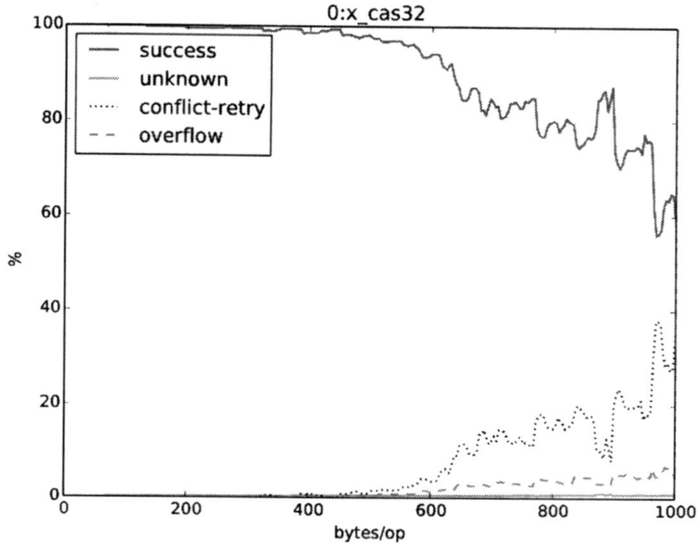

Figure 14. Success and failure rates for transactional *cas* operations competing against untransactional *cas* operations.

transactional memory writes. Presumably this is because the area of interference from the untransactional thread is always limited to the size of a cache line.

Finally, Figure 14 explores the competition of transactional compare-and-swap operations with untransactional memory reads. As with the previous example success rates are only shown for the transactional thread. This figure suggests that reads can interfere with active transactions. As read performance is greater than write performance it is possible for the reading thread to race ahead, thus the true extent of interfere may in practice be greater than that indicated in the figure. However, Figure 15 shows the performance of the same threads and suggests that the memory bus is saturated (convergence) with operations of 400 bytes and greater. Thus interference indicators for 400 to 1000 bytes should be broadly accurate.

2.7. *Summary*

The setup and teardown cost of a transaction is approximately 40 processor clock cycles. For common atomic operations such as compare-and-swap this cost can be amortized within approximately two to three words of memory access.

Transactions can realistically be up to 16KiB in size, but are probably far from optimal at this maximum size. For optimal performance transactions should probably be less than 1KiB.

There is no observable overhead on memory writes within a transaction, although there may be upto 20% overhead on memory reads. Thus the larger the transaction, the more effectively the transaction setup costs are amortized, at increased risk of failure.

Under contention, transaction failure (through conflict) becomes common when more than three or four cache lines are involved in the transaction. With high contention performance is equivalent to that of existing atomic operations such as compare-and-swap. There appears to be a degree of bias toward particular processor cores; however, this requires further investigation.

Transaction aborts are expensive, costing at least 150 processor clock cycles. This cost is in addition to the operations performed within the transaction. We infer that this cost or a

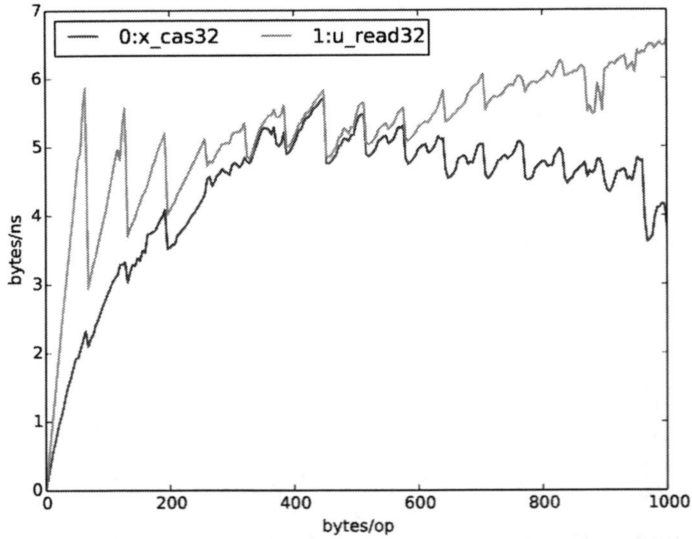

Figure 15. Performance of competing transaction *cas* and untransactional *read* operations.

higher one is also paid when a transaction is aborted due to interference. Where worst-case performance matters, for example within wait-free algorithms, traditional atomic operations (untransactional compare-and-swap) may be more appropriate.

3. Applications

This section describes how RTM can potentially be applied to the implementation of CPAs and their associated run-time systems. Subsections describe either novel techniques made possible using RTM or potential strategies for applying RTM to existing problems in the implementation of CPAs.

3.1. Fallback

A critical observation based on the performance details in section 2, is that a restricted transactional memory is not simply a CAS replacement. Replacing an algorithm which deterministically requires only one or two atomic operations with a transactional variant will likely reduce performance. Additionally, an existing algorithm embedded in a transaction is not guaranteed to make progress, as a transaction may encounter a conflict even when repeated if contention is present. This concern is common to lock-free and non-blocking algorithms [19] and where progress is to be guaranteed a fallback algorithm, perhaps with a lock, must be integrated. This fallback must also interact correctly with operating transactional code.

Fallback from transactional code to locked code can be achieved in a number of ways. Any read during a transaction adds the relevant cache line to the transaction's read set. Thus at the beginning of a transaction a lock word can be tested adding it to the read set. If non-transactional code takes this lock (changing the lock word) then the transaction will be aborted due to conflict.

Alternatively the transactional code could write the lock word at the beginning of the transaction. This will conflict with any other processor accessing the same lock word simultaneously regardless of other memory access within the transaction. Non-transactional code will still be able to override transactions by setting the lock word and aborting any running transaction, giving non-transactional code priority over transactional code. This is essentially the approach used by Intel's *Hardware Lock Elision* (HLE).

3.2. Conflict Synchronisation

The fallback mechanism detailed in section 3.1 highlights the use of transactional conflict as a form of thread interlock. The same mechanism can be used to implement synchronisation between a number of active threads.

One thread reads from a shared memory word and then spins performing no-op instructions. Other threads can then release the spinning thread by simply writing the shared memory word. Multiple threads could spin on the same word and be released simultaneously.

This mechanism has a great deal of similarity with a classic spin-lock; however, the spinning code is not required to read any memory. Whether the absence of memory operations makes a difference to performance needs to be established through experimentation, but we expect that there would be no observable difference due to existing caching mechanisms. The significant difference with RTM is that multiple memory locations can be monitored simultaneously with only marginally increased setup cost. Furthermore, using gather operations introduced with advanced vector extensions (AVX), also part of the Haswell microarchitecture, upto eight separate cache lines could be loaded simultaneously. This means that synchronisation on upto 512 bytes of memory could be implemented in just five machine instructions.

3.3. Context Switch Detection

As illustrated by the failure rates in section 2.4, operating system interference causes transactional abort. These aborts are detectable by the absence of error flags being set. In turn this means that user code can detect operating system context switches.

This effect could be applied in the implementation of timing critical code. It is worth noting that instructions which read the CPU cycle counter, e.g. `RDTSC`, are valid within a transaction. Thus transactions could be used to ensure accurate micro-benchmarking of instruction sequences. This may be useful during run-time system initialisation.

Broadly speaking operating system interference represents the same level of risk to run-time algorithms as data races with other processors. For example, operating system interference and scheduling is a key factor in why spin-locks are not appropriate for use in user code. Any user level locking primitive must account for operating system scheduling. Hence for algorithms with short and relatively deterministic execution times, RTM could greatly simplify implementation.

3.4. Exception Handling

Hardware exceptions which occur within a transaction are discarded once the transaction has been rolled back. This applies equally to events normally mediated by the operating system such as illegal memory accesses or instructions sequence. For example, if a *null* pointer is accessed within an transaction the transaction will abort, but the operating system will not be invoked. This allows very localised detection of exceptional conditions and mitigates the need for (program global) operating system signal handlers. Additionally, operating system signal propagation requires one or two orders of magnitude more time than a transactional abort.

3.5. Scheduling

Wait-free work-stealing algorithms such as those in the occam-π run-time system have demonstrated efficient and scalable scheduling of CPA workloads [20,13] and have also been used effectively with data parallel computations [21,22]. These algorithms are typically biased toward a work queue's owner, making stealing of work a costly operation. This cost is justified as the processor stealing work is otherwise idle. To ensure wait-freedom (bounded completion) work queues must also be of bounded size. It is also desirable to keep this bound low as it is directly related to maximum number of atomic operations the algorithm must perform.

RTM could be used to relax queue size constraints in common use cases and code paths. Existing operations which are efficiently implemented using compare-and-swap remain unchanged; however, operations which potentially traverse the queue, such as work-stealing, are performed transactionally. This maintains the existing bias as untransactional (local) operations will have priority.

As an extension of the above, a cleanliness (or bias) flag could be explicitly recorded within the work queue. When work is stolen from the queue this flag is updated at the beginning of a transaction to indicate the queue has been modified. The queue owner also tests the cleanliness flag when updating the queue in such a way that transactions conflict and abort. We speculated that using some form of this scheme it may be possible to implement owner updates without atomic operations, assuming all work-stealing is done transactionally.

As an optimisation, work-stealing schedulers often spin looking for work, before falling back to a use of condition variables or operating system primitives waiting for work. Using the method detailed in section 3.2 a transaction could instead be used to monitor work queues. This transaction can be constructed while searching for work to steal. If a monitored work queue is modified then the transaction will abort and the scheduler can again search for work.

3.6. Channel Communication

Previous work has demonstrated how channel communication in CPAs can be implemented efficiently using a single atomic operation per communication [13,23]. Therefore the more heavy-weight RTM does not provide an immediate benefit to implementation of point-to-point communication. However, there are often operations closely related to channel communication which may benefit from transactions or from being encompassed in a transaction that subsumes the communication.

In occam-π a shared channel has its reference copied when communicated. To support this behaviour each shared channel (or channel bundle) has an associated reference count. This reference count must be incremented when the shared channel is communicated. In principle this update could be made into a transaction along with the update of the channel state. While this would not represent any improvement in overhead, as only two atomic operations are involved, it could be of significant benefit when applied to more complex memory and object models than those used for occam-π.

3.7. Barriers

In order to support very large numbers of processes, priority and various forms of processor affinity, barriers within the occam-π run-time have a complex implementation and data structure [24]. We speculate that operations on these structures or the structures themselves could be simplified by use of a transactional model. Such changes would however need to efficiently handle transactional failure and guarantee progress as contention on barriers is common in many agent-based simulations.

3.8. Choice

Choice or alternation is the area of CPAs in which RTM may have the most impact. Here we briefly mention a few possible implementation starting points.

Brown has previously demonstrated the use of general and unrestricted *software transactional memory* (STM) to implement generalised choice over events [18]. This would be one possible route to use of RTM, although allowances would need to be made for the limited size of transactions.

For implementing choice over barriers an oracle mechanism similar to that proposed by Welch [25] could be adapted for transactional use. An oracle data structure common to multiple threads would be transactionally manipulated. This strategy could be particularly effective when compared to a lock-based implementation of the same data structure. This implementation might also be simpler than generalised alternation as proposed by Lowe [26].

The existing wait-free implementation of alternation for occam-π by Ritson, Sampson and Barnes [13] could be placed within a transaction. Consideration would need to be given to the relationship between the number of channels in the alternation and maximum transaction size. Additionally, the occam-π compiler does not guarantee that channels will not share cache lines and thus false sharing of channel cache lines could cause excessive transaction failures. If channels did occupy separate cache lines then in principle the transaction could only conflict as many times as there are channels involved, as each conflict represents a process committing to communication.

An alternative to the above strategy for occam-π would be to use a pre-enabling step as proposed by Barnes [27]. A transaction is initiated and channels are tested to see if there are waiting processes. This is done in priority order such that higher priority channels are added to the read set of the transaction first. If a channel is ready then communication is initiated with that channel, this can complete within the transaction. If no channels are ready then standard enabling and alternation wait is initiated using existing algorithms. The effect of this is that a channel being made ready by another process during the transaction will cause the transaction to abort and be restarted. This accelerates alternation over sets of channels where processes are already waiting. A lengthy enabling and disabling sequence is avoided and this method is multi-processor safe unlike Barnes original method.

4. Conclusions and Future Work

In this preliminary exploration we have establish baseline performance indicators for Intel's *Restricted Transactional Memory* (RTM) extensions as implemented on their Haswell microarchitecture. These performance indicators suggest that RTM is well suited for use on both relatively small (three memory words) and relatively large (upto 16KiB) transactions. RTM is available in the majority of 4th generation Core i7 and Core i5 processors. Thus will likely be widely available in laptop, desktop and server computers within two to three years.

Going forward further experimental work is required to explore the application of RTM to existing and emerging CPA run-time systems and languages. Additionally there are number of RTM characteristics which require further exploration:

- *Contention performance*: our results only evaluated a very small class of contention scenarios. A more intensive assessment is required. In particular, the observed bias to particular processor cores should be further explored.
- *Failure semantics*: our results (section 2.4) suggest that either the processor may not always accurately report the cause of a transactional abort, or that that there are other factors in addition to L1 cache size which affect the maximum size of a transaction.

- *Atomic operations within transactions*: instructions prefixed with `LOCK` are valid within a transaction, but we have not assessed their performance. It may be that processor implementations ignore the `LOCK` prefix, safely accelerating existing algorithms.
- *Nested transactions*: we have not tested nested transactions or established the nesting limit supported by the Haswell micro-architecture.
- *Performance counters*: our measurements do not include data from processor performance counters. These could be used to provide enhanced data on transactional failure causes and the impact of transactions on L1 and L2 caches.
- *Scheduler quantum*: an obvious experiment would be to change the operating system scheduling quantum and observe its impact on unknown transactional failures.
- *Transactional spin-locks*: our conflict based synchronisation mechanism as proposed in section 3.2 should be benchmarked to establish its applicability to real-world code.

Acknowledgements

This work was made possible by support from the EPSRC-funded MirrorGC project (EP/H026975/1) [2], and the University of Kent's Faculty of Sciences' research fund.

References

[1] Intel. *Intel® 64 and IA-32 Architectures Software Developer's Manual*, June 2013.

[2] Intel. *Intel® Architecture Instruction Set Extensions Programming Reference*, February 2012.

[3] K. Fraser. *Practical lock-freedom*. PhD thesis, University of Cambridge, King's College, September 2003.

[4] M. Herlihy. Wait-free synchronization. *ACM Trans. Program. Lang. Syst.*, 13(1):124–149, 1991.

[5] S. Owens, S. Sarkar, and P. Sewell. A better x86 memory model: x86-tso. In *Theorem Proving in Higher Order Logics*, volume 5674 of *Lecture Notes in Computer Science*, pages 391–407. Springer Berlin Heidelberg, 2009.

[6] S. Mador-Haim, L. Maranget, S. Sarkar, K. Memarian, J. Alglave, S. Owens, R. Alur, M.M.K. Martin, P. Sewell, and D. Williams. An axiomatic memory model for POWER multiprocessors. In *Computer Aided Verification, 24th International Conference, CAV 2012*, volume 7358 of *Lecture Notes in Computer Science*, pages 495–512. Springer, July 2012.

[7] P.H. Welch and F.R.M. Barnes. Communicating mobile processes: introducing occam-pi. In *25 Years of CSP*, volume 3525 of *Lecture Notes in Computer Science*, pages 175–210. Springer Verlag, April 2005.

[8] B. Sufrin. Communicating Scala Objects. In *Communicating Process Architectures 2008*, pages 35–54, sep 2008.

[9] J.B. Pedersen and B. Kauke. Resumable Java Bytecode - Process Mobility for ProcessJ targeting the JVM. In *Communicating Process Architectures 2009*, pages 159–172, November 2009.

[10] J. Gray. The transaction concept: virtues and limitations (invited paper). In *Proceedings of the seventh international conference on Very Large Data Bases - Volume 7*, VLDB '81, pages 144–154. VLDB Endowment, 1981.

[11] T. Haerder and A. Reuter. Principles of transaction-oriented database recovery. *ACM Comput. Surv.*, 15(4):287–317, December 1983.

[12] M. Herlihy and J.E.B. Moss. Transactional memory: architectural support for lock-free data structures. *SIGARCH Comput. Archit. News*, 21(2):289–300, May 1993.

[13] C.G. Ritson, A.T. Sampson, and F.R.M. Barnes. Multicore scheduling for lightweight communicating processes. *Science of Computer Programming*, 77(6):727 – 740, 2012.

[14] R. Haring, M. Ohmacht, T. Fox, M. Gschwind, D. Satterfield, K. Sugavanam, P. Coteus, P. Heidelberger, M. Blumrich, R. Wisniewski, A. Gara, G. Chiu, P. Boyle, N. Chist, and C. Kim. The IBM Blue Gene/Q Compute Chip. *IEEE Micro*, 32(2):48–60, March 2012.

[15] D. Dice, Y. Lev, M. Moir, and D. Nussbaum. Early experience with a commercial hardware transactional memory implementation. *SIGPLAN Not.*, 44(3):157–168, March 2009.

[2] `http://www.cs.kent.ac.uk/projects/gc/mirrorgc/`

[16] N. Shavit and D. Touitou. Software transactional memory. In *Proceedings of the fourteenth annual ACM symposium on Principles of distributed computing*, PODC '95, pages 204–213, New York, NY, USA, 1995. ACM.

[17] T. Riegel, C. Fetzer, and P. Felber. Snapshot isolation for software transactional memory. In *Proceedings of the First ACM SIGPLAN Workshop on Languages, Compilers, and Hardware Support for Transactional Computing*. June 2006.

[18] N.C.C. Brown. *Communicating Haskell Processes*. PhD thesis, University of Kent at Canterbury, 2011.

[19] M.M. Michael and M.L. Scott. Simple, fast, and practical non-blocking and blocking concurrent queue algorithms. In *Proceedings of the fifteenth annual ACM symposium on Principles of distributed computing*, PODC '96, pages 267–275, New York, NY, USA, 1996. ACM.

[20] C.G. Ritson, A.T. Sampson, and F.R.M. Barnes. Multicore Scheduling for Lightweight Communicating Processes. In *Coordination Models and Languages, 11th International Conference, COORDINATION 2009, Lisboa, Portugal, June 9-12, 2009. Proceedings*, volume 5521 of *Lecture Notes in Computer Science*, pages 163–183. Springer, June 2009.

[21] R.D. Blumofe, C.F. Jöerg, B.C. Kuszmaul, C.E. Leiserson, K.H. Randall, and Y. Zhou. Cilk: an efficient multithreaded runtime system. In *PPOPP '95: Proceedings of the fifth ACM SIGPLAN symposium on principles and practice of parallel programming*, pages 207–216, New York, NY, USA, 1995. ACM.

[22] G. Cong, S. Kodali, S. Krishnamoorthy, D. Lea, V. Saraswat, and T. Wen. Solving large, irregular graph problems using adaptive work-stealing. *Parallel Processing, 2008. ICPP '08. 37th International Conference on*, pages 536–545, September 2008.

[23] K. Vella. *Seamless Parallel Computing on Heterogeneous Networks of Multiprocessor Workstations*. PhD thesis, University of Kent, December 1998.

[24] C.G. Ritson. *Scalable Support for Process-Oriented Programming*. PhD thesis, University of Kent, July 2013.

[25] P.H. Welch, N.C.C. Brown, J. Moores, K. Chalmers, and B.H.C. Sputh. Alting Barriers: Synchronisation with Choice in Java using JCSP. *Concurrency and Computation: Practice and Experience*, 22:1049–1062, March 2010.

[26] G. Lowe. Implementing Generalised Alt. In *Communicating Process Architectures 2011*, pages 1–34, June 2011.

[27] F.R.M. Barnes. *Dynamics and Pragmatics for High Performance Concurrency*. PhD thesis, University of Kent at Canterbury, June 2003.

Communicating Process Architectures 2013
P.H. Welch et al. (Eds.)
Open Channel Publishing Ltd., 2013

Life of occam-Pi

Peter H. WELCH

School of Computing, University of Kent, UK

p.h.welch@kent.ac.uk

Abstract. This paper considers some questions prompted by a brief review of the history of computing. Why is programming so hard? Why is concurrency considered an *"advanced"* subject? What's the matter with *Objects*? Where did all the Maths go? In searching for answers, the paper looks at some concerns over fundamental ideas within *object orientation* (as represented by modern programming languages), before focussing on the concurrency model of communicating processes and its particular expression in the occam family of languages. In that focus, it looks at the history of occam, its underlying philosophy (*Ockham's Razor*), its semantic foundation on Hoare's CSP, its principles of *process oriented design* and its development over almost three decades into occam-π (which blends in the concurrency dynamics of Milner's π-calculus). Also presented will be an urgent need for rationalisation – occam-π is an experiment that has demonstrated significant results, but now needs time to be spent on careful review and implementing the conclusions of that review. Finally, the future is considered. In particular, is there a future?

Keywords. process, object, local reasoning, global reasoning, occam-pi, concurrency, compositionality, verification, multicore, efficiency, scalability, safety, simplicity

Introduction

Let's try a few questions to which the answer might be: *"occam, obviously"*. Which language most grievously irritates programmers who want to hack code quickly? Quite possibly – though it might also answer the question about which language *really* supports programmers who want to hack code quickly. Which language has concurrency built into its core design with a formal semantics founded on process algebra? Definitely – to the extent that it almost *is* a process algebra, with laws and proof rules and a model checker to enable formal verification to be managed as part of the program writing experience (by programmers who are not mathematicians and who do this as a matter of course).

Which language has concurrency safety rules built into its design that enable the compiler to prohibit data race hazards (even when the concurrency is highly dynamic, with process networks forming and breaking up dynamically)? Which language has a concurrency model that is deterministic *by default*, but which can build non-deterministic systems (when necessary) *by design*? Which language has the fastest and most scalable multicore scheduler on the planet (and which is proving highly effective in large-scale complex system simulations and the study of emergent behaviour)? Which language can be introduced to undergraduate CS students in *Fresher* week so that they can be programming LEGO robots to navigate an environment (reacting to bumps and following tracks) using concurrent processes for sensor data gathering, brain logic and motor drivers ... in 90 minutes? Which language had large-scale industrial use in embedded applications twenty years ago ... but has now been mostly forgotten? Yes, yes, yes, yes and (sadly) yes. Enough! Let's look at some history.

1. Some History

1.1. A Long Long Time Ago[1] ...

Computers were invented: calculating machines that were programmable and *Turing complete* ... meaning they could do anything any future computer could do (which was pretty cool).

Babbage's *Analytical Engine* [1] (1837) was probably the first, but was too far ahead of its time and handicapped by lack of funding and a bad (in retrospect) engineering decision over number representation (decimal). For working machines, we had to wait for a better understanding of the theory (Russell's *Principia Mathematica* [2,3] (1910-13), Gödel's *Incompleteness Theorem* [4] (1931), Church's *λ-calculus* [5] (1936) and Turing's *Computable Numbers* [6] (1936)) and seriously better funding (the *Second World War*). Built, programmed and operated in secret, they arrived: the Zuse Z3 [7] (Germany, 1941), Colossus [8] (Britain, 1943) and, probably, several others. Colossus, in particular, had significant impact on war outcomes.

Following the war, the pace picked up as the commercial potential became obvious – in the USA, at least (ENIAC [9], 1946). Ideas for hardware and software developed at a rate unprecedented for any other technology. A new and unfortunate race, concerning the phenomenon known as *Moore's Law*, started and continues to this day. Paraphrasing David May [10]: *"Hardware capabilities double every 18 months. Unfortunately, software overheads respond by matching this ..."*. So, we seem to be making progress – but is anything useful happening? There is cause for concern.

Edsger Dijkstra made a similar point: *"In the beginning when we had no computers, we had no problems. Then when we had small computers, we had small problems. Now that we have big computers, we have big problems ..."* [11]. Of course, now (2013) we have tiny computers again – mass-produced and in most everyone's pockets. Thanks to the ingenuity of our hardware engineers, their astonishing power is barely related to their physical size; and thanks to the ingenuity of our software engineers, we have astonishing problems. Our laptops/tablets/phones are millions of times more powerful than the computers that guided the Apollo astronauts to the moon. But what are they doing most of the time ... why, so often, won't they respond to us ... why the spinning wheel (or hour-glass) of death?

1.2. Bad News on the Doorstep ...

Something went wrong. A message in the preceding section is that mathematics is probably important. The pioneers of computing were mathematicians and engineers, who relied on (or were inventing) established mathematical models of the materials they were engineering. Such foundations seem largely ignored today. Walk into a class of (say) second-year Computer Science undergraduates in any (maybe, almost any) university and ask them what a *loop invariant* is ... or *recursion invariant* (if they are into functional programming) or *class invariant* (if they are into object-orientation). For the past three years, I have done that in my university and received mostly blank stares. This is not an uncommon reaction, but gentle further questioning revealed that they really didn't know. Some, thank goodness, did express curiosity though. I fear the result would be the same in most programming rooms (bedrooms/garages) of industry where most of the software we (have to) use gets written.

Nailing *invariants* in a system is one of the most powerful forms of analysis, whether in mathematics or programming. We must be forming invariants in our head whenever we write a loop (or recursive function or class); otherwise we would not know what to write and our programming would be guesswork. So, why not declare those invariants *explicitly*

[1] With special acknowledgements to Don McLean.

in our code, where they could also play a crucial role in verification (of which there is much more to say later)? We don't need to be mature mathematicians to write invariants – just good engineers. And we need to be good engineers to develop computer systems.

Perhaps invariance *is* taught in a *theory* course somewhere? But that would not be good enough: invariance is a fundamental concept in programming and should be taught alongside all the other fundamentals (variables, assignment, messages, events, sequence, parallel, conditionals, choice, loops, functions, parameters, recursion, ...). Missing invariance, however, is only one symptom of things that are wrong.

Dijkstra suggested that only well-trained mathematicians should be allowed to program computers. That is not my opinion ... but I believe that in failing to engineer certain crucial mathematics into the tools provided for programming, we have failed as engineers and should not be surprised when things go very wrong. Addressing such failure is long overdue.

1.3. A Generation Lost in Space ...

Where does concurrency fit in? A computer system – to be of use in this world – needs to model that part of the world for which it is to be used. In our world, things happen *at the same time*. If our modelling reflects that natural concurrency, the system should be *simpler*.

Yet concurrency is thought to be an "advanced" topic, much harder than sequential computing which, therefore, needs to be taught and mastered first. This *sequential-computing-first* tradition makes no sense. Concurrency is a powerful tool for *simplifying* the description of systems. Performance spins out from this, but is not the primary focus. This has radical implications on how we educate people for computer science *and* on how we practice it.

Of course, we need a model of concurrency that is mathematically rich and clean, yields no engineering surprises and scales well with system complexity. *Communicating Process Architectures*, based on the ideas and mathematics of Hoare's CSP [12,13,14,15,16] and Milner's π-calculus [17], fulfills this need. Traditional approaches to concurrency (e.g. shared data and locks) conflict with some basic assumptions of, and our intuition for, sequential programming. They are, not surprisingly, very difficult to use. The *'S'* in CSP, on the other hand, stands for *Sequential* and there is no conflict.

Matters have now reached a crisis because the nature of modern architecture (e.g. instruction-reordering processors, multicore/manycore nodes and networks thereof) *requires* concurrency in the systems that ride on them to be effective. This has forced the issue.

A key insight, which goes against much received wisdom, is that concurrency is a necessary concept for *all* stages of specification, design, verification, implementation and maintenance of computer systems. Concurrency ideas are as fundamental as, for example, those of sequence, looping or procedural abstraction and we must be confident in taking advantage of them. If concurrency remains treated as an advanced concept, too difficult to be used as a matter of routine and only reached for to raise system performance (e.g. supercomputing applications) or responsiveness (e.g. embedded systems), then our systems will be based on a model of the world that does not match reality and will be unnecessarily complicated ... and slow ... and wrong.

1.4. Some Happy News ...

Concurrency raises a few issues not present in sequential logic and they must be addressed. The ones we are, perhaps, most frightened by are race hazard, deadlock, livelock and process starvation.

The first of these concerns concurrent update and observation of shared data, which leads to data corruption and unplanned non-determinism. This is what *locks* of various forms were invented to prevent ... but for which, in all but the simplest of scenarios, super-hero programmers are required for their correct and efficient application. In the communicating process model, only immutable data may be shared and so the problems simply do not arise. Of course, it requires well-engineered programming languages to ensure that this rule of the model is enforceable and enforced. Providing that assurance, without unduly limiting their areas of application or compromising performance, is an engineering challenge that is solvable (and to which the original designers of occam [18,10] have made a massive contribution). Finding the super-hero programmers on the scale needed for programming future systems with locks is, however, not possible.

A classic example of overlooked deadlock is given by Dijkstra's *Dining Philosophers* [19]. It illustrates a common problem in concurrency: how to share resources safely between competing consumers. The problem arises when there are not enough resources to go around and consumers sometimes need exclusive use of two or more. Simple *process oriented* solutions can be found in Hoare's 1985 book [13]. The story of the dining philosophers was derived from a real deadlock that happened, very occasionally, in an early multiprocessor operating system. The resources being shared were a limited supply of memory buffers, used for smoothing the reading and writing of files. Today – almost 30 years later – memory is not so scarce! Yet, operating system (or specific application) deadlock is rampant. We have been made to grow tolerant of our laptop/tablet/phone (or one of its apps) freezing on us. We have been, and still are, making the same mistakes again and again. Will we accept this forever?

Livelock – where a process demands processing resource but refuses to deliver or accept anything (e.g. a kill signal) – is also sadly common. Why does my laptop sometimes go into a spasm of 100% core activity, refuse to respond to keyboard or mouse events and force a powerdown? As I type these words, why does my web browser insist on sustained consumption of 50% of one of my cores, even though the pages open are all static, and only a full system reboot restores sanity (for a while)? Yet, livelock is addressed by CSP. We do not have to flounder in uncertainty. Systems can, and should, be designed to be free from livelock and that absence can, and should, be formally verified (Sections 3.2.2 and 3.4.5).

Process starvation is where some parts of the system fail to progress compared with other parts. This is always due to unfair servicing of their requests for resources, either by application processes managing those resources or by the run-time system scheduler managing access to the processors. The former is a particularly easy trap to fall into when using locks (e.g. the *"Wot, No Chickens?"* example, [20,21], in which the approved way for programming Java monitors leads to permanent starvation of one process despite a total processor loading of less than 1%). Under the communicating process model, the former is simply managed by sharing the ends of the application service channels (i.e. strict FIFO policing of *interleaved* use of the channels) or by well-known *fair* choice algorithms by the servers. The latter is rarely a problem for communicating processes, so long as the communication channels are *synchronised* or, occasionally, have small finite buffers that block when full. This is because processes that need resource (either to receive or deliver information) will block, leaving the run-time scheduler no choice but to schedule other processes including, eventually, those that will supply or accept the information that will unblock the first set – unless, of course, the system is mis-designed or mis-programmed. In these circumstances, process scheduling is self-driven, leaving little for the scheduler to consider. *[Note:* if the communication primitives are *asynchronous* and supported by unlimited buffering (e.g. Erlang mailboxes), process scheduling to avoid starvation will have to be more clever ... and expensive.*]*

Summary: communicating processes with synchronised message passing avoids, just by its nature, the dangers of race hazard and process starvation. Deadlock and livelock are easy to design against and formally verify that the design achieves their absence. Why, then, is this not state of the art? Why do we accept these repeated failures in concurrent systems, especially as modern processor architecture and our ambition for systems with ever more complex behaviours and scale mean that all systems will have to be concurrent?

2. We Need to Talk About Objects

Originally, objects were supposed to be autonomous entities, each containing a cohesive collection of data and algorithms for managing that data. A system is a collection of objects interacting by sending each other messages. Objects are typed through a hierarchy of classes, where sub-classes represent specialisation (or, if you prefer, extension) from the super-class.

The first two sentences of the preceding paragraph actually describes the model of communicating processes discussed in this paper. We claim that processes and channels fulfills the original aims for objects in ways that the objects and methods of modern "object-oriented" languages fail.

A serious problem with modern objects is that they are not *autonomous* – i.e. they cannot, in general, be understood on their own; their semantics are system sensitive. We sometimes tell our students that objects are water-tight containers of data, interacting with other objects through formal public interfaces. Unfortunately, objects leak information.

One problem is that an object does not declare the external resources its implementation needs (e.g. other classes of object, specific methods, monitor locks); its public interface only guides how it may be used by others (which does not necessarily cover how it is *affected* by others). This *information hiding* is sometimes a good thing – why should the user of an object know what other resources it uses? But that information is needed to understand how the whole system is put together. UML class entity relation diagrams show calling dependencies between classes, which is marginally interesting, but does not say anything about the dynamic network of objects at runtime. To understand a system, we need that information and, to find it, must hunt through the whole source code.

Of course, processes and channels give us concurrent systems; whereas objects and methods, in their modern guise, execute sequentially (ignoring, for now, special mechanisms for launching threads). The issues raised in the following subsections relate just to sequential programming with objects.

2.1. *Hidden References ("not a word was spoken")*

Suppose an object, `A`, has reference to two other objects, `B` and `C`. Consider somewhere in `A`'s (Java) coding:

```
B.reset ();
C.something ();
```

After the `C.something` method returns, can we rely on `B` still being in its *reset* state? `C` may have a reference, or knows someone who knows someone who knows someone who has a reference, to `B` and the call to `C` may have altered `B` – so no! Did the documentation of `C`'s class note that it might change something in an object of `B`'s class? If that was viewed as an implementation detail, then no. So, we have to search source code to find the network of objects in our system to discover if `B` might have been kicked! Local analysis of just the above two lines is insufficient. Such hidden side-effects happen often in OO code and the result is mystery.

An alarming postscript to the above scenario is that the side-effect on `B` can happen *even if* `B` *is held in a (supposedly) private field of* `A`*, was constructed by* `A` *and has never had its reference given away by* `A`! The last of the 'someones' in the second sentence of the previous paragraph may be `A` itself, who certainly knows about `B` (i.e. the side-effect was by a *callback*, on which more is said in Sections 2.2, 2.3 and 2.4).

Given the semantics of heap objects and multiple aliasing (which is part of the power of object-orientation, albeit a dangerous one), the above is, perhaps, not surprising and good engineers will discipline their designs to limit and control its occurrence. When it happens though, strong documentation is needed to guide the reader through the relevant trail of source code and unmask the mystery of those two lines. This is evidence of weak engineering in our programming technology – we should not have to be doing this.

2.2. *Globalisation and Loss of Privacy ("the day the music died")*

This example was produced more than ten years ago by Tom Locke – to whom I remain very grateful. I continue to be shocked each time I see it.

Suppose an object, `A`, has reference to an object `C` and a *private* integer field, `count`. The latter does not live on the heap and cannot be shared with other objects. Surely local reasoning about `count` will work? Consider somewhere (else) in `A`'s coding:

```
count = 42;
C.something ();
```

After the `C.something` method returns, can we rely on `count` still having the value `42`? Alas, no! Although no object other than `A` may directly change the value of `count`, `C` may have a reference, or knows someone who knows someone who knows someone who has a reference, back to `A` – who can change it for them. Again, we have to start searching a lot more code than the above two lines.

The conclusion is that local fields of objects must be regarded as global, regardless of any notions of *privacy* declared for them. In the past, proliferation of global variables meant that local reasoning about code was unsound. *Structured programming* led us safely out of that mire. *Object orientation* seems to have taken us back in.

2.3. *Global Reasoning Required (the Counter Example)*

This section gives a short complete example of a class whose semantics cannot be determined from its code, although it looks as they they should! The code seems perfectly innocent, obeying simple principles of good object orientation:

```
public class Counter {

  private int count = 0;

  private Logger logger;

  public Counter (Logger logger) {
    this.logger = logger;
  }

  public int getCount () {
    return count;
  }

  public void increment () {      // adds 1 to count (are you sure?)
    count++;
    logger.log (count);           // log this event
  }

}
```

Inspecting this code, we would hope that an invocation of the `increment` method would increment the value of the `count` field of the invoked object by one (as indicated by the comment). For example, if `counter` is a `Counter` object and we see the code:

```
x = counter.getCount (); counter.increment (); y = counter.getCount ();
```

would we not expect `y` to get the value `x + 1`? We should not need to look up what a `Logger` object does with its `log` method: primitive data types are passed by *value* in Java, so that call cannot change `count`. Unfortunately, none of these sentiments are reliable!

Let's look at the `Logger` class:

```
public class Logger {

  public void log (int count) {
    System.out.println ("Log " + System.currentTimeMillis () + ": " + count);
  }

}
```

Well, that seems mostly harmless ... so, are we safe?

Unfortunately, no! Because of the *inheritance* mechanism – one of the most promoted features of object-orientation – we have to look further. `Logger` may not be the class of the object given to `Counter`. It could be a sub-class and that could do anything ... like a callback! For example:

```
import java.util.Random;

public class Blogger extends Logger {

    private Counter counter = null;

    private boolean awake = true;  // flag to ignore callbacks

    public void setCounter (Counter counter) {
      this.counter = counter;
    }

    private Random random = new Random ();

    public void log (int count) {
      if (awake) {
        if (counter != null) {
          awake = false;            // callbacks will be ignored
          for (int i = 0; i < random.nextInt (100); i++) {
            counter.increment ();
          }
          awake = true;             // callbacks accepted again
        }
        super.log (count);
      }
    }

}
```

Now, if `Counter` is constructed with a `Blogger`, rather than a `Logger`, its behaviour will become somewhat unexpected *(to a reader of just its code)* should the `Blogger`'s `setCounter` method ever get invoked with a reference back to the `Counter`. For example:

```
Blogger blogger = new Blogger ();
Counter counter = new Counter (blogger);

for (int i = 0; i < 10; i++) {
  counter.increment ();               // adds 1
}
```

```
blogger.setCounter (counter);

for (int i = 0; i < 10; i++) {
  counter.increment ();               // adds a random amount!
}
```

The point is that the `Counter` class seems very simple, clean and innocent. Yet its semantics cannot be determined in isolation – we have to be aware of the network of objects in which it is placed. This need for non-local analysis is a serious trap for the unwary object orienteer. If a class as simple as `Counter` can have its semantics corrupted, what about everything else? If a system does not decompose into parts that can be understood in isolation and that understanding not violated when those parts are put together, then complex systems will always be unsafe. And we have not begun to talk about concurrency and objects.

Of course, `Counter` could be coded to protect itself from callbacks, using the same *ignore flag* trick employed by `Blogger`. But that is something of a distraction to its purpose and we would rather not have to do it.

2.4. Class Invariants ("do you believe in rock & roll?")

In Section 1.2, class *invariants* were mentioned as something that might be important. They are not in many textbooks. At my institution, we do not currently teach them ... although there are plans! I fear they may not be currently much used in practice.

Class invariants are predicates expressing properties of, and relationships between, data held within an object that are always true before and between method calls. All constructors and initialisation code must establish them. The implementor of a method can rely on them at the start of the method and must re-establish them before the end.

Unfortunately, the above description is incomplete – another trap for the unwary! The implementor has to work harder to ensure that invariants are maintained. The problem, again, is callback. *Any* callout from a method to another object has the potential for callback. Discovering if this is so for a particular system requires non-local analysis. Discovering if this so for a class whose objects are to be used in systems not known to the designer of the class (e.g. `Counter`) is not possible. Precautions, therefore, must be taken.

Should a callback happen and a class invariant be invalid at the point of callout, the assumptions of the method called back will not be honoured and it will be operating on a corrupt object – with unknown consequences. If the implementor of the method calling out does not know whether this will happen, the class invariants must be re-established *before* calling out. One way to do this is to program the *ignore flag* trick used by the `Blogger` class in Section 2.3 – the invariants then become extended through being implied by that flag. However, if there is a callback, there is probably a reason for it and simply doing nothing may not be an option; so the original invariants must be properly re-established. How often in this care taken?

So, we are faced with either non-local analysis (which makes complex system development and maintenance ultimately impossible) or non-trivial added complexity to any method that invokes other objects (which has the same result).

In Section 3.3, we argue that the communicating process model for computation does not suffer from these problems. Local-only reasoning is sufficient to establish the semantics of processes, with no complex trickery needed. Those semantics are not violated by sequential or concurrent composition. Large scale and complex systems can be built and maintained with confidence. And, of course, their natural concurrency means they can directly take advantage of distributed and multicore platforms. So, there will be correctness and there will be performance.

3. Process Oriented Design and occam-π

We call the programming discipline underlying Communicating Process Architectures *Process Orientation (PO)*. A process is a *component* that encapsulates a cohesive collection of data and algorithms for managing that data. Unlike an object, its data and algorithms are strictly *private* – the outside world can neither see its data nor execute its code. Each process is *alive*, its algorithms executed by threads of control that stay always inside the process.

A process *variable* changes its value if and only if that process makes that change. None of its state can change behind its back, which gives a simple and intuitive semantics for sequential operation. A major, but largely unremarked, advantage of this is that PO code directly and safely exploits modern instruction-reordering processors – no special care needs to be taken by PO programmers.

Processes interact through synchronised message-passing over named channels, or through multiway synchronisation on named barriers. Networks can be built dynamically through the runtime construction of processes, channels and barriers; and by the communication of channel and barrier *ends* through channels. These mechanisms have direct representation in CSP and the π-calculus and (can) have extremely lightweight implementation – a few tens of cycles – on modern architectures.

Synchronised communication has several benefits. Section 1.4 mentions the force it provides for scheduling, simplifying the work of the process scheduler to enable *fair* progression of all parts of the system. A semantic power is the knowledge it gives to the sender of a message: if the message has gone, the sender knows the receiver has received. A corollary is the semantic power it gives to a receiver: if the receiver is not in a state to take a message from some source, it may refuse *(simply by not having the code to accept)*. There is no need for complex logic involving *condition variables* [22] or Java *monitors* (`wait` and `notify`) [21] or for runtime expensive logic such as *busy polling*.

For these reasons, we claim that processes are *autonomous reusable components* and, of course, they plug into and build concurrent systems.

3.1. A Diagram Language (in 4 + 2 Pictures)[2]

The starting point for process oriented design is to sketch some process networks. Networks have structure - networks within networks - and the design can be done from any level and in any direction. Figure 1 shows the small diagram language for process-oriented design.

Figure 1(a) shows part of a network, with three processes connected by 3 *point-to-point* channels to each other and 2 channels to the rest of the network. Figure 1(b) shows a single *client* process with access, through a channel, to a bank of *servers*. The client does not care which server takes any request it sends. The servers *share* the reading end of the channel and, when not busy with an earlier task, form a queue to receive new instructions (*interleaving*, in CSP terms, their use of the channel). Figure 1(c) shows the opposite: multiple clients sharing the writing end of a channel, with a single server at the other. Again, clients who seek access interleave their use of the channel (disciplined by a queue) to reach the server. Channels may be shared at both ends (not shown), where (needy) clients queue up to seek service and (idle) servers queue up, separately, to provide. In these examples, a client must not care which server it gets and a server must not care which client.

The channels in these diagrams are unidirectional. However, they may also represent *bundles* of channels, so that communication may be two-way. If bundle-*ends* are shared, a process acquiring them (by reaching the front of the queue) has exclusive use of all the channel-ends inside, so that an extended two-way conversation can be made. In the case of such sharing, the arrowhead points to the *server* end.

[2]Four paragraphs in this section are based on paragraphs from Sections 3.1.1 and 3.1.2 of [23].

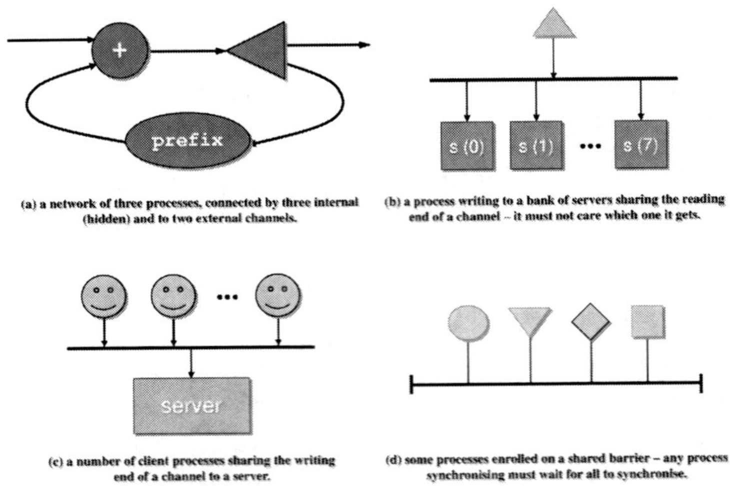

(a) a network of three processes, connected by three internal (hidden) and to two external channels.

(b) a process writing to a bank of servers sharing the reading end of a channel – it must not care which one it gets.

(c) a number of client processes sharing the writing end of a channel to a server.

(d) some processes enrolled on a shared barrier – any process synchronising must wait for all to synchronise.

Figure 1. Process oriented design: components and connectors (*from [24]*). There are no rules for the shapes of component icons (*processes*) – whatever is meaningful to the designer is allowed.

Figure 1(d) shows four processes synchronising on a barrier. This is a multi-process event: if one process offers to synchronise (i.e. tries to get over the barrier), it must wait for *all* processes connected to the barrier to do the same. The last process to synchronise releases all the others.

occam-π [25,26,27,28,29,24,30,23]) has a simple syntax that directly captures these network structures. Figure 1(a) maps to:

```
CHAN REAL64 a, b, c:              -- channels are strongly typed
PAR
  adder (in?, c?, a!)             -- these processes
  delta (a?, b?, out!)            -- can be listed
  prefix (b?, c!)                 -- in any order
```

where only the 3 completely used channels (`a`, `b` and `c`) are declared. The 2 remaining channels (`in` and `out`) connect to the rest of the network and so must have wider scope (i.e. they are declared global to this piece of code). *Notation:* the expression `c!` is the writing-end of the channel named `c` and `c?` is its reading-end.

We can abstract those missing channels into parameters for a complete process component. The parameter channels represent external connections that must be wired into the network using this component; the internally declared channels represent internal wiring that is hidden from the external network:

```
PROC integrate (CHAN REAL64 in?, out!)
  ...  above five lines
:
```

Visually (figure 2), this abstraction of a network into a reusable component corresponds to embedding the network (with loose ends) on to a named chip (with I/O pins). The `integrate` component can now be used in forming higher-level networks – in the same way as `adder`, `delta` and `prefix` were used in its own construction.

For completeness, we show occam-π coding for the other three diagrams of Figure 1. The server-bank network in (b) may be set up by:

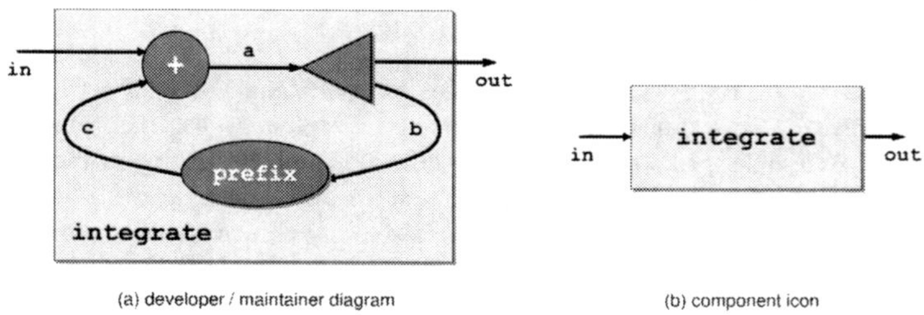

Figure 2. Process oriented design: a network of processes is also a process.

```
SHARED ? CHAN SOME.SERVICE c:          -- only the reading-end is shared
PAR
  triangle (c!)                        -- one writer process
  PAR i = 0 FOR 8
    s (i, c?)                          -- many servers, sharing the reading-end
```

The many-client system in (c) is:

```
SHARED ! CHAN ANOTHER.SERVICE c:       -- only the writing-end is shared
PAR
  PAR i = 0 FOR n.clients
    smiley (i, c!)                     -- many clients, sharing the writing-end
  server (c?)                          -- one reader process
```

The network of processes in (d), synchronising on a shared barrier is:

```
BARRIER b:
PAR ENROLL b
  circle (b)
  triangle2 (b)
  diamond (b)
  square (b)
```

We have given these examples of occam-π syntax because they show how network structure is explicitly declared. There are no equivalent mechanisms in *threads and locks* programming. Erlang networks are represented implicitly by process variables holding references to (the mailboxes of) other processes. Discovering the networks involved means searching disparate pieces of code. Discovering *structure* in the networks is harder still. The concepts are one-dimensional – there is no such thing as *threads-within-a-thread* nor, in Erlang, *processes-within-a-process*. Yet, without network structure, there would be no universe.

The iconography in Figures 1 and 2 reflects hardware design: *chips* and *wires*. The physical constraints implied by hardware mechanics reflect needed semantic rules underlying process autonomy (layered construction, localisation and inviolability of state, interaction only if explicitly connected). Software has no need for such constraints: we can let threads of logic quietly interfere with and depend upon each other's state – just because we can! Even though we may try to retrofit controls (e.g. *locks*) to try and avoid the worst consequences, the freedom afforded means mistakes will be made and we exploit it at our peril.

The hardware discipline that process-oriented design imposes on software has important benefits for simplicity and performance. Engineers – whatever their field[3]– have long experience of network (or *circuit*) diagrams, with intuition for their semantics long established. The underlying formal semantics of CSP and π-calculus align with that intuition so that we get no surprises, at least at this low-level, from our concurrency.

[3]Sadly, it seems that software engineering may be an exception to this.

3.2. CSP Semantics (in one page)

This section presents no CSP syntax or examples, but outlines the key properties of its three most common semantic models (*traces, failures* and *divergences*) together with the concept of *refinement*. Why this is important for the specification and verification of future computer systems – and why *verification* is important – is left for the reader to consider.

We note that the FDR model-checker [31] enables machine-checked verification of the refinements described below, so long as the systems being tested do not have too many states. Model checking is an art that needs to become accessible to systems designers and programmers as normal practice. Assistance for this is proposed in Section 3.4.5. By setting up and asking the right questions, surprisingly large systems (e.g. those with more states than atoms in the universe [32], though not infinitely many) can be verified.

3.2.1. Safety (traces refinement)

An *event* (e.g. channel communication, barrier synchronisation) happens when, and only when, all processes relevant to it choose to engage. A process trace is a finite sequence of events in which a process *may* engage.

`P` *trace-refines* `Q` means the traces of `P` are also traces of `Q` – anything `P` may do, so may `Q`. Turning this round, if there is something `Q` cannot do, `P` cannot do it either. Now, if `Q` is a *specification*, then `P` is safe in the sense that `P` cannot exhibit behaviour (presumably 'bad') disallowed by `Q`.

This is not enough – e.g. the `STOP` process trace-refines anything, since it does nothing (not even terminate); but it's probably not what the specifiers had in mind!

3.2.2. Liveness (failures refinement)

A process *state* is what a process has become after executing one of its traces (and may be represented by that trace). An event is *external* to a process if other processes may engage on it. A state is *stable* if there is no *internal* (i.e. hidden) event on which it may engage. A stable *resolution* of a state is a stable state reached by zero or more internal events only. A process *failure* is a state paired with a set of external events on which a stable resolution of that state refuses to engage. A *deadlocked* system has reached a stable state where *all* pairings of that state (i.e. the trace that led it there) with external events are failures.

`P` *failure-refines* `Q` means (`P` trace-refines `Q`) *and* (the failures of `P` are also failures of `Q`). So, if a *<trace, event-set>* is not a failure of `Q`, it is not a failure of `P` either. Now, if `Q` is a *specification*, then `P` fulfills its liveness conditions: if the specification (`Q`) says that *in this state you will react to one of these events* (i.e. there is no failure here), the implementation (`P`) *will react* (and the reaction will be safe because of trace refinement).

3.2.3. Liveness (failures-divergences refinement)

A process state is *divergent* if an infinite sequence of *internal* events is possible from that state. This is usually a bad thing. A divergent state is, of course, unstable but *may* recover to a stable state. However, a divergent state is considered so dangerous that further consideration of process behaviour is not worth pursuing. All divergent process states, together with all subsequent states (even if stable), are written off and deemed to be the same. This means that everything is a failure with them and slightly changes the meaning of *failure-refines* in the following definition.

`P` *failure-divergence-refines* `Q` means (`P` failure-refines `Q`) and (the divergences of `P` are also divergences of `Q`). Now, if `Q` is a *specification* with no divergences (which would be usual), then the implementation (`P`) also has no divergences.

If divergence-freedom is verified, *failures* and *failures-divergences* refinement are the same – so only the former need be used for further analysis.

3.3. Local-only Reasoning (the Counter Example)

This section reviews issues raised in Section 2 as they relate to the communicating process model.

3.3.1. Expect Processes to Change (They are Alive)

Section 2.1 refers to a sequence of two method invocations on two different objects, where the second one has a hidden side-effect on the object reset by the first. If we replace the objects with processes running concurrently and the method invocations with channel communications, the surprise does not arise:

```
SEQ
  to.B ! reset
  to.C ! something
```

The process at the other end of the `to.B` channel (presumably `B`) is alive. Having been reset, it is free to move on – so we have no expectation that it is still in its reset state, even before the execution of the second line.

If the action in the second line causes process `C` (at the other end of the `to.C` channel) to interact (directly or indirectly) with `B`, that can be seen as a possibility by looking at the network diagram (or its explicit coding) for this part of the system. Only if connections are visible and we are concerned, need we consider the behaviour of `C`.

So, if the `B` process changes since we reset it, there is no surprise. The problem with the object code is that the system is purely sequential and the `B` object – which cannot do anything on its own – changes without any visible sign to the reader of that code.

3.3.2. Local Reasoning for Local Data

Section 2.2 shows a disturbing inability to secure control over the value of a supposedly *private* integer field. If that integer is data local to a process, control is assured: the value of a variable changes if and only if the process owning it makes a change:

```
SEQ
  count := 42
  to.C ! something
```

Local reasoning on just these three lines of code is sufficient. They are executed in sequence. The value of `count` is set once and not changed subsequently. Therefore, its value after the sequence ends is `42`.

No further analysis is needed. If the `C` process (after receiving `something`) tries to communicate back to this process so that `count` will be changed, that cannot yet have happened simply because there is no code in the above lines to accept such a communication – so, if one is being attempted, it will be blocked.

3.3.3. Counter as a Process

Section 2.3 develops this further, presenting a simple `Counter` class whose semantics cannot be determined from its code – they depend on the context in which it is used.

If `Counter` becomes a *process*, its semantics are locally determined. Figure 3 shows an icon for this `Counter` process. The `answer` and `log channels` carry integers (the value of the internal `count`). The channel `ask` carries two kinds of question and, in occam-π, is best described with a *protocol*:

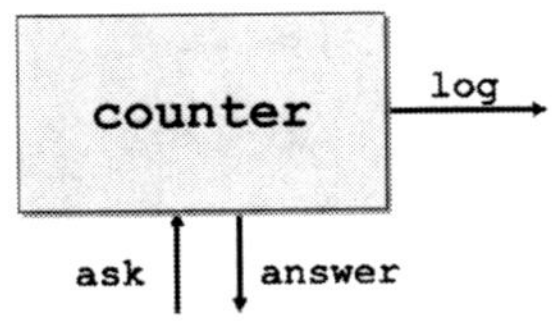

Figure 3. Process oriented design: process semantics does not depend on context.

```
PROTOCOL COUNTER.ASK
  CASE
    increment
    get.count
:
```

The process almost writes itself:

```
PROC counter (CHAN COUNTER.ASK ask?, CHAN INT answer!, log!)
  INITIAL INT count IS 0:
  WHILE TRUE
    ask ? CASE
      increment                         -- adds 1 to count
        SEQ
          count := count + 1
          log ! count                   -- log this event
      get.count
        answer ! count
:
```

Unlike the case for the object code (Section 2.3), we can immediately deduce that the response to the `increment` request is to add 1 to the internal `count`. The behaviour of the surrounding processes (Figure 4) does not matter.

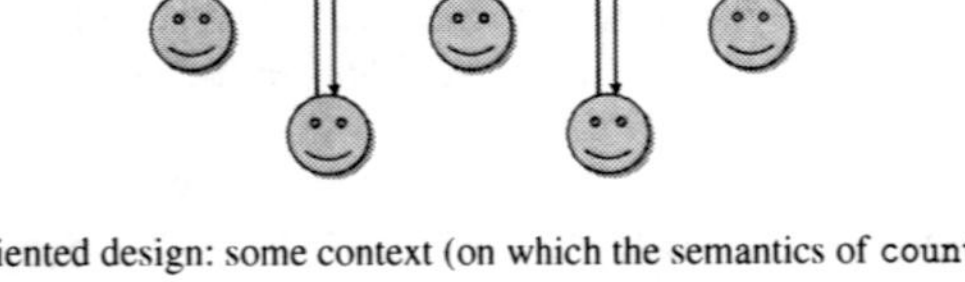

Figure 4. Process oriented design: some context (on which the semantics of `counter` do not depend).

If the `logger` process (on the end of the `log` channel) is, mysteriously, one of the *smiley faces* and calls back, its request would be queued until this process can take it. The `increment` request that triggered the `log` message will complete *before* any callback is serviced.

Of course, should `logger` never accept the `log` message, the servicing of the `increment` request would also never complete. Also, if `logger` held the `log` message in an occam-π *extended input* and tried to call back (which breaks a design rule for extended input), there would be deadlock. Neither of these scenarios is the concern of the `Counter` process. Its semantics are that the `ask` messages are processed correctly and that their servicing continues so long as its responses (on the `answer` and `log` channels) are accepted. Ensuring the latter is the responsibility of the network builder.

3.3.4. Process Invariants

No new concept is needed. Data in processes have no special semantics (cf. object fields, class fields, method variables). We just need the standard notions of *loop* and *recursion* invariants.

3.4. Rationalising occam-π (Unfinished Business)

occam-π [25,26,33,34] has been an experiment in language design and implementation for the communicating process model, starting around the mid-1990s from the occam2.1 language developed by INMOS (later SGS-Thomson Microelectronics, who kindly open-sourced their compiler and libraries for the Transputer). In parallel, libraries enabling this model for a range of standard languages (including CTJ [35,36,37,38] and JCSP [39,40, 41,42] for Java, CCSP [43,44,45,46,47] for C, C++CSP [48,49] for C++, CHP [50,51] for Haskell, CSP [52] for Scala and PyCSP [53], python-csp [54] and Hydra [55] for Python) have been and are being developed. The work started at Kent but has spread to researchers at many institutions around the world. Numerous papers – too many to reference more than a tiny fraction – have been written and published, not least in the CPA conference series.

Applications are an essential driver for the research. For occam-π, these have recently included small platform targets (e.g. the Arduino [56,57,58]), operating systems (RMoX [59, 60]), the engineering of nano-scale machines (TUNA [61,62,63]) and the modelling and control of complex systems with emergent behaviours (CoSMoS [23,64,65,66,67,68]).

occam-π is a careful blending of the dynamics of Milner's π-calculus (mobile channels and processes) with Hoare's CSP underpinning of classical occam, together with significant extension of CSP capabilities (e.g. barriers [27], shared channel-ends [28]). Supporting this, the lightest (by some margin) multicore scheduler available today has been researched and developed, on which applications automatically and effectively scale with the number of cores supplied [46,45]. We also have a fast algorithm for the resolution of general CSP choice (between barriers, input/output guards and timeouts) [69,70], though this has only so far been exploited by JCSP (the CSP library for Java) and an experimental compiler for CSP [71].

However, the occam-π extensions are showing signs of strain and it is now time to look to rationalise and integrate the lessons learnt into a leaner and improved design. Whether this should lead to a new language family or one that is obviously still occam should be considered by its community of stakeholders. The rest of this section outlines a few ideas.

3.4.1. Recursive Union Types

These are an old idea – long overdue for occam. Pascal and C had them back in the 1970s, although they were somewhat unsafe in that they allowed data to be written as one of the variant types and interpreted as another. Securely typed versions underpin modern functional programming languages and were proposed (in 1991) for the (sadly unimplemented) occam3 language [72]. More recently, they are a key element of the proposed *Guppy* [73] follow-up language from occam-π.

Here is described one of the *occam Enhancement Proposals*: OEP-156 (2006) [74]. Its syntax reflects that of *variant (*`CASE`*) protocols* in occam-π, whose purpose it is intended to replace (Section 3.4.4). Two examples:

```
DATA TYPE FOO
  CASE
    sugar, BOOL, REAL64, []BYTE
    salt, BYTE, BYTE
    pepper
:
```

```
DATA TYPE COLOUR
  CASE
    red
    green
    blue
:
```

Each element type of the union is introduced by a user-chosen *tag* word, followed by zero, one or more component fields (i.e. a *record*). The tags must be different. In the second example, there are only tags and the union type corresponds to an *enumerated* type.

Union types are first class – their values may be freely assigned, communicated, abbreviated, passed as parameters and returned from functions, subject to the normal rules for strong typing. Example literals have the same syntax as those for RECORD types:

```
[sugar, TRUE, 99.99, "Krakatoa"]              [red]
[salt, 42, 'A']                               [green]
[pepper]                                      [blue]
```

In case of ambiguity arising from different union types having identical variants (same-named tags and same-typed fields), the type of the literals must be resolved with the standard occam syntax for type decoration. For example:

```
[pepper] (FOO)                                [blue] (COLOUR)
```

Processing values of a union variable requires a CASE process to determine the variant. For example, if x if a variable of type FOO:

```
CASE x
  sugar, BOOL b, REAL64 r, []BYTE s
    ...  b, r and s abbreviate the component fields
    ...  optional follow-up (can change x variant, abbreviations not in scope)
  salt, BYTE m, n
    ...  m and n abbreviate the component fields
    ...  optional follow-up (can change x variant, abbreviations not in scope)
  pepper
    ...  can change x variant (no field abbreviations)
:
```

Thus, values of one variant cannot be processed as another. The optional process following the response to some variants is only needed if the variant has to be changed. This is not allowed in the first response to the variant if there are field abbreviations, since those would be invalidated. The abbreviations are not in scope for the follow-up process.

Recursive union types are allowed. For example:

```
DATA TYPE TREE
  CASE
    node, TREE, FOO, INT, TREE      -- left, data, count, right
    empty
:
```

Here is a process for inserting an element of FOO, which the code assumes has a (user-defined) total ordering with the usual comparison operators, in the above tree:

```
PROC insert (VAL FOO x, TREE t)
  CASE t
    node, TREE left, FOO y, INT count, TREE right
      IF
        x < y
          insert (x, left)
        x = y
          count := count + 1
        x > y
          insert (x, right)
    empty
      t := [node, [empty], x, 1, [empty]]      -- variant changed
:
```

This is standard recursive tree walking. Note, however, that in line with all dynamically allocated occam-π elements, no *"null pointer exceptions"* can occur at runtime – a major win. This is because the *definedness tracking* by the compiler either ensures that all variables and fields have assigned values before they can be observed or demands the explicit insertion (by the programmer) of runtime checks.

For convenience of recursive declarations generally, we propose to change the historic occam language rule that defined names do not come into scope until *after* their declaration (i.e. after the closing colon). Instead, we adopt the standard practice for modern languages that all names in the same block of declarations are in each other's scope. This requires the compiler to make two passes through each declaration block, but that is not a problem.

A more serious issue for recursive data structures is whether to allow *branch-and-join* or *cycles*. Both these introduce *aliasing* (i.e. multiple references to the same data), which is otherwise not present in occam-π. Indeed, this is essential for the static analysis (by the compiler) that eliminates data race hazards in occam-π systems (Section 1.4) – and must not be lost! *Separation logic* [75] may rescue us here. Otherwise, we may need to allow only *tree* structures to be built, which do not introduce aliasing, through a combination of compiler and (hopefully not) runtime checks. Trees also avoid unpredictable overheads of garbage collection, which would remain linearly proportional to the size of the structure being recovered.

Recursive union types make new fields of application practical for occam-π systems. For example, the compiler for the revised language could be written in itself. Novel and simple parallel approaches to compiling are open to research – to be efficiently consumed by multicore processors.

3.4.2. Unify Static and Dynamic Allocation

occam-π introduces *mobile* data types. Like any occam-π data, there can only be one owner at a time. When assigned or communicated, non-mobile data is copied but mobile data is *moved* – meaning that the source variable loses it to the target (i.e. only one copy ever exists). Mobile data is constructed through dynamic allocation from heap storage. If source and target are in the same memory space (i.e. have access to the same heap), movement is by copying the reference to the target and *undefining* the source variable (meaning that it cannot be looked at again until it is *redefined* by assignment or input). Otherwise, movement has to be by copying (to the heap in the target's memory space), but still with *undefining* the source variable (so that the semantics of movement are the same whether it is within or between memory spaces).

Arrays with unspecified sizes can be declared mobile and allocated to runtime determined sizes (which may change during the life of a variable). Currently, the only way to get dynamically sized arrays in occam-π is to make them mobile. The syntax for declaring and constructing such arrays is, therefore, different and somewhat more complex than the syntax for declaring known-sized arrays. The position is even more complicated for runtime sized arrays of channels!

We propose to simplify this, leaving the question of static versus dynamic allocation to be handled invisibly by the compiler and runtime. Data will just be data. Some data, if the logic in the application demands that at most one copy exists, can be declared `MOBILE`.

For non-recursive processes, stack allocation can be arranged – as at present – by the compiler *before* runtime. Recursive process stack allocation happens dynamically, when any level is started. The same is true for processes replicated in parallel by a runtime found value.

Items with compiler-known *small* size (less than 8 or, perhaps, 16 bytes) are allocated on their process stack. Everything else is dynamically allocated on the heap, with references on the stack. *[Note:* currently, compiler-known large items are pre-allocated in *Vectorspace*, following the mechanism from the Transputer implementation of occam. Putting all large items on the heap, regardless of whether their sizes are pre-known, simplifies things.*]*

The programmer is blind to the above. In particular, there is the same syntax for arrays, regardless of whether its size is known to the compiler:

```
[n]THING t:         -- n may be a run-time value
```

Array size is no longer part of the type. An array variable declared with one size may be assigned to an array with another size (same type, of course). An array variable may be declared without size, but must then be assigned (either by assignment or incoming communication) to an actual array value before being used.

Assignment and communication are handled in the most efficient way. Regardless of any declared mobility, stack items – always small – are assigned/communicated by *copying* and heap items by *reference*. In the latter case, this would normally be by copying the reference held on the stack. However, if compiler usage analysis of the sending process shows the assigned/communicated data is used later by that process, a reference to a (deeply) cloned copy is sent. This is Brown's algorithm from his paper at CPA 2009 [76].

If data is explicitly declared MOBILE, the above decision is different: if compiler usage analysis of the sending process shows the assigned/communicated data is used later by that process, this is a semantic error and the compilation fails (reporting the error). This is the current algorithm for occam-π mobiles.

3.4.3. Remove occam2.1 *Protocols*

Having proposed extensions to the language, the principles of *Ockham's Razor* suggest there may be some things to remove. We propose removing occam2.1 PROTOCOLs. These are not real *protocols* in the usual meaning of that term for computer networks. They define a flexible range of *message structures* that are no longer needed, thanks to more powerful types.

The *sequential protocol* is simply replaced by occam2.1 RECORD types. A reason this was not done before is that there was one semantic win for this protocol over record types: a sequential protocol message is a *sequence* of communications, so that an earlier item of received data could be used to address the location of a later one. For example:

```
in ? i; A[i]
```

Fortunately, this capability is won back through *session protocols* (see Section 3.4.4), with no loss of syntactic clarity or runtime efficiency.

The *counted array protocol* is replaced by the revised array type (Section 3.4.2), whose size is not part of the type but carried as part of its value.

The *variant (*CASE*) protocol* is replaced by *union (*CASE*)* types. Before the decision to allocate all large items of data on the heap, there was one pragmatic win for variant protocols over such unions. Where the receiving process logic meant that only small varieties of message were to be accepted, space for the larger variants did not have to be (pre-)allocated – whereas a variable of a union type would have needed space reserved for its largest variant. Now, however, union types with large variants will always be on the heap, so that only space for its reference needs to be reserved in the receiving process.

These proposals mean that communication consists of copying only standard-sized amounts of data (8 or, perhaps, 16 bytes for references and primitive types) – except, of course, when communicating across a memory frontier. An additional benefit from this is that buffered channels that are *generic* across all types become easy to build – a memory frontier is not a problem (since the buffering can be placed on either side or both). Currently, implementing a buffered channel carrying (say) a variant protocol requires special code for each variant, which is tedious to write and inefficient to run (compared with the new proposal).

3.4.4. Introduce Session Protocols

These were proposed by Adam Sampson [77,78]. They are communication protocols in the sense normally understood (i.e. *patterns* of communication). They are associated with a single channel, which may have *shared* ends. The channel is directed (in the same sense as a current channel record is directed), but may be used in both directions (possibly at the same time). We call them *session* protocols, reflecting the idea of *session types* [79,80,81,82]. As the channel is used, the type of message it carries (and its direction) may change.

The simplest session protocol is one data type, sent one way, once. This corresponds to a classical channel. The very simple example in Section 3.4.3 becomes syntactic sugar for:

```
SEQ
  in ? i
  in ? A[i]
```

where the `in` channel is engaged in part of a session specifying that an integer is followed by whatever type is held by elements of the `A` array.

Structured sessions consist of separately typed messages flying in (nested) `SEQ`, `ALT` and/or `PAR`, including repetition (through recursion or looping). This declared structure is the session protocol – syntactic details are not yet settled. The compiler checks that all code operating on the channel conforms, tracking use across all processes and procedures. Channel parameters carrying a session protocol will have to declare which (named) part of the protocol their `PROC` implements. The intention is to eliminate all protocol errors through compiler checks.

Session protocols *may* let us drop channel records (i.e. channel bundles) from the language. These are mainly used for two-way conversations that would be more safely handled by a session protocol (and with less syntactic clutter).

3.4.5. Verification within the Language

This is a proposal to make formal verification of occam-π programs manageable within the language [83,84,32].

The language is extended with qualifiers on types and processes (to indicate relevance for verification and/or execution) and assertions about refinement (including deadlock, livelock and determinism). The compiler abstracts a set of CSP equations and assertions, delegates their analysis to the FDR model checker and reports back in terms related to the occam-π source. The full (FDR) range of CSP assertions is accessible, with no knowledge of CSP syntax required by the programmer (but who will need at least the semantic understanding outlined in Section 3.2). Programs are proved just by writing and compiling programs.

Model-checking is a skilled art. There is more to be done than simply asking the compiler to verify, for example, deadlock-freedom (although that may be a useful starting point). The skill lies in *asking the right questions* and setting up the necessary auxiliary processes needed to frame them. By enabling this to be done as part of normal programming, the intention is to let formal verification become everyday practice [30,83,84]. We suggest that this is urgently needed.

3.4.6. Call Channels and Process Mobility

`CALL` channels were proposed for occam3 [72]. They are syntactic sugar for a simple and common session protocol: *lend-return*, where a process lends some of its resources (i.e. anything in its scope) to a process (at the end of the call-channel) and waits for that process to signal it has finished. The borrowing process may change some of the resource it has been lent. This may be implemented by *lend* and *return* communications over the channel *(the session)*, though obvious and safe shortcuts can be made if lender and borrower are in the same memory space. Syntax for declaring and calling the channel reflects that for declaring and calling a procedure (PROC). The syntax for accepting the call reflects the implementation body of a procedure, except that it happens at the places of acceptance and can be different each time. The semantics are those for classical *extended rendezvous* (e.g. Ada task entries).

A proposal extending CALL channels to allow *variant* interfaces has been made [85,86]. The syntax is different from the occam3 proposal, being expressed as a *protocol* with which channels may be typed. The interfaces in the protocol still have procedural syntax, but using them gives explicit reminder of the synchronisations involved (!, by the lender, and ?? by the borrower). The *double question-mark* is to be consistent with the notion of *extended input* [87] (see Section 3.4.7) in occam-π, which already provides the semantics of extended rendezvous (but without the return of resources). Variant call channels have long been available in the JCSP library.

We have highlighted variant call channels because they provide a direct way to show the *duality* between mobile channels and mobile processes in a future occam-π (or derivative). But we need to re-consider the form of mobile processes currently implemented in occam-π. At present, they work efficiently and allow whole networks (and networks of networks) to be moved, but are somewhat tedious to program because they can only have a *single* interface and that interface allows only channel-end parameters. Also, the mobile network (or solitary process) must be explicitly programmed to *freeze* before it can be moved. We are proposing a new model, [86], that gives mobile processes *variant* interfaces, allows *any* kinds of parameter, greatly simplifies their use and can be moved at any time – even when executing (except during rendezvous with a current host). It is these that are the *dual* of (mobile) variant call channels. A similar proposal for mobile processes has been made for the ProcessJ language by Pedersen and Sowders [88] and is further developed by Pedersen and Smith this year [89].

Almost all of the large dynamic models built so far with occam-π have used *channel*, rather than *process*, mobility. These have a difficulty for migrating processes across memory boundaries, since it is not easy for the runtime system to decide whether a process needs to be moved across such a boundary to follow a migrating *channel-end* whose other end it is holding (and we don't try)! Instead, the programmer must intervene (e.g. [90,91]). However, if a *process* is moved across a memory boundary, that's a pretty good hint to the runtime that it ought to move it (and we think we know how this may be arranged).

3.4.7. and More ...

There are many more ideas that need attention. Not least are several from the occam3 language (in addition to the CALL channels just discussed). There are an interesting new range of process declarations (INITIAL, FINAL, RESOURCE and SERVER), parametrised MODULE types and *libraries*[4]. The reader is invited to look again at the reference manual [72].

The *extended input* process of occam-π was mentioned in Section 3.4.6. This holds the sender of a message while the receiving process responds, rescheduling the sender only when the receiver chooses – the sender is unaware of the extended rendezvous. It has some surprising and powerful uses (e.g. for inserting processes into a channel for monitoring/black-boxing traffic *without* changing the semantics of synchronised communication over the channel). Implementation has been achieved with *zero* overhead for communications that do not use it (i.e. no checking has to be made by a sending process).

Brown and Ritson have proposed the converse: *extended output*, where the receiver process – once committed – is held while the sender decides what to send [93,94]. This also has

[4]For large software projects, package support built into the language would provide great assistance and is more-or-less expected in modern languages. Currently, occam-π uses traditional mechanisms (pre-processor [92], conditional compilation, separate compilation, makefiles).

interesting applications for hiding processes. But it also allows *fishing* by processes looking after sensors in real-time applications, so that they send only the latest available data once a receiver has been *caught*. It is also a useful primitive for modelling/implementing higher level synchronisation mechanisms (such as Teig's *x-channels* [95,96]).

The *fast* algorithm for resolving choice between arbitrary CSP events (e.g. barriers), mentioned at the start of this Section (3.4), is linear with respect to the number of events offered by each process (and needs no *back-tracking*). JCSP provides barrier and output *guards* with a cost that does not impact choices that restrict themselves to the set classically allowed by occam (inputs, timeouts and skips). In occam-π, we might also like to be able to write:

```
ALT
  SYNC bar
    ...  over the barrier, carry on
  out ! n
    ...  message taken, continue
  in ? x
    ...  message arrived, process it
  tim ? AFTER timeout
    ...  timed out, respond
```

However, the fast algorithm works for shared-memory multicore, but does not easily span multiple memory spaces. There may be a way to combine McEwan's *two-phase commit* protocol [97] (for distributed memory synchronisation) with our fast resolution (for shared memory) – more research is needed.

Finally, generic types have long been needed within the language. The code for buffering INTs is no different from that for REAL64s, nor for anything else! With the replacement of classical occam PROTOCOLs by session protocols (Section 3.4.4), union types (Section 3.4.1) and the revised array types (Section 3.4.2), this will be simpler to devise and achieve. Whether this should extend to a hierarchy of *types-with-associated-operations* (e.g. the data type FOO with operators <, = and > from Section 3.4.1), though not *classes*, should be considered.

4. Conclusions

Much received opinion is that object orientation and/or functional programming has solved the problems of management of state and that the problems of concurrency can be finessed with better use of traditional techniques. We do not subscribe to this view.

There is exciting, interesting and important work to do in language engineering for concurrent systems – by which we mean, of course, *all* future systems. Whether this takes the form of a rationalised and extended occam-π or completely new families (such as Guppy and ProcessJ) does not matter. All and more may happen. The important element is that lessons from the occam family are carried forward, with particular regard to *engineered-in* safety against avoidable concurrency errors (such as race hazard), safety against avoidable sequential errors (such as misplaced aliasing and callbacks), semantics that are composable and not context-sensitive (so that only local reasoning is needed), the ability to reason about systems intuitively as well as formally, and simplicity and (therefore) performance. Such engineering cannot happen without the close involvement of those developing the theory and those responsible for applications. We need to keep talking.

The future for software systems rebuilt on simple, secure and stable foundations, so that our energies as designers and programmers can focus on creative ideas, is there to be taken. For the future of our communities, environment and curiosity, it needs to be taken. Unless we take it, which means *work* and *funding*, it will not be there.

Acknowledgements

The opinions reported here are those of the author. The ideas, tools and applications around which they formed are the result of decades of work by highly skilled, insightful and respected colleagues all round the world. Special thanks are owed to Fred Barnes, Carl Ritson and Adam Sampson at Kent, to our partners on the TUNA and CoSMoS projects [67], to all past, present and future members of the the occam-π [33] and KRoC [34] (or derivatives) development teams and to the IFIP Working Group 2.4 for their patient listening, constructive criticism and valuable advice over the past ten years. Finally, some giants of the field (on whose shoulders we gratefully stand) must be acknowledged: Tony Hoare and Bill Roscoe (CSP), Robin Milner (π-calculus) and David May (classical occam).

References

[1] Wikipedia. Analytical Engine, 1837. http://en.wikipedia.org/wiki/Analytical_engine.

[2] Alfred N. Whitehead and Bertrand Russell. *Principia Mathematica (vols. I-III)*. Cambridge University Press, 1910-1913.

[3] Wikipedia. Principia Mathematica, 1910-1913. http://en.wikipedia.org/wiki/Principia_Mathematica.

[4] Wikipedia. Gödel's incompleteness theorems, 1931. http://en.wikipedia.org/wiki/Godel%27s_incompleteness_theorems.

[5] Wikipedia. Lambda Calculus, 1936. http://en.wikipedia.org/wiki/Lambda_calculus.

[6] Alan M. Turing. On Computable Numbers, with an Application to the Entscheidungsproblem. *Proc. London Math. Soc.*, s2-42(1):230–265, October 1937. doi:10.1112/plms/s2-42.1.230.

[7] Wikipedia. Z3 Computer, 1941. http://en.wikipedia.org/wiki/Z3_(computer).

[8] Wikipedia. Colossus Computer, 1943. http://en.wikipedia.org/wiki/Colossus_computer.

[9] Wikipedia. ENIAC, 1946. http://en.wikipedia.org/wiki/ENIAC.

[10] M. D. May. CSP, occam and Transputers. In Ali E. Abdallah, Cliff B. Jones, and Jeff W. Sanders, editors, *25 Years of CSP*, volume 3525 of *Lecture Notes in Computer Science*, pages 75–84. Springer Verlag, April 2005.

[11] Ngi. Nieuwe Ontwikkelingen voor Software Development, 2005. https://www.ngi.nl/Afdelingen/Informatie-Systemen/Verslagen/Nieuwe-ontwikkelingen-voor-Software-Development.html.

[12] Charles A. R. Hoare. Communicating Sequential Processes. *CACM*, 21(8):666–677, August 1978.

[13] Charles A. R. Hoare. *Communicating Sequential Processes*. Prentice-Hall, 1985.

[14] Andrew W. Roscoe. *The Theory and Practice of Concurrency*. Prentice Hall, 1997.

[15] Steve Schneider. *Concurrent and Real-time Systems — The CSP Approach*. Wiley and Sons Ltd., UK, Baffins Lane, Chichester, UK, 1999. ISBN: 0-471-62373-3.

[16] Andrew W. Roscoe. *Understanding Concurrent Systems*. Springer, 2010.

[17] Robin Milner. *Communicating and Mobile Systems: the π-Calculus*. Cambridge University Press, 1999. ISBN-10: 0521658691, ISBN-13: 9780521658690.

[18] Dick Pountain and David May. *A Tutorial Introduction to occam Programming*. McGraw-Hill, Inc., New York, NY, USA, 1987.

[19] E.W. Dijkstra. Hierarchical Ordering of Sequential Processes. *Acta Informatica*, 1(2):115–138, June 1971. http://www.cs.utexas.edu/users/EWD/ewd03xx/EWD310.PDF.

[20] P.H. Welch. Wot, No Chickens? Java Threads mailing list archive, September 1996. http://www.cs.kent.ac.uk/projects/ofa/java-threads/0.html.

[21] P.H. Welch. Java Threads in the Light of occam/CSP. In P.H. Welch and A.W.P. Bakkers, editors, *Proceedings of WoTUG 21*, volume 52 of *Concurrent Systems Engineering*, pages 259–284, Amsterdam, The Netherlands, April 1998. WoTUG, IOS Press. ISBN: 90-5199-391-9. http://www.cs.kent.ac.uk/pubs/1998/702/content.pdf.

[22] Charles A. R. Hoare. Monitors: An Operating System Structuring Concept. *Communications of the ACM*, 17(10):549–557, October 1974.

[23] Peter H. Welch, Kurt Wallnau, Adam T. Sampson, and Mark Klein. To Boldly Go: an occam-π Mission to Engineer Emergence. *Natural Computing*, 11(3):449–474, September 2012. http://dx.doi.org/10.1007/s11047-012-9304-2.

[24] Peter H. Welch. Concurrency Design and Practice. `http://www.cs.kent.ac.uk/projects/ofa/sei-cmu/`, November 2007. A course on concurrency and process-oriented design, based on occam-π.

[25] P. H. Welch and F. R. M. Barnes. Communicating Mobile Processes: introducing occam-π. In Ali E. Abdallah, Cliff B. Jones, and Jeff W. Sanders, editors, *25 Years of CSP*, volume 3525 of *Lecture Notes in Computer Science*, pages 175–210. Springer Verlag, April 2005.

[26] F. R. M. Barnes and P. H. Welch. Communicating Mobile Processes. In Ian East, Jeremy Martin, Peter H. Welch, David Duce, and Mark Green, editors, *Communicating Process Architectures 2004*, volume 62, WoTUG-27 of *Concurrent Systems Engineering Series, ISSN 1383-7575*, pages 201–218, Amsterdam, The Netherlands, September 2004. IOS Press. ISBN: 1-58603-458-8.

[27] P. H. Welch and F. R. M. Barnes. Mobile Barriers for occam-π: Semantics, Implementation and Application. In Jan F. Broenink, Herman W. Roebbers, Johan P.E. Sunter, Peter H. Welch, and David C. Wood, editors, *Communicating Process Architectures 2005*, volume 63, WoTUG-28 of *Concurrent Systems Engineering Series*, pages 289–316, Amsterdam, The Netherlands, September 2005. IOS Press. ISBN: 1-58603-561-4.

[28] P. H. Welch and F. R. M. Barnes. A CSP Model for Mobile Channels. In *Communicating Process Architectures 2008*, volume 66, WoTUG-31 of *Concurrent Systems Engineering Series*, pages 17–33, Amsterdam, The Netherlands, September 2008. IOS Press. ISBN: 978-1-58603-907-3.

[29] P. H. Welch. *An occam-π Quick Reference Guide*. Programming Languages and Systems Research Group, University of Kent, `https://www.cs.kent.ac.uk/research/groups/plas/wiki/OccamPiReference/`, 2011.

[30] Peter H. Welch and Jan B. Pedersen. Santa Claus: Formal Analysis of a Process-Oriented Solution. *ACM Trans. Program. Lang. Syst.*, 32(4):14:1–14:37, April 2010. `http://doi.acm.org/10.1145/1734206.1734211`.

[31] Formal Systems (Europe) Ltd. *Failures-Divergence Refinement (FDR2 User Manual)*. Oxford University Computing Laboratory. `http://www.cs.ox.ac.uk/projects/concurrency-tools/`, 2.94 edition, May 2012. `http://www.cs.ox.ac.uk/projects/concurrency-tools/download/fdr2manual-2.94.pdf`.

[32] Peter H. Welch and Neil C.C. Brown. Self-Verifying Dining Philosophers. Presentation to IFIP Working Group 2.4, September 2011. `https://www.cs.kent.ac.uk/research/groups/plas/wiki/IFIP_WG24`.

[33] A. T. Sampson, C. G. Ritson, M. C. Jadud, F. R. M. Barnes, and P. H. Welch. *occam-π Home Page*. Programming Languages and Systems Research Group, University of Kent, `http://occam-pi.org/`, 2010.

[34] Frederick R. M. Barnes, Peter H. Welch, Jim Moores, and David C. Wood. *The KRoC Home Page*. Programming Languages and Systems Research Group, University of Kent, `http://www.cs.kent.ac.uk/projects/ofa/kroc/`, 2010.

[35] G.H. Hilderink, J.F. Broenink, and A. W. P. Bakkers. Communicating Java Threads. In *Parallel Programming and Java, Proceedings of WoTUG 20*, volume 50, pages 48–76, University of Twente, Netherlands, 1997. IOS Press, Netherlands.

[36] G.H. Hilderink, J.F. Broenink, and A. W. P. Bakkers. Communicating Threads for Java. In *Proceedings of WoTUG-22: Architectures, Languages and Techniques for Concurrent Systems*, pages 243–261. Proceedings of the 22nd World Occam and Transputer User Group Technical Meeting, Keele, United Kingdom, 1999.

[37] A. W. P. Bakkers, G.H. Hilderink, and J.F. Broenink. A Distributed Real-time Java System Based on CSP. In *Proceedings of WoTUG-22: Architectures, Languages and Techniques for Concurrent Systems*, pages 229–241. Proceedings of the 22nd World Occam and Transputer User Group Technical Meeting, Keele, United Kingdom, 1999.

[38] M. M. Bezemer, R. J. W. Wilterdink, and J.F. Broenink. LUNA: Hard Real-Time, Multi-Threaded, CSP-Capable Execution Framework. In P.H. Welch, A. T. Sampson, J. B. Pedersen, J. M. Kerridge, J.F. Broenink, and F. R. M. Barnes, editors, *Communicating Process Architectures 2011, Limmerick*, volume 68 of *Concurrent System Engineering Series*, pages 157–175, Amsterdam, November 2011. IOS Press BV. WoTUG-33.

[39] P. H. Welch and N. C. C. Brown. The JCSP (CSP for Java) Home Page, 2011. `http://www.cs.kent.ac.uk/projects/ofa/jcsp/`.

[40] P.H. Welch. Process Oriented Design for Java: Concurrency for All. In P.M.A.Sloot, C.J.K.Tan, J.J.Dongarra, and A.G.Hoekstra, editors, *Computational Science - ICCS 2002*, volume 2330 of *Lecture Notes in Computer Science*, pages 687–687. Springer-Verlag, April 2002. Keynote Tutorial.

[41] P. H. Welch, N. C. C. Brown, J. Moores, K. Chalmers, and B. H. C. Sputh. Integrating and Extending JCSP. In Alistair A. McEwan, Steve Schneider, Wilson Ifill, and Peter Welch, editors, *Communicating Process*

Architectures 2007, volume 65 of *Concurrent Systems Engineering Series*, pages 349–370, Amsterdam, The Netherlands, July 2007. IOS Press. ISBN: 978-1-58603-767-3.

[42] Peter H. Welch, Neil C. C. Brown, James Moores, Kevin Chalmers, and Bernhard H. C. Sputh. Alting Barriers: Synchronisation with Choice in Java using JCSP. *Concurrency and Computation: Practice and Experience*, 22:1049–1062, March 2010. The DOI should redirect to `http://onlinelibrary.wiley.com.chain.kent.ac.uk/doi/10.1002/cpe.1471/abstract`.

[43] J. Moores. CCSP – a Portable CSP-based Run-time System Supporting C and occam. In B.M. Cook, editor, *Architectures, Languages and Techniques for Concurrent Systems*, volume 57 of *Concurrent Systems Engineering series*, pages 147–168, Amsterdam, The Netherlands, April 1999. WoTUG, IOS Press. ISBN: 90-5199-480-X. `http://www.cs.kent.ac.uk/pubs/1999/753`.

[44] James Moores. *The Design and Implementation of occam/CSP Support for a Range of Languages and Platforms*. PhD thesis, The University of Kent at Canterbury, Canterbury, Kent. CT2 7NF, December 2000.

[45] Carl G. Ritson, Adam T. Sampson, and Frederick R. M. Barnes. Multicore Scheduling for Lightweight Communicating Processes. In John Field and Vasco Thudichum Vasconcelos, editors, *Coordination Models and Languages, COORDINATION 2009, Lisboa, Portugal, June 9-12, 2009. Proceedings*, volume 5521 of *Lecture Notes in Computer Science*, pages 163–183. Springer, June 2009. `http://www.cs.kent.ac.uk/pubs/2009/2928/`.

[46] Carl G. Ritson, Adam T. Sampson, and Frederick R. M. Barnes. Multicore Scheduling for Lightweight Communicating Processes. *Science of Computer Programming*, 77(6):727–740, 2012. Revised from LNCS (5521) paper with same name.

[47] Carl G. Ritson. *Scalable Support for Process-Oriented Programming*. PhD thesis, School of Computing, University of Kent, March 2013. (Awarded July, 2013).

[48] N. C. C. Brown and P. H. Welch. An Introduction to the Kent C++CSP Library. In J.F. Broenink and G.H. Hilderink, editors, *Communicating Process Architectures 2003*, WoTUG-26, Concurrent Systems Engineering, ISSN 1383-7575, pages 139–156, Amsterdam, The Netherlands, September 2003. IOS Press. ISBN: 1-58603-381-6.

[49] N. C. C. Brown. *C++CSP Home Page*. Programming Languages and Systems Research Group, University of Kent, `http://www.cs.kent.ac.uk/projects/ofa/c++csp/`, 2010.

[50] N. C. C. Brown. *Communicating Haskell Processes Home Page*. Programming Languages and Systems Research Group, University of Kent, `http://www.cs.kent.ac.uk/projects/ofa/chp/`, 2010.

[51] Neil C. C. Brown. Communicating Haskell Processes: Composable Explicit Concurrency Using Monads. In Peter H. Welch, Susan Stepney, Fiona A.C. Polack, Frederick R.M. Barnes, Alistair A. McEwan, Gardner S. Stiles, Jan F. Broenink, and Adam T. Sampson, editors, *Communicating Process Architectures 2008*, volume 66 of *Concurrent Systems Engineering*, pages 67–83, Amsterdam, The Netherlands, September 2008. WoTUG, IOS Press.

[52] Bernard Sufrin. Communicating Scala Objects. In *Communicating Process Architectures 2008*, volume 66, WoTUG-31 of *Concurrent Systems Engineering Series*, pages 35–54, Amsterdam, The Netherlands, sep 2008. IOS Press.

[53] Brian Vinter, John Markus Bjrndalen, and Rune Mllegard Friborg. PyCSP Revisited. In *Communicating Process Architectures 2009*, volume 67, WoTUG-32 of *Concurrent Systems Engineering Series*, pages 263–276, Amsterdam, The Netherlands, nov 2009. IOS Press.

[54] Sarah Mount, Mohammad Hammoudeh, Sam Wilson, and Robert Newman. CSP as a Domain-Specific Language Embedded in Python and Jython. In *Communicating Process Architectures 2009*, volume 67, WoTUG-32 of *Concurrent Systems Engineering Series*, pages 293–309, Amsterdam, The Netherlands, nov 2009. IOS Press.

[55] Waide B. Tristram and Karen Bradshaw. Hydra: A Python Framework for Parallel Computing. In *Communicating Process Architectures 2009*, volume 67, WoTUG-32 of *Concurrent Systems Engineering Series*, pages 311–324, Amsterdam, The Netherlands, nov 2009. IOS Press.

[56] M.C. Jadud. Parallel Programming for the Rest of Us, 2013. `http://concurrency.cc/`.

[57] Christian L. Jacobsen, Matthew C. Jadud, Omer Kilic, and Adam T. Sampson. Concurrent Event-driven Programming in occam-π; for the Arduino. In *Communicating Process Architectures 2011*, volume 68, WoTUG-33 of *Concurrent Systems Engineering Series*, pages 177–193, Amsterdam, The Netherlands, June 2011. IOS Press. ISBN: 978-1-60750-773-4.

[58] Ian Armstrong, Michael Pirrone-Brusse, A. Smith, and Matthew C. Jadud. The Flying Gator: Towards Aerial Robotics in occam-π. In *Communicating Process Architectures 2011*, volume 68, WoTUG-33 of *Concurrent Systems Engineering Series*, pages 329–340, Amsterdam, The Netherlands, June 2011. IOS Press. ISBN: 978-1-60750-773-4.

[59] F.R.M. Barnes. RMoX: an occam-pi Operating-System, January 2005. Available at: `http://rmox.net`.

[60] F.R.M. Barnes and C.G. Ritson. Process-Oriented Device Driver Development. *Concurrency and Computation: Practice and Experience*, 22(8):995–1006, June 2010.

[61] S. Stepney, P.H. Welch, F.A.C. Pollack, J.C.P. Woodcock, S. Schneider, H.E. Treharne, and A.L.C. Cavalcanti. TUNA: Theory Underpinning Nanotech Assemblers (feasibility study), January 2005. EPSRC grant EP/C516966/1. Available from: `http://www.cs.york.ac.uk/nature/tuna/index.htm`.

[62] C. G. Ritson and P. H. Welch. A Process-Oriented Architecture for Complex System Modelling. *Concurrency and Computation: Practice and Experience*, 22:965–980, March 2010.

[63] C. Ritson and P.H.Welch. 3D Blood Clotting (TUNA), 2006. `https://www.cs.kent.ac.uk/research/groups/sys/wiki/3D\_Blood\_Clotting/`.

[64] P. H. Welch, A. T. Sampson, and D. C. Wood. *"To Boldly Go: an occam-π Mission to Engineer Emergence" – supplementary materials*. Programming Languages and Systems Research Group, University of Kent, `http://projects.cs.kent.ac.uk/projects/naco-boldly/`, 2011.

[65] Paul S. Andrews, Adam T. Sampson, John Markus Bjørndalen, Susan Stepney, Jon Timmis, Douglas N. Warren, and Peter H. Welch. Investigating Patterns for the Process-oriented Modelling and Simulation of Space in Complex Systems. In S. Bullock, J. Noble, R. Watson, and M. A. Bedau, editors, *Artificial Life XI: Proceedings of the Eleventh International Conference on the Simulation and Synthesis of Living Systems*, pages 17–24. MIT Press, Cambridge, MA, August 2008.

[66] Fiona A.C. Polack, Paul S. Andrews, and Adam T. Sampson. The Engineering of Concurrent Simulations of Complex Systems. In *2009 IEEE Congress on Evolutionary Computation (CEC 2009)*, pages 217–224. IEEE Press, 2009.

[67] Susan Stepney, Peter H. Welch, Fiona Polack, Jon Timmis, Fred R. M. Barnes, and Andy Tyrrell. CoSMoS Home Page, 2010. EPSRC EP/E053505/1 and EP/E049419/1. `http://www.cosmos-research.org/about.html`.

[68] S. Stepney, P.H. Welch, J. Timmis, C. Alexander, F.R.M. Barnes, M. Bates, F.A.C. Polack, and A. Tyrrell. CoSMoS: Complex Systems Modelling and Simulation infrastructure, April 2007. EPSRC grants EP/E053505/1 and EP/E049419/1. URL: `http://www.cosmos-research.org/about.html`.

[69] P.H. Welch, F.R.M. Barnes, and F.A.C. Polack. Communicating Complex Systems. In Michael G. Hinchey, editor, *Proceedings of the 11th IEEE International Conference on Engineering of Complex Computer Systems (ICECCS-2006)*, pages 107–117. IEEE, August 2006.

[70] P. H. Welch, N. C. C. Brown, J. Moores, K. Chalmers, and B. H. C. Sputh. Alting Barriers: Synchronisation with Choice in Java using CSP. *Concurrency and Computation: Practice and Experience*, 22:1049–1062, 2010.

[71] F.R.M. Barnes. Compiling CSP. In P.H. Welch, J. Kerridge, and F.R.M. Barnes, editors, *Communicating Process Architectures 2006*, volume 64 of *Concurrent Systems Engineering Series*, pages 377–388, Amsterdam, The Netherlands, September 2006. IOS Press. ISBN: 1-58603-671-8.

[72] Geoff Barrett. *occam3 Reference Manual*. INOS Ltd., March 1992. `http://www.wotug.org/occam/documentation/oc3refman.pdf`.

[73] F.R.M. Barnes. Towards a New Language for Concurrent Programming. *CPA 2011 Fringe:* slides at `http://www.wotug.org/papers/CPA-2011/Barnes11/Barnes11-slides.pdf`, August 2011. Guppy Home Page: `http://frmb.org/guppy.html`.

[74] P.H. Welch. Union Data Types. *occam Enhancement Proposal 156*, April 2006. `https://www.cs.kent.ac.uk/research/groups/plas/wiki/OEP/156`.

[75] J.C. Reynolds. Separation Logic: a Logic for Shared Mutable Data Structures. In *Proceedings of the 17th. Annual Symposium on Logic in Computer Science*, pages 55–74, Copenhagen, Denmark, July 2002. WoTUG, IEEE Press. ISBN: 0-7695-1483-9. `http://dx.doi.org/10.1109/LICS.2002.1029817`.

[76] Neil C.C. Brown. Auto-Mobiles: Optimised Message-Passing. In *Communicating Process Architectures 2009*, volume 67, WoTUG-32 of *Concurrent Systems Engineering Series*, pages 225–238, Amsterdam, The Netherlands, nov 2009. IOS Press.

[77] Adam T. Sampson. Two-Way Protocols for occam-π. In Peter H. Welch, S. Stepney, F.A.C Polack, Frederick R. M. Barnes, Alistair A. McEwan, G. S. Stiles, Jan F. Broenink, and Adam T. Sampson, editors, *Communicating Process Architectures 2008*, volume 66, WoTUG-31 of *Concurrent Systems Engineering Series*, pages 85–97, Amsterdam, The Netherlands, sep 2008. IOS Press.

[78] Adam T. Sampson. *Process-Oriented Patterns for Concurrent Software Engineering*. PhD thesis, University of Kent, October 2010.

[79] K. Honda. Types for Dyadic Interaction. In *CONCUR93: Proceedings of the International Conference on Concurrency Theory*, page 509523. Springer-Verlag, 1993. LNCS 715.

[80] Kaku Takeuchi, Kohei Honda, and Makoto Kubo. An Interaction-based Language and its Typing System. In *PARLE 94: Parallel Architectures and Languages Europe*, pages 398–413. Springer-Verlag, 1994. LNCS 817.

[81] Nobuko Yoshida and Vasco T. Vasconcelos. Language Primitives and Type Discipline for Structured Communication-based Programming. In *Proceedings of the European Symposium on Programming 1998*, pages 122–138. Springer-Verlag, 1998. LNCS 1381.

[82] Simon Gay, Vasco Vasconcelos, and Antonio Ravara. Session Types for Inter-Process Communication. Technical report, Department of Computing Science, University of Glasgow, 2003. TR-2003-133.

[83] Peter H. Welch, Jan Baekgaard Pedersen, Frederick R. M. Barnes, Carl G. Ritson, and Neil C.C. Brown. Adding Formal Verification to occam-π. In *Communicating Process Architectures 2011*, volume 68, WoTUG-33 of *Concurrent Systems Engineering Series*, pages 379–379 *(Endnote Presentation)*, Amsterdam, The Netherlands, jun 2011. IOS Press. ISBN: 978-1-60750-773-4. `http://www.wotug.org/papers/CPA-2011/Welch11/Welch11-slides.pdf`.

[84] Peter H. Welch, Jan B. Pedersen, Frederick R.M. Barnes, and Carl G. Ritson. Self-Verifying Concurrent Programming. Presentation to IFIP Working Group 2.4, September 2011. `https://www.cs.kent.ac.uk/research/groups/plas/wiki/IFIP_WG24`.

[85] P.H. Welch and E. Bonnici. Variant Call Protocols. *occam Enhancement Proposal 178*, August 2010. `https://www.cs.kent.ac.uk/research/groups/plas/wiki/OEP/178`.

[86] Eric Bonnici and Peter H. Welch. Mobile Processes and Call Channels with Variant Interfaces (a Duality). In *Communicating Process Architectures 2011*, volume 68, WoTUG-33 of *Concurrent Systems Engineering Series*, pages 377–377 *(Fringe Presentation)*, Amsterdam, The Netherlands, June 2011. IOS Press. ISBN: 978-1-60750-773-4.

[87] F.R.M. Barnes. Extended Rendezvous. *occam Enhancement Proposal 110*, December 2001. `https://www.cs.kent.ac.uk/research/groups/plas/wiki/OEP/110`.

[88] Jan Baekgaard Pedersen and Matthew Sowders. Static Scoping and Name Resolution for Mobile Processes with Polymorphic Interfaces. In Peter H. Welch, Adam T. Sampson, Jan Baekgaard Pedersen, Jon Kerridge, Jan F. Broenink, and Frederick R. M. Barnes, editors, *Communicating Process Architectures 2011*, pages 71–85, jun 2011.

[89] J.B.Pedersen and M.L.Smith. ProcessJ: A Possible Future of Process Oriented Design. In *Communicating Process Architectures 2013*. Open Channel Publishing, August 2013.

[90] John Markus Bjørndalen and Adam T. Sampson. Process-Oriented Collective Operations. In *Communicating Process Architectures 2008*, volume 66 of *Concurrent Systems Engineering*, pages 309–328, Amsterdam, The Netherlands, September 2008. WoTUG, IOS Press. `http://www.cosmos-research.org/docs/cpa2008-poco-slides.pdf`.

[91] Adam T. Sampson, John Markus Bjørndalen, and Paul S. Andrews. Birds on the Wall: Distributing a Process-Oriented Simulation. In *2009 IEEE Congress on Evolutionary Computation (CEC 2009)*, pages 225–231. IEEE Press, May 2009.

[92] Frederick R. M. Barnes. occam-π Pre-processing Support, 2006. `http://frmb.org/occ21-extensions.html#preproc`.

[93] N.C.C. Brown and C.G. Ritson. Extended Outputs. *occam Enhancement Proposal 142*, March 2006. `https://www.cs.kent.ac.uk/research/groups/plas/wiki/OEP/142`.

[94] Neil C.C. Brown. How to Make a Process Invisible. In *Communicating Process Architectures 2008*, volume 66, WoTUG-31 of *Concurrent Systems Engineering Series*, pages 445–445 *(Fringe Presentation)*, Amsterdam, The Netherlands, sep 2008. IOS Press.

[95] Oyvind Teig. XCHANs: Notes on a New Channel Type. In *Communicating Process Architectures 2012*, pages 155–170. Open Channel Publishing, August 2012.

[96] Peter H. Welch. An occam Model of XCHANs, 2013. `https://www.cs.kent.ac.uk/research/groups/plas/wiki/An_occam_Model_of_XCHANs`.

[97] A.A. McEwan. *Concurrent Program Development*. DPhil thesis, The University of Oxford, 2006.

Communicating Process Architectures 2013
P.H. Welch et al. (Eds.)
Open Channel Publishing Ltd., 2013

Mutually Assured Destruction (or the Joy of Sync)

Fringe Presentation

Peter H. WELCH [a], Jan Bækgaard PEDERSEN [b] and Frederick R.M. BARNES [a]

[a] *School of Computing, University of Kent, UK*

[b] *Department of Computer Science, University of Nevada Las Vegas, NV, USA*

P.H.Welch@kent.ac.uk, matt.pedersen@unlv.edu, F.R.M.Barnes@kent.ac.uk

Abstract. In contrast to the *Client-Server* pattern, *Mutually Assured Destruction (MAD)* allows the safe opening of communication by either of two processes with the other. Should both processes commit to opening conversation together, both are immediately aware of this and can take appropriate action – there is no deadlock, even though the communications are synchronous. A common need for this pattern is in real-time control systems (e.g. robotics), artificial intelligence, e-commerce, model checking and elsewhere. A typical scenario is when two processes are given a problem to solve; we are satisfied with a solution to either one of them; whichever process solves its problem first kills the other and makes a report; the one that is killed also reports that fact. In this case, the opening communication between the two processes is the only communication (a *kill* signal). If both solve their problem around the same time and try to kill each other, *MAD* is the desired outcome (along with making their reports). A simple and safe solution (verified with FDR) for this pattern with occam *synchronous* communications is presented. This is compared with solutions via *asynchronous* communications. Although these avoid the potential deadlock that a naive strategy with synchronous communications would present, they leave a mess that has to be tidied up and this tidying adds complexity and run-time cost. The *Joy of Sync* arises from the extra semantic information provided by synchronous communications – that if a message has been sent, the receiving process has taken it. With this knowledge comes power and with that power comes the ability to implement higher level synchronisation patterns (such as *MAD* and *Non-blocking Syncs*) in ways that are simple, verifyably safe and impose very low overheads.

Keywords. mutually assured destruction, non-blocking synchronisation, occam-pi, concurrency, synchronisation, simplicity, verification, performance

Communicating Process Architectures 2013
P.H. Welch et al. (Eds.)
Open Channel Publishing Ltd., 2013

FDR3: the Future of CSP Model Checking

Fringe Presentation

Thomas GIBSON-ROBINSON

Department of Computer Science, University of Oxford, UK

thomas.gibson-robinson@cs.ox.ac.uk

Abstract. Over the last couple of years a brand new version of FDR, FDR3, has been under development at Oxford. This includes greatly improved performance (typically 2.5 times faster than FDR2 on a single core) and, for the first time, a parallel model-checking mode that scales almost linearly with the number of available cores. In this talk I will give a demonstration of FDR3, including the new debug viewer, the integrated version of ProBE, and the enhanced error messages. FDR3 is due for general release this October.

Keywords. model checking, concurrency, CSP, FDR, FDR3, Probe, simplicity, scaling, performance

Communicating Process Architectures 2013
P.H. Welch et al. (Eds.)
Open Channel Publishing Ltd., 2013

Using FDR to Model Check CSP-Like Languages

Fringe Presentation

Thomas GIBSON-ROBINSON

Department of Computer Science, University of Oxford, UK

`thomas.gibson-robinson@cs.ox.ac.uk`

Abstract. This talk presents a demonstration of `tyger`, a tool that has been constructed to allow a large class of *CSP-like languages* to be simulated in CSP. In particular, FDR will be used to demonstrate model checking of a simple CCS version of the classic Dining Philosophers problem. The theory behind this demonstration will be presented in [1].

Keywords. CSP-like, tyger, simulation, FDR, Dining Philosophers

References

[1] T. Gibson-Robinson. Efficient Simulation of CSP-Like Languages. In *Communicating Process Architectures 2013*. Open Channel Publishing, August 2013.

Communicating Process Architectures 2013
P.H. Welch et al. (Eds.)
Open Channel Publishing Ltd., 2013

An Introduction to Go

Fringe Presentation

Richard BETON

Big Bee Consultants Ltd., UK

`www.bigbeeconsultants.co.uk`

Abstract. Go is a new open-source programming language from a team of Google's developers. It provides a modern garbage-collected runtime that has particular support for concurrency. The pattern for this is based on Hoare's CSP, using channels for synchronisation and communication, and discouraging the direct sharing of variables. Processes are light-weight co-operative threads that the runtime time-slices onto underlying operating system threads; this allows an application's natural concurrency to be expressed as fully as needed by the natural concurrency of the application domain. The presentation provides an overview of Go's main features, covering in particular the unusual interfaces and duck-typing, the reflection and the concurrency. Focussing on concurrency in detail, as an example, a simple web-crawler application is converted step by step to operate concurrently, and at each stage this allows a particular feature of Go's concurrency to be demonstrated.

Keywords. concurrency, language, Go, CSP, light-weight

Communicating Process Architectures 2013
P.H. Welch et al. (Eds.)
Open Channel Publishing Ltd., 2013

The Guppy Language: an Update

Fringe Presentation

Frederick R.M. BARNES
School of Computing, University of Kent, UK
f.r.m.barnes@kent.ac.uk

Abstract. As a possible successor and perhaps replacement language for occam-π, we have been working piecemeal for the past few years on a new language – Guppy. Rather than being a completely new language, it aims to rationalise and simplify the concurrent programming concepts and idioms that we have explored whilst developing occam-π. This short talk will look at the current state of the language and its implementation, that is now able to compile and run a simple *"commstime"* benchmark, scheduled by the existing occam-π run-time system (CCSP).

Keywords. concurrency, language, occam-pi, Guppy

Communicating Process Architectures 2013
P.H. Welch et al. (Eds.)
Open Channel Publishing Ltd., 2013

An occam Model of XCHANs

Fringe Presentation

Peter H. WELCH
School of Computing, University of Kent, Canterbury, UK
P.H.Welch@kent.ac.uk

Abstract. Øyvind Teig, in [1], proposed a higher level channel construct (XCHAN) that attempts to reconcile those wedded to asynchronous message passing with the synchronous form in CSP. Sending a message does not block the sender, but the message may not get sent: the sender receives a success/fail result on each send. The XCHAN provides a conventional feedback channel on which it signals when it is ready to take a message. Being ready means that it has space (if it is buffered) or a reading process has committed to take the message (if it is not buffered). Sending to a ready XCHAN always succeeds; sending to an XCHAN that is not ready always fails. The sender can always wait for the signal from the XCHAN (whilst ALTing on, and processing, other events) before sending. We can model an XCHAN by a process in occam-π. Buffered XCHANs are easy. Zero-buffered XCHANs are a little harder, because we need to maintain end-to-end synchronisation. However, occam-π's extended input (??) and output (!!) primitives enable the process implementing the XCHAN to be hidden from its users. Unfortunately, extended outputs are not yet in the language, but their semantics can be simulated by making the receiving process read twice and ignore the first (which is just a signal whose taking must commit the reader to its second read). An important message is that sane higher level synchronisation mechanisms are usually not hard to implement efficiently via the low level CSP primitives offered by occam-π. Although not yet measured for XCHANs, it is likely that such simulation in occam-π will have competitive performance with direct implementation elsewhere. occam-π code and documentation for this model is available from [2], together with the slides from this presentation.

Keywords. XCHAN, occam-pi, concurrency, synchronisation, simplicity

References

[1] Oyvind Teig. XCHANs: Notes on a New Channel Type. In *Communicating Process Architectures 2012*, pages 155–170. Open Channel Publishing, August 2012.
[2] Peter H. Welch. An occam Model of XCHANs, 2013. https://www.cs.kent.ac.uk/research/groups/plas/wiki/An_occam_Model_of_XCHANs.

Communicating Process Architectures 2013
P.H. Welch et al. (Eds.)
Open Channel Publishing Ltd., 2013

Names of XCHAN Implementations

Fringe Presentation

Øyvind TEIG[1]

Autronica Fire and Security AS, Trondheim, Norway

Abstract. Two names for possible XCHAN implementations are suggested. The original presentation [1] describes the *"classic"* scheme where the `xchan-ready` channel is used only if the original sending fails. The occam-π model [2,3] uses the *"pre-confirmed"* scheme, where a signal on `xchan-ready` is a necessary precondition to any communication. It is believed that *"feathering"* [4] seems to be possible only with the classic scheme.

Keywords. XCHAN, implementation, classic, pre-confirmed, occam-pi, feathering,

References

[1] Ø. Teig. XCHANs: Notes on a New Channel Type. In *Communicating Process Architectures 2012*, pages 155–170. Open Channel Publishing, August 2012.
[2] P. H. Welch. An occam Model of XCHANs. In *Communicating Process Architectures 2013*. Open Channel Publishing, August 2013. *Fringe presentation.*
[3] P. H. Welch. An occam Model of XCHANs, 2013. `https://www.cs.kent.ac.uk/research/groups/plas/wiki/An_occam_Model_of_XCHANs`.
[4] Ø. Teig. Selective Choice "Feathering" with XCHANs. In *Communicating Process Architectures 2013*. Open Channel Publishing, August 2013.

[1]The author works with concurrent software for fire detection systems, but this "industrial paper" does not necessarily reflect views taken by the company. See: `http://www.teigfam.net/oyvind/work/work.html` and `http://www.autronicafire.com`.

Subject Index

Subject Index

Author Index

CPSIA information can be obtained at www.ICGtesting.com
Printed in the USA
LVOW07s1144071213

364281LV00001B/10/P

9 780956 540973